PENGUIN BOOKS

THE WAY TO GO

Kate Ascher serves on the faculty of Columbia University's Graduate School of Architecture, Planning, and Preservation and is a partner at BuroHappold overseeing the firm's U.S. Cities practice. Prior to taking up her current positions, she worked at the Port Authority of New York and New Jersey, the New York City Economic Development Corporation, and Vornado Realty Trust. Her former books include *The Works: Anatomy of a City* and *The Heights: Anatomy of a Skyscraper.*

THE WAY TO GO

MOVING BY SEA, LAND, AND AIR

Kate Ascher

Art direction by Design Language
Research by Rob Vroman

Penguin Books

PENGUIN BOOKS
An imprint of Penguin Random House LLC
375 Hudson Street
New York, New York 10014
penguin.com

First published in the United States of America by The Penguin Press,
a member of Penguin Group (USA) LLC, 2014
Published in Penguin Books 2015

Credits for contributing artists appear on page 198.

Photography credits appear on page 199.

ISBN 978-1-59420-468-5 (hc.)
ISBN 978-0-14-312794-9 (pbk.)

Printed in the United States of America
1 3 5 7 9 10 8 6 4 2

Set in Tisa, Chaco, and Forza
Designed by Design Language

Cover design by Design Language
Cover illustration by Ken Batelman

A NOTE TO READERS

Transportation moves fast. The way people move from place to place today bears little resemblance to the way they did a century ago, when planes and cars were still in prototype phase. One can only imagine how the world will travel a century from now, when new types of materials, more sophisticated computers, and greater understanding of aerodynamics will undoubtedly provide for faster, safer, and more fuel-efficient journeys.

Although research into many of these areas is well under way, few of today's cutting-edge technologies are covered in this book. Instead, it focuses on those engineering concepts that have stood the test of time and remain relevant and central to the major transportation modes: sea, rail, road, and air. Inventions like the wheel, the rudder, the combustion engine, and the wing continue to lie at the heart of moving people and cargo, though their operation in modern vehicles has grown increasingly complex.

Trying to explain to lay readers how these core transportation technologies work has proved no easy task, and we will undoubtedly have fallen short on many counts. While every effort has been made to ensure that the explanations on the pages that follow are broadly accurate, neither the publisher nor the author assumes responsibility for omissions, errors, or changes that occur after publication.

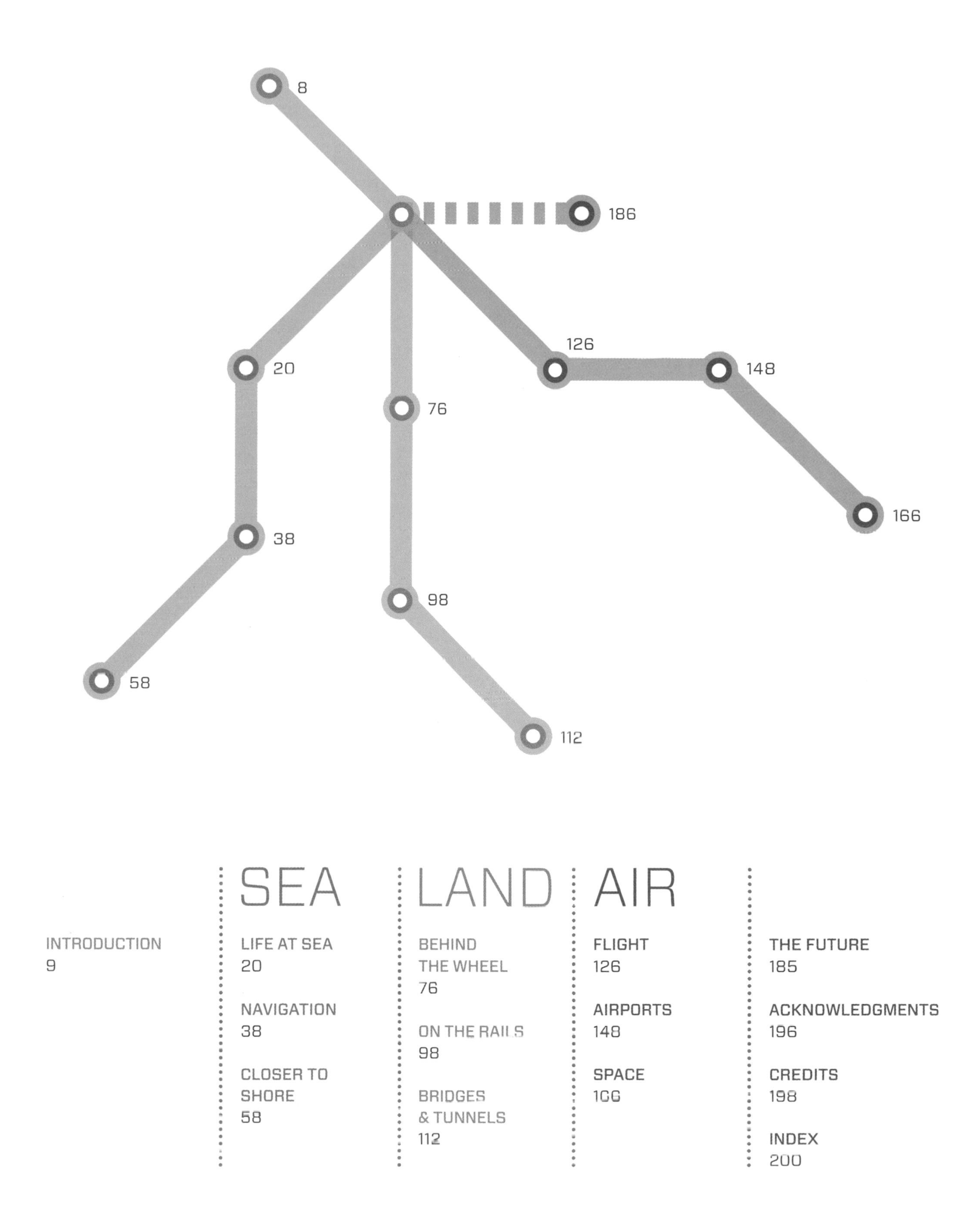

INTRODUCTION
9

SEA

LIFE AT SEA
20

NAVIGATION
38

CLOSER TO SHORE
58

LAND

BEHIND THE WHEEL
76

ON THE RAILS
98

BRIDGES & TUNNELS
112

AIR

FLIGHT
126

AIRPORTS
148

SPACE
166

THE FUTURE
185

ACKNOWLEDGMENTS
196

CREDITS
198

INDEX
200

INTRODUCTION

In late-eighteenth-century France, the diligence—a fairly primitive form of stagecoach—carried intrepid travelers from Paris to London. On a typical journey, as recounted by Jonathan Conlin in his book *Tales of Two Cities*, a handful of customers embarking in Paris braved tight quarters and rutted roads to reach first Abbeville, where they spent the night, and then Boulogne or Calais, where they passed a second. Once across the Channel, passengers were put up in Dover for a third night—before departing the next morning for Canterbury and Rochester, the jumping-off points for the final leg of the journey to London.

Today, the journey from the heart of Paris to London is counted in minutes rather than days. Eurostar trains leave every hour from the Gare du Nord, reaching speeds of just under 200 mph (335 km/h) and completing the trip in two hours and fifteen minutes. The underwater crossing of the Channel is fleeting and almost indiscernible, the arrival on English soil marked only by the noise of mobile phones picking up signals from British telecom providers.

It is not just people who are going places faster today. Cargo that once took months to sail from its origin to destination can be loaded into the belly of a plane and arrive in hours—or in days, if time is less critical and it travels by ship. The cost of sending that cargo has dropped dramatically as well, helping to create the global economy as we know it today: products that could never have shouldered the historic costs of long-distance transport now appear regularly on supermarket shelves to compete with locally sourced ones.

From trains and cars to boats to planes, transportation has united the world in novel and unprecedented ways—touching virtually every corner of the earth. Long-distance, cross-border travel is now utterly routine, leading to greater interchange among places, cultures, and people than ever before. International trade has likewise been transformed, with people on opposite sides of the globe sharing footwear, alcohol, and even produce from the same factories or farms.

But the revolution ushered in by the transportation advances of the last two centuries has come at a price. Growing dependence on fossil fuels has led to alarming rises in both water and air pollution, with environmental consequences that are only now beginning to be understood. This hunger for fuel has given birth to the notion of "energy security," a term never far from the top of the geopolitical agenda and responsible for more than one war. Even the growth of tourism and recreational travel has been a mixed blessing, having led to the desecration of some of the world's most beautiful spots and to the marginalization of many native cultures.

For better or worse, however, one thing is certain: transportation shapes our existence. It helps determine what we wear and eat, where we work and play, and how we maintain our social networks. Yet most of us take it largely for granted. We assume that the car will run and the highways will be moving, that the trains and planes will get us to our destinations more or less on time, and that the products we buy at the supermarket from farms thousands of miles away will be available year-round. Rarely if ever do we stop to wonder at the miracle of modern transport that underpins our daily lives.

The following chapters pay homage to that miracle. In explaining "the way to go," they lift the hood and shine light on the complex, sophisticated systems that have revolutionized the transport of people and goods over the course of the last century. From vehicles and technologies to facilities and people, they explore the often invisible networks that keep the world on the move—at sea, on land, and in the air.

The first section of the book looks at transport on water. The earliest "way to go," it is also the one that has changed least over the course of mankind's recorded existence. Many of the inventions that first allowed ships to move without human power—hulls, rudders, and sails—are still in use today. So too are inventions borne of military combat at sea—from submarines to sonar to aircraft carriers. Today, these historic innovations are accompanied by myriad modern technologies—for navigation, for loading cargo, and for protecting life at sea.

The second section of the book deals with transportation on land—on roads, along rails, and via tunnels and bridges. Though most people in the developed world ride in a car or train every day, only a small number understand how they work. Even fewer know much about the infrastructure that supports them: paved roadways and rail track, signal systems and stoplights, tunnels and bridges. How these support systems are constructed and operated is as much a part of the land transport story as the science behind the magnetic-levitation train or electric car.

The final section of the book looks at aviation—both traditional air transport and space travel. The newest way of getting somewhere, travel by air is also the most global—with standards for aviation promulgated on a worldwide basis and international cooperation in the development of space technology now commonplace. One need only step onto an airplane destined for a foreign capital to be reminded of just how thoroughly advances in air transport have transformed and strengthened relations among people and nations.

Across all three modes of travel, the book touches on historic and current forms of transport—both civilian and military. Though its reach is broad, it does not attempt to be comprehensive with respect to the wide range of technologies and systems in use around the globe. Nor does it serve as a guide to how people will move in the future: rapid innovation has been the hallmark of the industry since the invention of steam power and remains so today. *The Way to Go* serves instead as a panoramic snapshot of transportation as we know it now—a tribute to how people and things move at the dawn of the twenty-first century.

The story of modern transportation is in many ways the story of modern society. At its heart, it's the story of inventions that underpinned leaps forward in the transport of people and goods—powering economic growth and development around the globe and transforming a world made up of empires and colonies into one with almost two hundred independent nation-states.

Two centuries of transportation innovations set the stage for this book. Things like the steam engine, the internal combustion engine, the railroad, and the jet plane fundamentally and forever changed the world around them. While they were products of the Industrial Revolution, they also created their own revolutions—in the way people worked, traded, and lived. Those revolutions continue to resonate today both in developing countries, where the growth of new cities is increasingly dependent on cars and trains, and in the developed world, where established cities would grind to a halt without their public transit and aviation infrastructures.

Some of the most dramatic inventions—the car, the plane, the rocket—are the most recent. But to start with them would overlook thousands of years of mankind's history. For before there were machines and fossil fuels to move people and cargo there were animals for transport on land and sailboats for transport by water. Both forms of carriage were effective, if slow—supporting trade, agriculture, and communications.

HISTORY

Although transportation is no more than a means to an end for people and goods, the means of movement has shifted frequently in size and shape and from mode to mode—providing a compelling time line of efforts to "get there" more efficiently over the last two centuries.

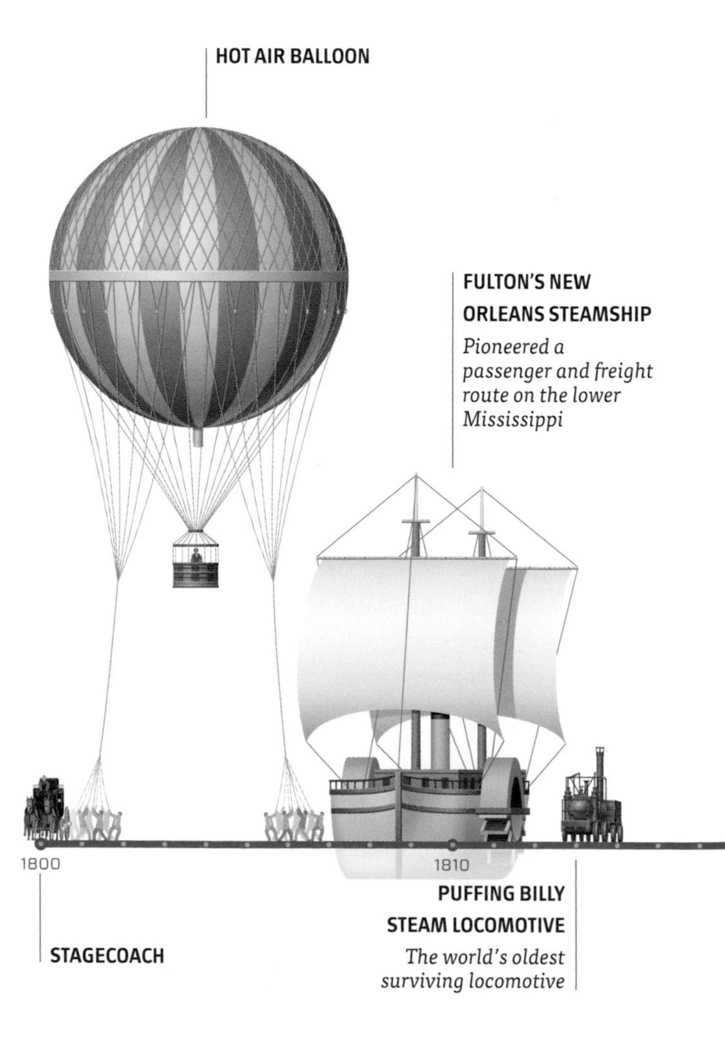

There was also infrastructure, some of it sophisticated, for these animals and boats. Much of the basic technology that supports transport on land and sea today had its inspiration in ancient times. Roman roads tied together that far-flung empire as it spread across the Mediterranean and to points east. Likewise, the Grand Canal of China permitted trade and communication between that empire's political center in the north and its agricultural belt in the center and south—knitting together by water, and thus stabilizing, a vast expanse of territory.

Indeed water is where transportation began. Though today we think of ocean or river transport as among the slowest means of moving people or goods, it was once the fastest. Overland speeds by horse were historically slow—typically from five to ten miles per hour—on roads ranging from nonexistent to muddy. With the possible exception of a small handful of private turnpikes developed in Britain, travel by boat was almost universally easier and safer than travel by land until the nineteenth century.

Waterborne transport would lead Europeans to the New World, to colonies whose initial purpose was to produce commodities for trade—from furs to cotton to grain. City after city was founded on the coast, often along waterways that served as conduits for inland trade. In some places, new waterways were created to facilitate that trade and the manufacturing it ultimately supported—most notably in Britain, where an extensive system of canals was developed to serve the burgeoning Industrial Revolution, and in the United States, where the Erie Canal led to New York State's prominence in international trade for over a hundred years.

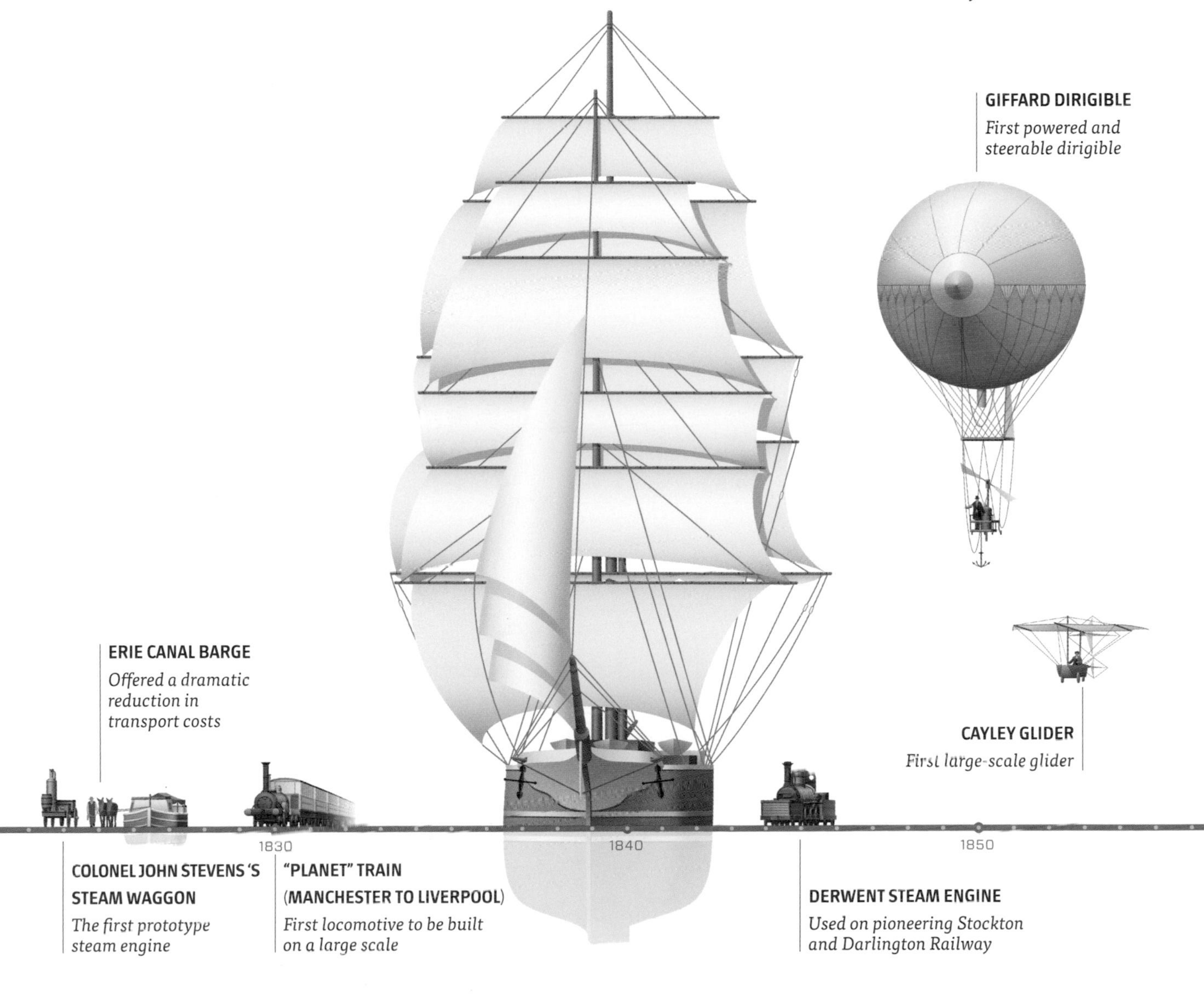

Not surprising, perhaps, the first manifestation of the Industrial Revolution in transportation came on the water—in the form of the steam engine. Though originally devised to remove water from mines, steam power soon found a profitable place on the high seas, where the predictability of travel by steamer was a major advantage over the vagaries of sailboat schedules. Within a relatively short period after its debut in the early nineteenth century, steam power would all but eclipse wind power in maritime trade—permanently transforming the nature of moving goods and people by water.

Railway technology embraced steam at roughly the same time. The first steam-powered train, conceived to haul coal, appeared in 1825 in Britain. Rail transport quickly proved far more efficient than road, due to the poor condition of roads and to the fact that trains could take heavier loads. The invention of the railroad also dramatically lowered long-distance travel times: journey time across the United States dropped from several months to one week with the opening of a transcontinental rail line in the 1860s. Similar efficiencies were associated with new transcontinental trains across Siberia and Canada.

During the second half of the twentieth century, experiments with rail technology moved into the city. Some of these experiments involved trains at grade level, running along tracks in the street. Others involved trains on overhead trestles, referred to in many cities of the United States as "the el"—short for "the elevated." Both relied on steam power—not necessarily the quietest of technologies, but an improvement from a sanitary perspective over the horse-drawn carriages that had dominated urban transport to that point.

A number of cities around the world began experimenting with steam-powered and electric trains underground, both for inner-city portions of commuter rides from the suburbs and for point-to-point transit within the city itself. These early subway lines—in places like London, New York, and Paris—would serve as the foundation for what would later become very sophisticated and extensive subway systems.

But no technology would revolutionize passenger transportation more than the internal combustion engine, which marked the shift from coal to oil-based energy production and would give rise directly to the automobile. Early experiments with the technology took place throughout the late nineteenth century, but it was not until 1888 that a motorcar prototype substantial enough to be put into commercial production

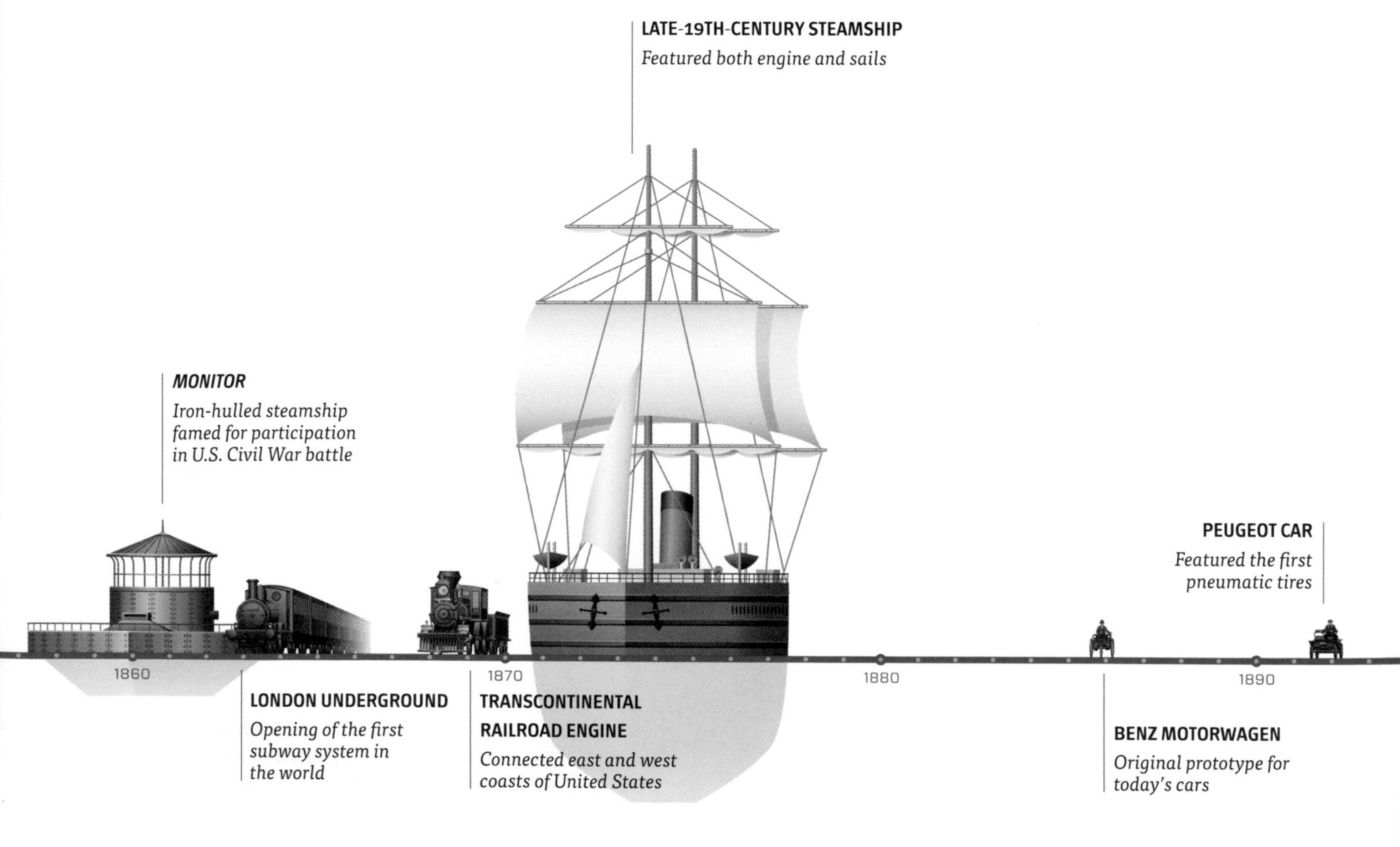

was developed. While much of the credit for the development of the car goes to German engineers like Karl Benz, it was an American—Henry Ford—who figured out how to produce it cheaply enough for the masses.

The combustion engine would also revolutionize cargo movement, giving rise to the trucking industry and making land-based transportation across large distances competitive with water and rail carriage. To this day, these three modes of carrying cargo—by truck, water, and rail—compete with one another for the lion's share of the world's trade—with many intermodal journeys starting with one mode of travel and finishing with another.

One more mode of transport would enter the competition for passengers and cargo in the first half of the twentieth century: air. Internal combustion technology, as the Wright brothers knew, could power engines in the air as well as on the ground. By the second decade of the twentieth century, the technology to reliably support human flight had been established. The earliest airports opened in the 1920s, initially for military purposes and often in coastal areas. Many were soon converted to civilian use as demand for recreational flight and airline capacity grew.

While the car had brought the freedom to move freely and in an unscheduled manner between different points, the airplane provided the ability to travel to places heretofore unreachable in less than a day's travel. At 600 miles per hour, roughly 60 to 100 times the speed of a horse-pulled cart, trips that once took weeks could be done by air in a day—knitting the world together in ways all but unthinkable fifty years earlier.

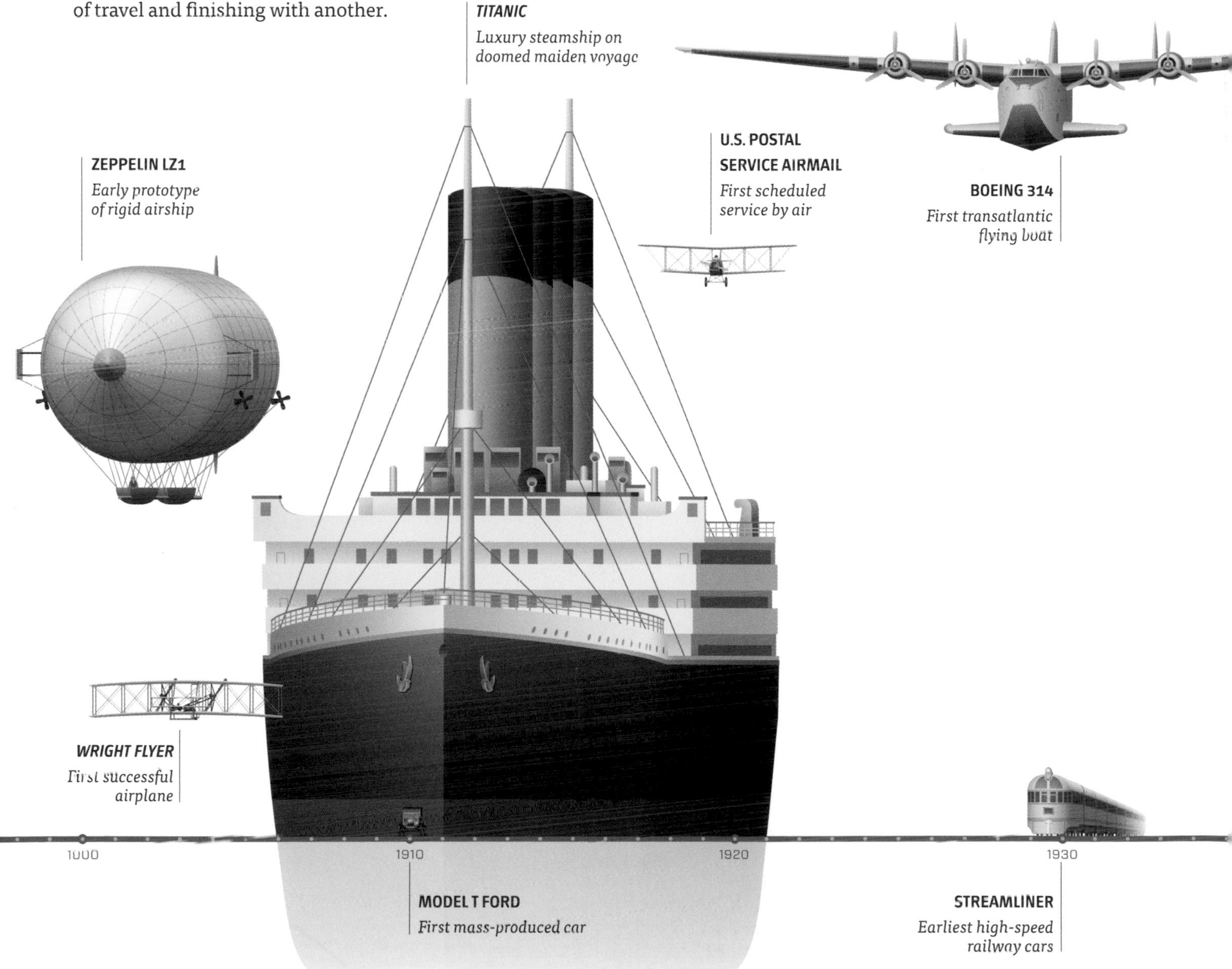

Transportation and trade would be redefined once again before the end of the twentieth century—by the maritime container. Until its commercialization in the 1960s and 1970s, cargo moving on and off ships was handled piecemeal—by longshore gangs assisted by hooks and pulleys. Malcom McLean's invention of a cargo container that could be loaded once—and then seamlessly moved from truck to ship and back again—led to a precipitous drop in the price of maritime trade. With shipping costs down to as little as a twentieth of their previous level, markets for international trade in commodities once not worth sending by ship emerged almost overnight—leading to an explosion in maritime trade.

Perhaps the last major innovation to shape the world of transport in the twentieth century was the rocket. The method of channeling explosive force devised largely by Robert Goddard allowed humanity to "escape its earthly bounds," and journey to places that

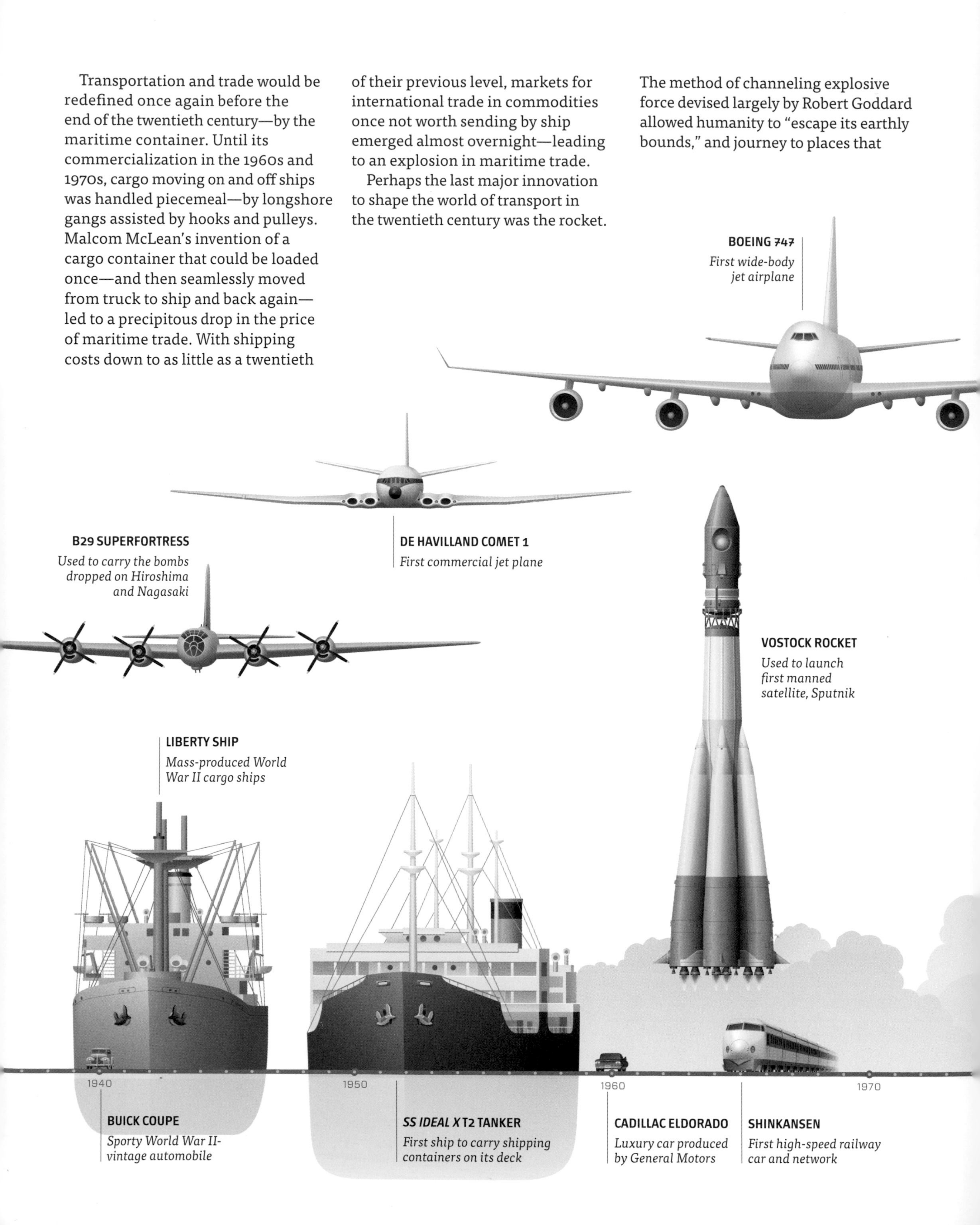

early astronomers could hardly have dreamed of reaching. While only a tiny number of humans have been able to take advantage of rocket technology to travel into space, a large number of cargoes have gone there—including animals, plants, industrial and medical products, and a wide range of telecommunication equipment.

The twenty-first century holds the promise of further changes in transportation—almost certainly with respect to automation and very likely in the fuels we rely on for travel. And although the race to the moon is now over, the goal of improving life on earth through space exploration is likely to remain. Each rocket launched to serve the space station now orbiting the earth or to visit other planets paves the way for more diverse activities in places once beyond the limits of man's imagination—and for faster, safer, and cheaper ways of getting there.

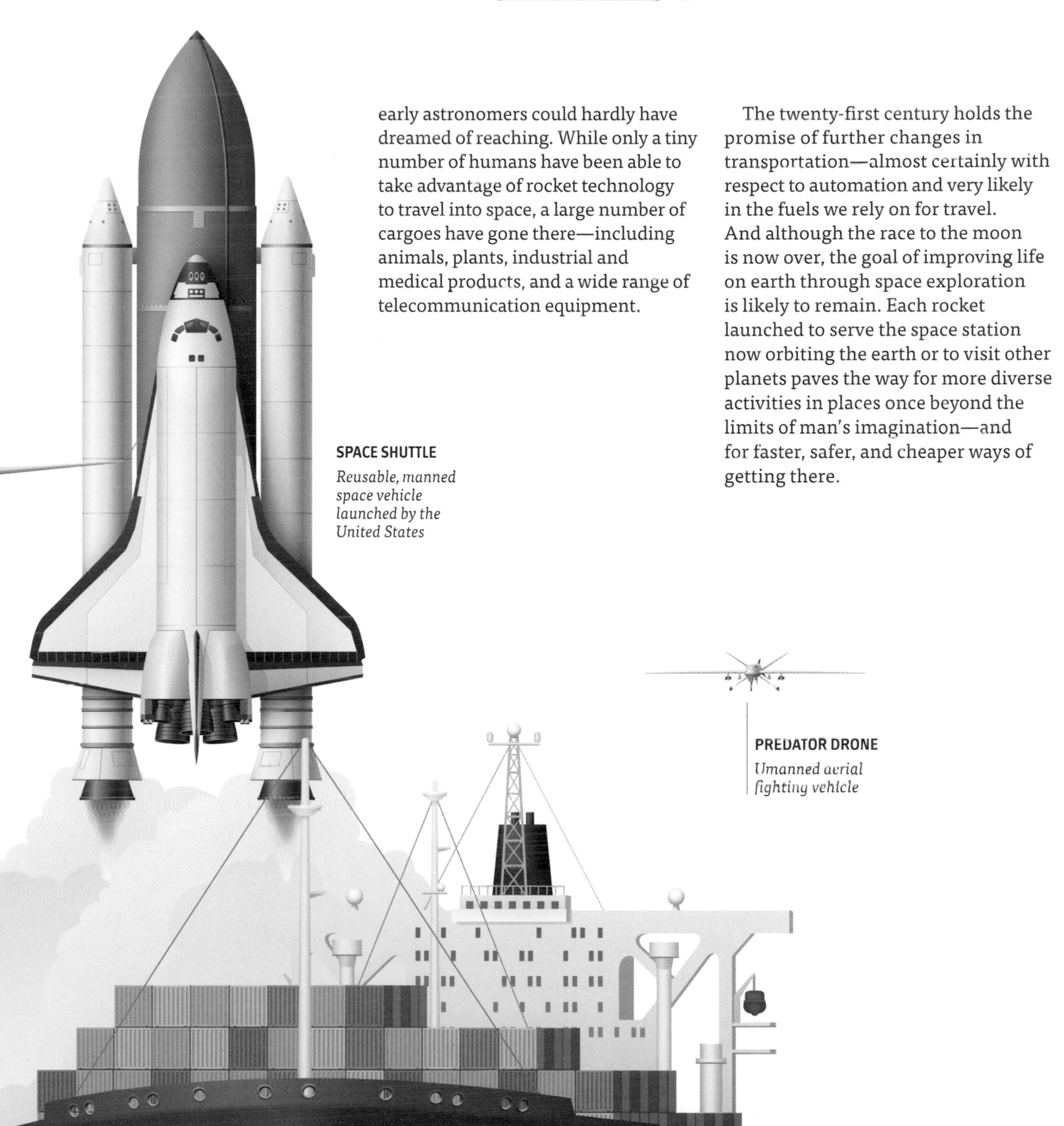

SEA

LIFE AT SEA

Man's fascination with the sea dates as far back as recorded history, and likely further. Though the oldest existing relic of a seaworthy boat was reputedly constructed around 8,000 B.C., historians have speculated that seaworthy boats might have existed tens of thousands of years earlier; indeed, there is no other explanation for the early population of Australia.

Primitive boats have existed for as long as people have lived on water. During the Stone Age, dugout canoes were a common method of travel along the coast and a useful way to supplement fishing on the shore. Inland, rivers such as Asia's Indus gave birth to the science of navigation over 5,000 years ago. Vikings, Egyptians, Polynesians, and other societies were renowned for their skill in constructing ships—some extremely large, for use in trade and commerce, and some much sleeker, for military purposes.

But the sea and navigation really came into their own in the fifteenth century, during what is often referred to as the Age of Sail or the Age of Discovery. Chinese junks, developed during the twelfth to fifteenth centuries, had proved the value of multiple masts in overseas trade. Over the next two centuries, multimasted ships became the vehicles for vast empire-building exercises that stretched from Europe to Asia and to the new continent on the western shore of the Atlantic.

Sail technology, which previously had evolved little, developed rapidly during this period. The small dhows, or simple rigged sailing vessels, that had been used for localized travel and trade gave way to the caravel, or carrack, sailing ship of Columbus' era—memorialized by his Nina, Pinta, and Santa Maria. Multiple sails and masts offered greater control and greater speed. The ships remained wooden, however, and thus susceptible to damage in rough weather.

The first iron vessels appeared in the late eighteenth century, primarily as experiments. By the first half of the nineteenth century, commercial and military sail-based vessels were becoming ironclad and more durable. However, the new steam technology would soon make commercial sailing a thing of the past. Not only were captains able to control, almost precisely, the direction in which their ship moved but they were also able to adhere to a schedule for the first time in history. With some famous exceptions, most notably the RMS *Titanic*, wind and weather no longer determined the timing or fate of ocean vessels.

The relative regularity and predictability that the steam engine offered prompted an explosion in maritime transport—initially for trade purposes but eventually for people as well. The great waves of immigration that fueled the growth of the United States in the nineteenth century relied heavily on the regular steamer services that called at a variety of European ports. These journeys also gave rise to much higher-class travel, for recreation and business, between Europe and the eastern seaboard of the United States.

With the invention of the internal combustion engine and the gas turbine, steam engines in ships began to disappear—and faster, larger, and more streamlined vessels would characterize twentieth-century travel by water. So too would a variety of purpose-built boats, including nuclear-powered submarines and icebreakers, aircraft carriers, hydrofoils, and hovercraft—to name just a few.

While the shape, size, and technologies associated with ships have changed significantly over the last several centuries, many of the core concepts that explain how ships work have changed little over time: for example, why boats float, how rudders and propellers work, how the shape of the hull affects the speed and stability of a boat. Each of these factors has an enormous influence on the safety and performance of almost any boat—and thus on life at sea.

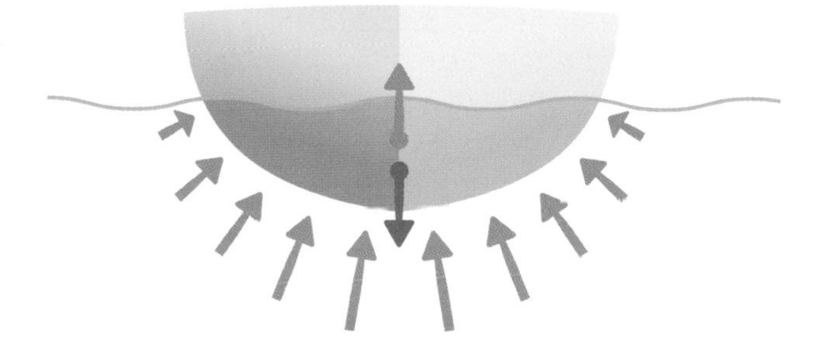

BUOYANCY

What makes something float? The "eureka" moment occurred when the ancient Greek mathematician Archimedes discovered that when an object (a gold crown in his case) is immersed, it will displace its own weight of fluid. In other words, when a boat displaces a weight of water (the buoyant force) equal to its own weight it floats. The boat is said to be positively buoyant if there is additional or reserve volume above the waterline. If the buoyant force is less than the object's weight (negatively buoyant), the object will drop lower in the water until enough reserve volume is submerged to reach a floating equilibrium or until it reaches bottom.

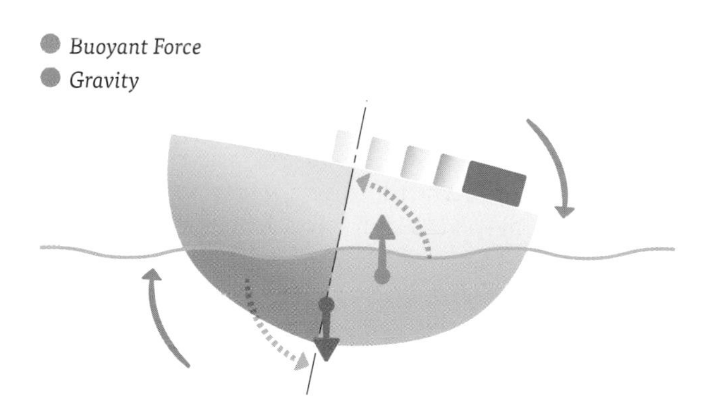

STABILITY

It is not enough to make a ship float—it must also be able to stay upright. The combined center of gravity of the vessel and all its contents should be as low as possible. Ships will often take on ballast weight in the form of lead or other metals to lower their center of gravity. Hull shape also factors heavily into stability—slender hull forms (like canoes and rowing sculls) tend to be less stable than ships with generous beams. Ancient shipbuilders improved stability by trial and error, but today's ship designers use sophisticated mathematics to predict static and dynamic stability under a variety of sea conditions.

SEA TRAVEL INVENTIONS

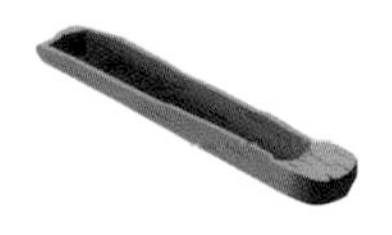

8000–7500 B.C.
The first evidence of manmade dugout canoes found in the Netherlands

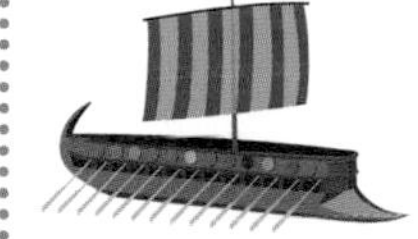

4000–3000 B.C.
Egyptians add sails to their oar-driven boats

1100 B.C.
Phoenicians build the first known warships

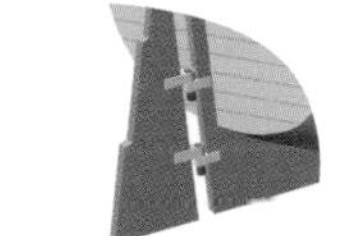

1st Century A.D.
The rudder is invented in China

12th–15th Centuries
Chinese junks introduce multiple masts and watertight bulkheads

1620
The first submarine is built in England

1700s
Various concepts for the steamship are introduced in England, France, and the United States

1775
The Turtle, the first military submarine, is built in the United States

1787
John Wilkinson of England builds a barge constructed of iron

1820s–1830s
The screw propeller is introduced in Europe

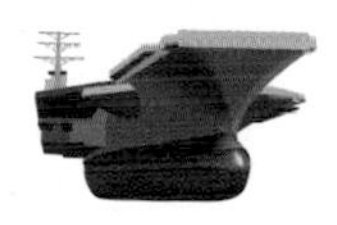

1910
The bulbous bow is invented by U.S. Rear Admiral David Taylor

BOTTOMS UP

Many things determine the speed of a boat in the water, including the power of its engine and the condition of the surrounding seas. But perhaps the most important factor in determining how a boat moves through the water is the shape of its hull.

Hulls are designed first and foremost to prevent a boat from overturning while allowing it simultaneously to move smoothly through the water. They come in many different shapes and sizes, based on the waters they frequent and the relative importance of stability, speed, and performance. For example, the slender hull of a rowing scull is designed to optimize speed but offers very little stability and is unreliable outside of calm rivers and lakes. By connecting two of these slender hulls together with a platform above the water, a catamaran is created—a vessel that is much more stable and nearly as fast, but still not well suited for the open ocean.

Ship designers must balance these trade-offs when matching the hull shape and design with a ship's mission. United States Coast Guard motor lifeboats are designed

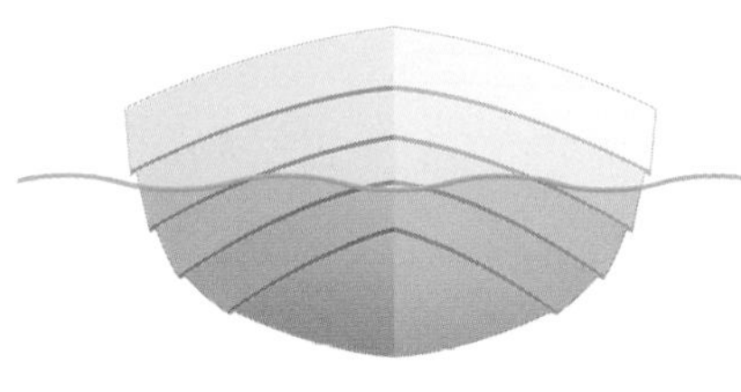

ROUND BOTTOM

Probably the most widely used hull shape. A rounded bilge provides for ample internal volume while providing good seakeeping and stability.

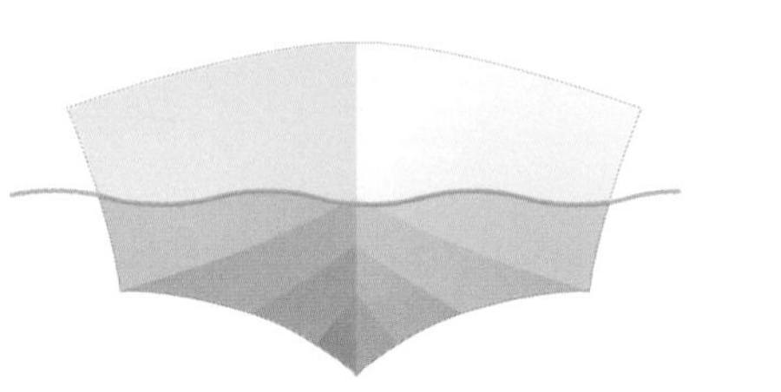

DEEP-V

A deep-V, or chined, hull facilitates planing at higher speeds as the boat lifts out of the water. This shape also is relatively easy to construct.

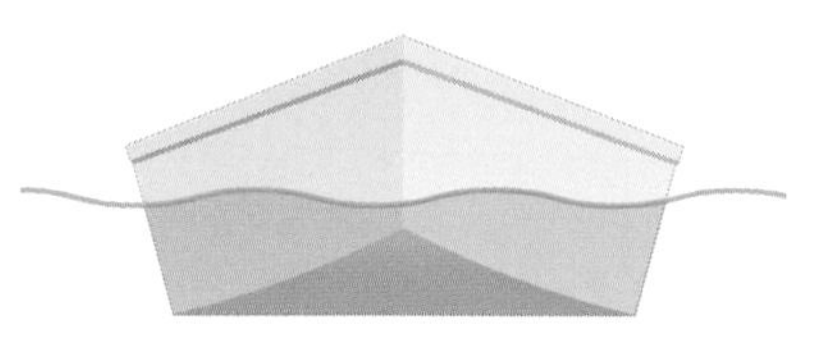

FLAT BOTTOM

Flat-bottomed boats allow operation in shallow waters and range in size from small fishing boats to large amphibious warships.

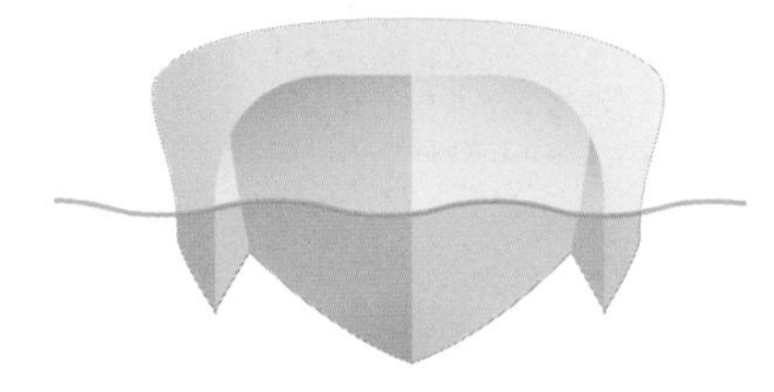

CATHEDRAL HULL

A cathedral hull is a hybrid between a deep-V hull and a trimaran (three hulls) and is principally used in pleasure boats. When inverted, it is said to look like a cathedral.

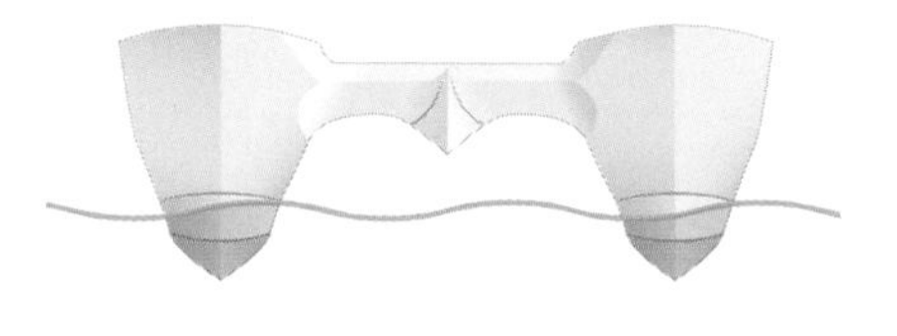

CATAMARAN

Actually two hulls connected by a structure above the waterplane, catamarans are known to be fast and stable (as long as both hulls stay submerged).

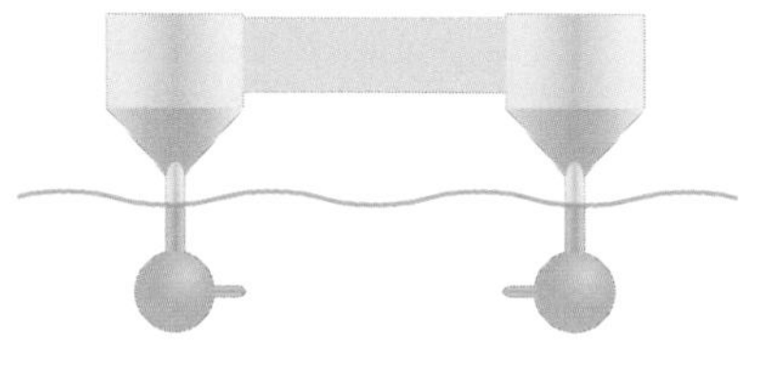

SWATH

An acronym for small waterplane area twin hull, SWATHs are very stable ships in rough seas and uses range from research vessels to casino boats.

BULBOUS BOW

Many large ships, including merchant and naval vessels, employ a strange-looking appendage at the bow just below the waterline. At first glance, this bulbous bow would seem to increase the ship's resistance—and indeed it does at low speeds. But at high speeds, the bulb actually improves performance by creating a destructive wave that helps offset wave-making resistance.

Not all bulbous bows look alike. The shape and location of the bulb is tuned, or specified more precisely, during tow-tank testing to optimize its performance for the ship's design cruising speed. Thanks to the savings from the bulbous bow, a ship may employ a smaller engine or burn less fuel (up to 15 percent less) to achieve top speeds.

to rescue stranded people in the roughest weather and surf and can roll a complete 360 degrees without capsizing. The trade-off for this extreme stability is an uncomfortable, or "snappy," ride. In contrast, cruise ships must be designed for passenger comfort and employ stabilizers to reduce dynamic accelerations but cannot roll to the same extent without capsizing.

Below the water surface, hulls are designed to minimize resistance and thus reduce the power required to move the vessel at a desired speed. The resistance of a ship is composed of two main components, which are largely a function of its speed in moving through the water. At low speeds, frictional (or viscous) resistance dominates and is largely influenced by hull shape and smoothness. At higher speeds, wave-making resistance takes over as more and more energy is used to plow the ship through the water, creating waves. During design, extensive modeling and tank testing are conducted to optimize hull resistance and ensure that enough power is installed to overcome this resistance and reach the boat's design speed.

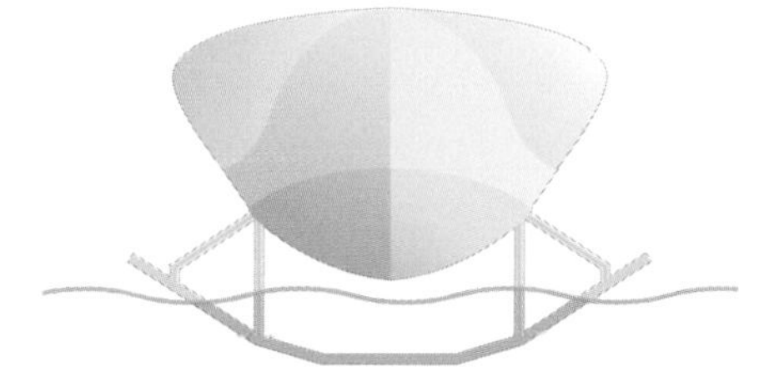

HYDROFOIL

Some vessel designs aim to reduce high-speed wave-making resistance by lifting the hull completely out of the water. Hydrofoils feature a small "wing," or foil, protruding on an angle from the boat's hull. As the boat gains speed, the movement of the water over the wing lifts the boat's hull out of the water—effectively eliminating wave-making resistance and providing a more efficient use of the boat's power at higher speeds.

The idea of the hydrofoil fascinated the engineers of the early nineteenth century. Alexander Graham Bell, father of the telephone, spent considerable time in the Italian Lake District developing his own version of an experimental hydrofoil. Bell's 1919 hydrofoil set a world marine speed record (71 mph or 114 km/hr) that stood for two decades, setting the stage for the commercial hydrofoil service that continues to serve the lake communities of northern Italy.

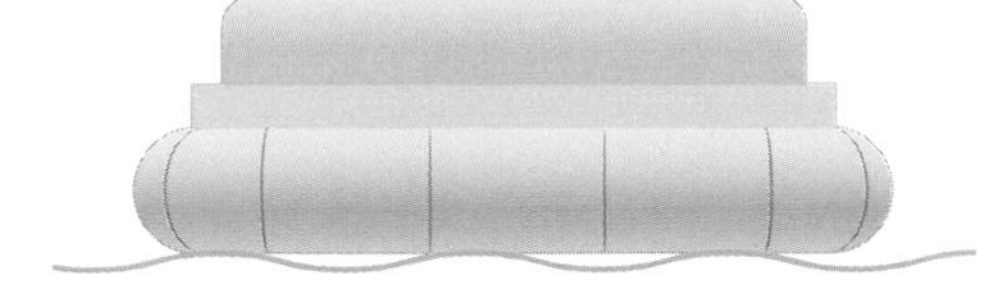

HOVERCRAFT

Hovercraft, which literally hover, or float, above the water on a layer of air, appeared in the mid-twentieth century in Great Britain as an alternative means of reducing water resistance as a boat moves forward. Powerful engines produce a cushion of air that the hovercraft rides along as it travels over the water. A "skirt" sitting under the vessel traps the air under the craft, creating a cushion of air above the water.

Originally dismissed as a plane by the British Navy and as a boat by the British Air Force, the hovercraft nevertheless proved a successful means of crossing the English Channel for decades—until competition from the Channel Tunnel put an end to cross-Channel hovercraft operations in 2000. Hovercrafts continue to be employed elsewhere, including military applications for amphibious landings and operations.

SHIPSHAPE

At any given point in time, thousands of ships are plying the world's oceans, coasts, and inland waters. These range from military vessels carrying troops in war or peace to ferry boats shuttling back and forth from one port to another to container and bulk ships moving cargo in international trade. Less heralded are the scores of workboats designed to support them: tugboats, coastal patrol vessels, and repair ships, among others. Nearly all of these vessels are somehow involved in supporting the transportation of people or goods.

This mission is central to understanding why there are so many types and sizes of ships. A ship's shape and design are largely driven by what it carries: container ships have deep holds and hatch covers to facilitate the stacking of

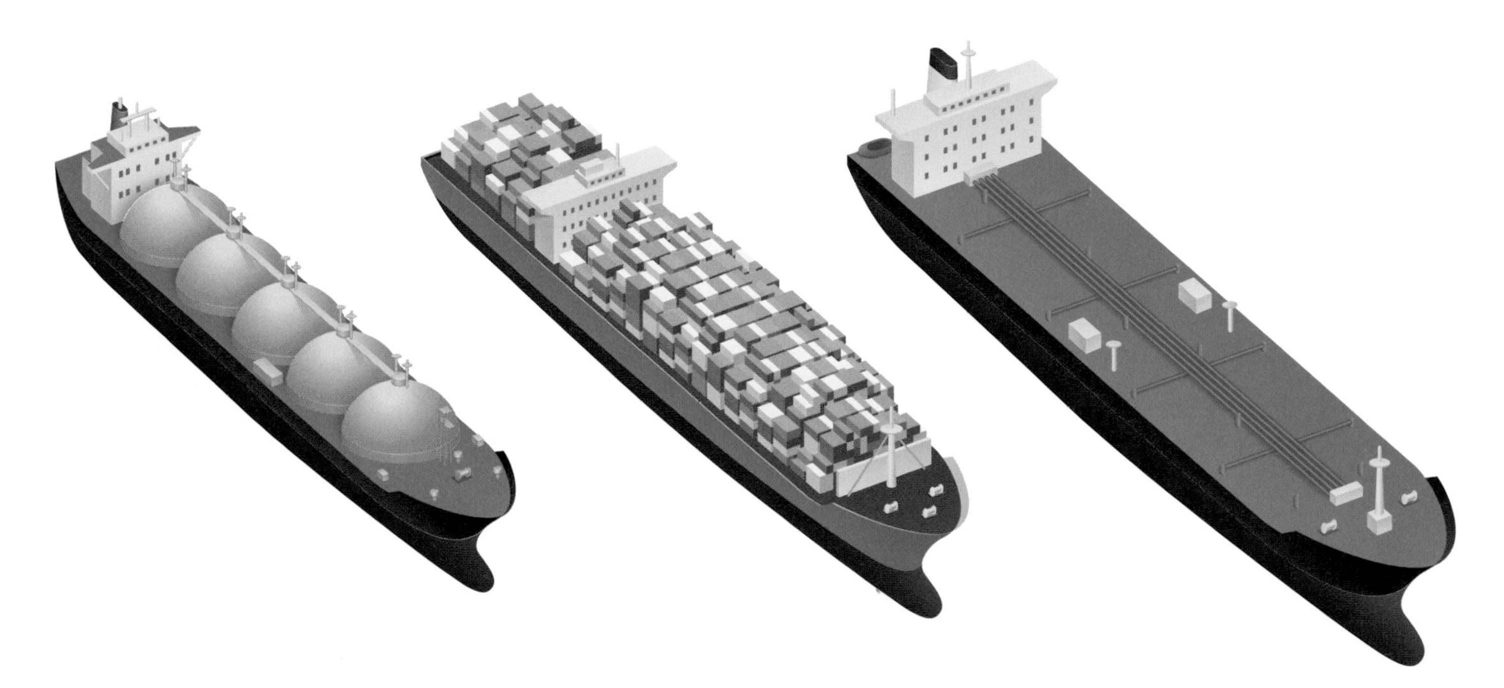

LNG CARRIER

Designed to carry liquefied natural gas, these ships may be over 1,100 feet long (335 m) and feature 4 to 6 distinctive spherical-shaped tanks.

CONTAINER SHIP

Capable of carrying up to 14,500 twenty-foot containers, the largest container ships are over 1,200 feet long (365 m).

OIL TANKER

Supertankers are capable of carrying over 2 million barrels of oil and can be as long as the world's tallest skyscrapers.

BULK CARRIER

Used to carry bulk commodities like iron ore and coal, these mammoth ships feature expansive deep holds covered by large watertight hatches.

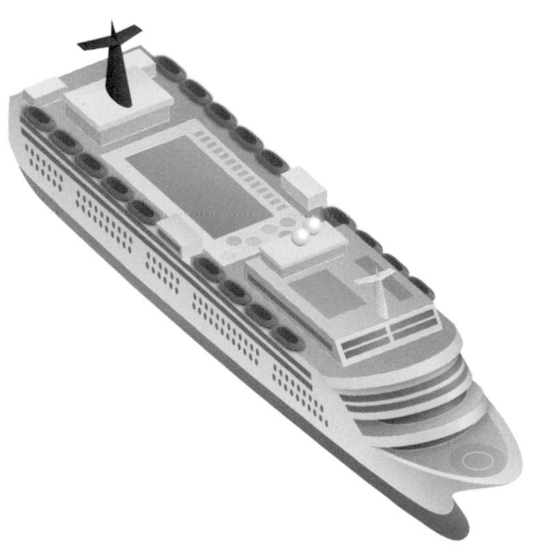

CRUISE SHIP

At nearly 1,200 feet (365 m), the world's largest cruise ships are floating cities able to carry over 6,000 passengers.

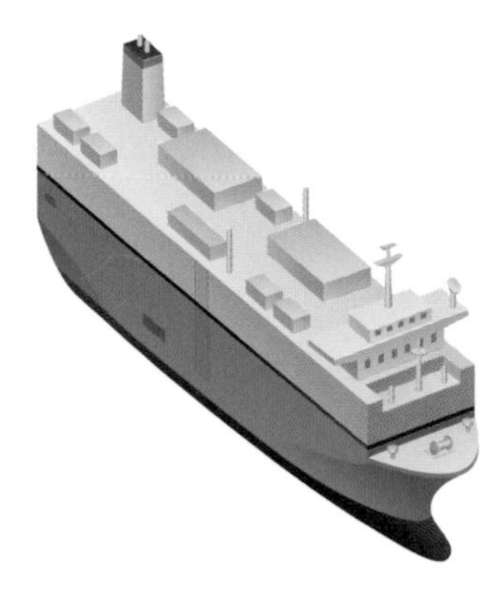

RO-RO

Roll-on/roll-off vessels are used to carry vehicles able to be loaded on their own wheels. These ships feature bow and/or stern ramps for ease in loading/unloading.

containers; ro-ro (roll-on/roll-off) ships are shaped like a box to accommodate the movement of wheeled vehicles; cruise ships rise high off the sea to maximize accommodation and views, and ferry boats are designed to move large numbers of people on and off swiftly.

The missions of naval ships have evolved over the centuries, but two are consistent—protection of trade and protection of people and property. Form follows function in warship design—sleek destroyers and submarines glide seamlessly through the ocean carrying their payload of missiles and torpedoes while aircraft carriers boast broad decks for airplane takeoffs and landings and amphibious ships feature flat bottoms for launching troops and equipment close to shore.

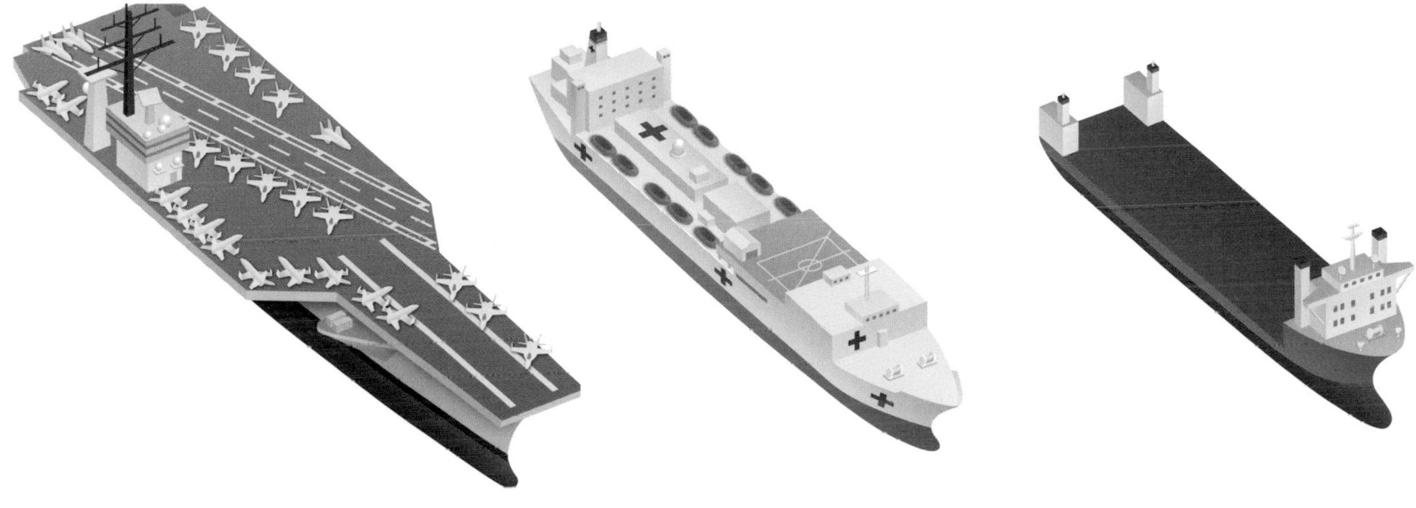

AIRCRAFT CARRIER

Aircraft carriers serve as the ocean base for dozens of aircraft and thousands of crew.

HOSPITAL SHIP

The floating hospitals support military and peacetime missions. Attacking a hospital ship is a war crime and all hospital ships must be clearly marked with a red cross or red crescent.

HEAVY LIFT

Heavy-lift ships are designed to move large loads like oil platforms and other ships. Many are semisubmersible, which allows the ship to float its load on and off.

COMMERCIAL FISHING

Oceangoing commercial fishing vessels feature extensive onboard facilities for processing and freezing their catch.

CORVETTE

Smaller surface combatants like the Swedish Visby class feature stealth technology to decrease their radar signatures to other combatants and missiles.

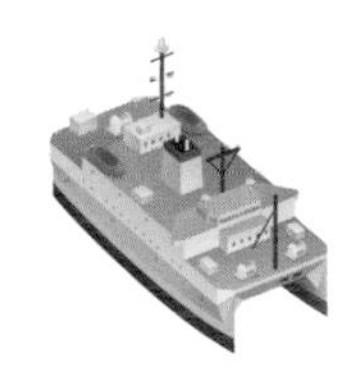

SURVEILLANCE/RESEARCH

Smaller military surveillance and civil research vessels utilize towed sonar systems for intelligence or scientific data collection.

SUBMARINE

At nearly 400 ft (120 m) long, modern nuclear-powered fast-attack submarines are built for speed and stealth.

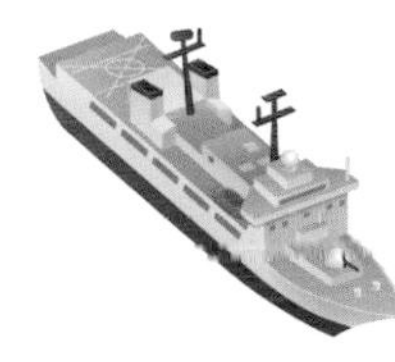

COAST GUARD CUTTER

The largest cutters used by the U.S. Coast Guard measure 400 feet in length (115 m) and enforce laws and treaties and perform search and rescue on the high seas.

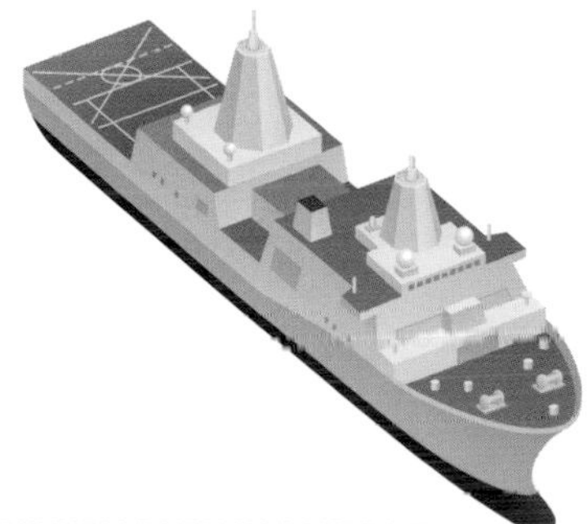

AMPHIBIOUS TRANSPORT DOCK

Measuring nearly 700 feet (215 m) long, these amphibious warships carry helicopters and air-cushioned landing craft to deliver up to 800 marines ashore.

SUBMARINE BASICS

Some of the most intriguing maritime vessels travel under, rather than over, the water. Just how they do that is one of the marvels of maritime engineering—fascinating mankind for hundreds of years. Indeed, the desire to go under the sea existed long before the Industrial Revolution provided engines to push submarines forward: the earliest submarines were lumbering, barrel-shaped craft powered by humans with oars or pedals.

Jules Verne's *Twenty Thousand Leagues Under the Sea*, written in 1870, galvanized interest in submarines at a time when the idea of one was far more advanced than it was in practice. Military uses proved central to the technological development of the submarine in the late nineteenth century: the ignoble premise of being able to attack enemy ships from an invisible position under the water would transform naval warfare.

Berthing and eating (also known as messing) compartments are located in the forward half of the submarine. Some submarines do not have enough beds for all personnel and some junior sailors are forced to "hot rack" (take turns using a bed).

The forward-most compartment is the sonar dome that contains hull-mounted passive arrays to listen for other ships and submarines.

The stern of the submarine contains the engine and machine rooms and can take up over half the overall length. Essential equipment includes propulsion systems, electrical-generating equipment, heating and ventilation, and atmosphere-control machinery.

Whether nuclear or diesel powered, a battery is an essential part of a submarine, providing a critical source of backup power in emergencies or while submerged.

STEERING A SUB

Submarines typically use three types of control surfaces to control their direction, attitude, and ascent/descent.

- *The stern planes, horizontally oriented just forward of the propeller, are used to control the angle of attitude of the submarine, allowing it to climb bow up and down like a plane.*
- *The rudder, also located forward of the propeller, controls the ship's direction. At higher speeds, the submarine can bank and the rudder may begin to act like a stern plane.*
- *The fair-water planes (on the "sail" that protrudes from the hull) or the retractable bow planes (near the bow) provide lift and help the submarine rise or dive without changing angle.*

In order to submerge, a submarine must intentionally sink itself. To do this, submarines rely on ballast tanks located between the vessel's inner and outer hulls. Adding water to these tanks creates negative buoyancy—that is, the weight of the submarine becomes greater than the buoyant force pushing it upward and so it sinks. Likewise, the expulsion of ballast water from these tanks reduces the submarine's weight, allowing it to rise (positive buoyancy).

While submerged, a submarine strives to maintain an equilibrium known as neutral buoyancy, in which its weight equals its buoyant force. Should it want to go deeper, it will undertake a dynamic dive using its speed and control surfaces, or will hydroplane to move through the water the way an aircraft uses wings and flaps. Submarines also rely on trim tanks located near the bow and stern to adjust the attitude of angle of the ship.

HOW SUBMERGING WORKS

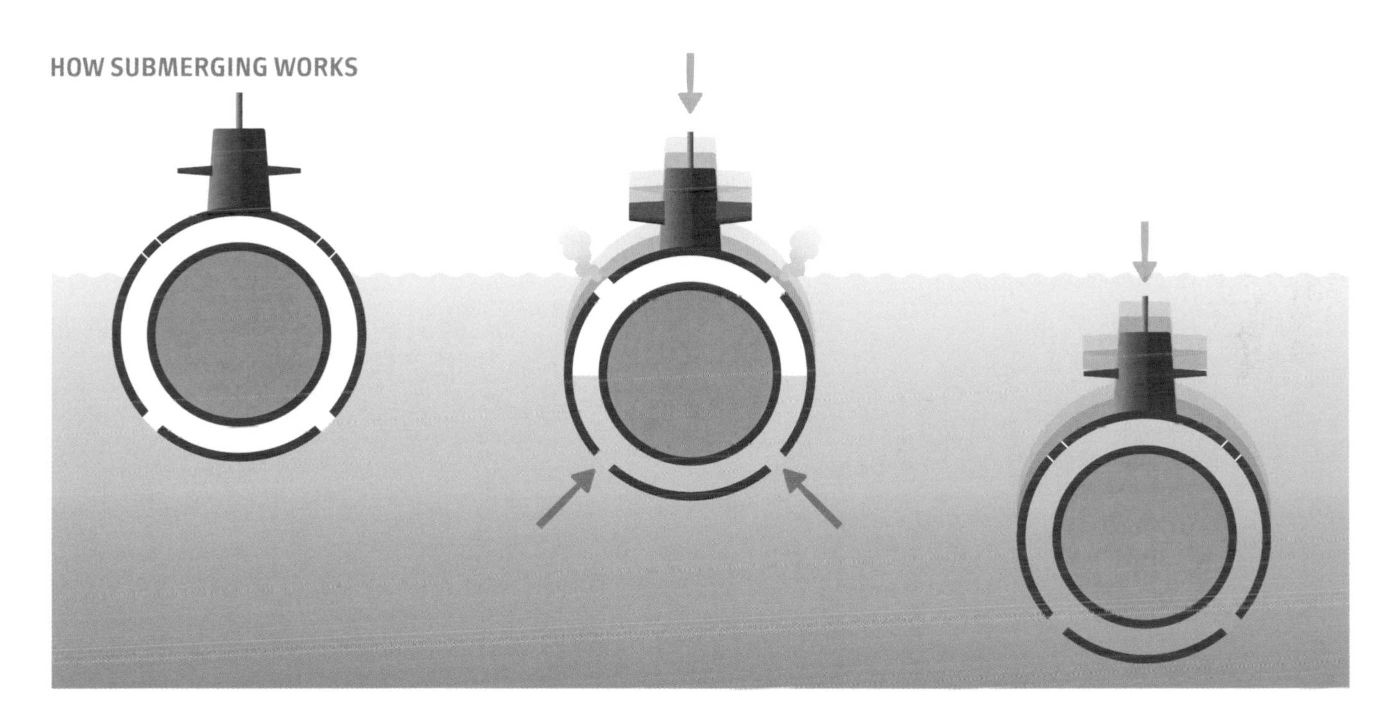

When the submarine is ready to submerge, the main ballast tank vents located on the top of the submarine are opened. High-pressure air is vented through the openings and water is allowed to enter the ballast tanks through grates near the bottom.

As the ballast tanks fill, the ship gets heavier and becomes neutrally or negatively buoyant and begins to sink. Once the stern planes and fair-water (bow) planes are below the water, they are used to drive the ship below the waves.

Once at a safe depth, the main ballast tank vents are cycled open and shut to ensure all the air has escaped. Seawater is pumped to and from onboard tanks to achieve neutral buoyancy.

EMERGENCY BLOW

As most moviegoers and naval buffs know, submarines can stay under the water for extended periods of time—months on end in some cases. But in case of emergencies, especially flooding, they may need to get to the surface quickly.

To do that, a submarine will "blow the tanks"—which involves rapidly clearing all the water from the ballast tanks aboard the vessel. High-pressure air is injected into the tanks to expel all the ballast water while the submarine's planes and propeller are used to help lift the craft toward the surface. Once on the surface, a low-pressure blower system is typically used to finish blowing any remaining ballast water and restore positive buoyancy.

Blowing the tanks can be a dangerous and not always successful operation. In 1963, the nuclear-powered submarine U.S.S. *Thresher* suffered flooding in the engine room on its initial dive following a maintenance overhaul. The ship's crew performed an emergency blow to get to the surface, but the rapid depressurization of the high-pressure air tanks caused moisture in their air to freeze and block the air lines to the main ballast tanks. With seawater rushing into the ship, the crew struggled to rise to the surface—but failed. Decades later, divers found that the ship had imploded into thousands of pieces under the crushing pressure of the ocean.

FULL STEAM AHEAD

Like all forms of transport, travel over water requires turning power into forward motion. Power can come from a number of sources: people (kayaks and canoes), nature (sailboats), or fuel. But the power source needs to be harnessed and channeled in a way that uses it most efficiently—by designing hulls, sails, or propellers that smoothly transmit that force into forward motion.

Nearly all ships engaged in commercial activity today rely on diesel reciprocating engines or gas turbines, which transmit power to a vessel's propellers via a propeller shaft—either directly or via some form of gearbox. These engines are classified in a variety of ways—by their power, operating cycle (two-stroke or four-stroke), layout of their pistons, or speed in revolutions per minute. Somewhat counterintuitively, the larger the vessel, the lower the speed of the engine: today's large-ship propellers are at their most efficient in the range of a slow-speed diesel engine (somewhere between 60 and 200 revolutions per minute).

Large ships have multiple engines for redundancy and operational flexibility. The power from these engines might be transmitted to a single propeller shaft through complicated gearing or electric motors or might be directly geared to it. At higher speeds, ships may employ reduction gears that reduce the speed of the rapidly spinning engine to slower rotational speeds more efficient for the ship's propellers.

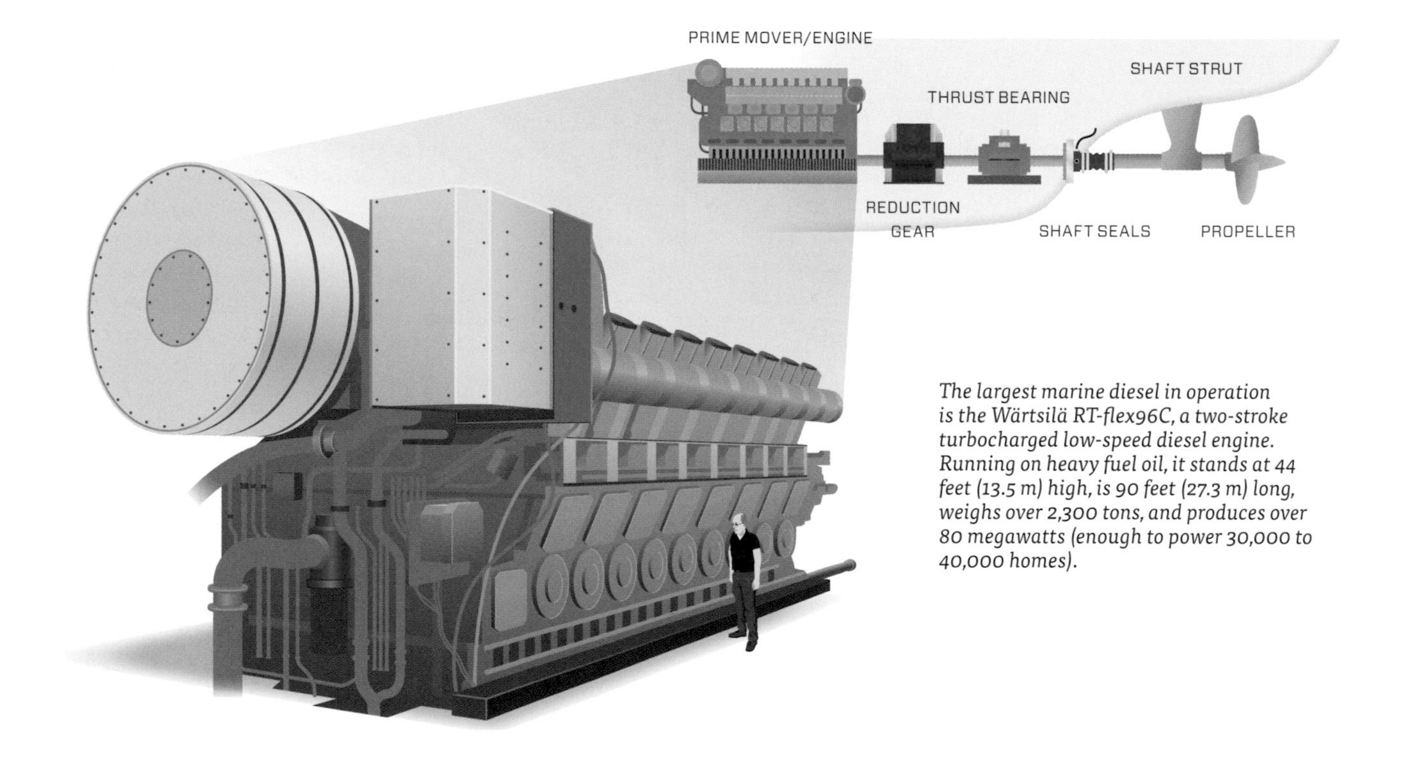

The largest marine diesel in operation is the Wärtsilä RT-flex96C, a two-stroke turbocharged low-speed diesel engine. Running on heavy fuel oil, it stands at 44 feet (13.5 m) high, is 90 feet (27.3 m) long, weighs over 2,300 tons, and produces over 80 megawatts (enough to power 30,000 to 40,000 homes).

NUKES AT SEA

Some oceangoing ships are powered by nuclear rather than fossil fuel. Heat from nuclear fission (splitting of atoms) produces tremendous amounts of energy in the form of heat that is used to create steam, which in turn drives turbines, providing power to the vessel. Vast amounts of power can be provided by relatively small amounts of enriched uranium fuel so that refueling is needed only after years at sea—making nuclear power particularly suitable for submarines. Nuclear fuel is also used to power the American Navy's fleet of aircraft carriers and the powerful fleet of Russian icebreakers that have opened up parts of the Arctic to something approaching year-round navigation.

PROPELLERS

VARIABLE PITCH

In the most common type, the pitch (like threads of a screw) varies from the blade root to the tip and results in greatly increased efficiency.

HIGHLY SKEWED

Skewing the blade to a scimitar shape helps reduce vibrations transmitted to the hull. These are commonly seen on cruise liners, ferries, and submarines.

SUPER-CAVITATING

Cavitation (the localized pressure-induced boiling of water on the blade) can destroy most propellers. These propellers, used by military and racing boats, use the cavitation to reduce friction.

AZIMUTH THRUSTER

Rather than being attached to a fixed longitudinal shaft, the azimuth thruster allows the propeller to turn in all directions to generate precise thrust.

CONTROLLABLE PITCH

Ideal for single-speed engines, the propeller blades are rotated along their longitudinal axis to change their pitch (and even direction of thrust).

VOITH SCHNEIDER

A series of vertical, rudderlike blades spin in circles. Precise thrust in any direction is computer generated by controlling the angle of each blade.

RUDDERS

The direction of forward movement in a ship is guided by a rudder, which, depending on the size and type of ship, might be hung from the stern or mounted inboard as an extension of the ship's keel or hull. A rudder, typically a flat sheet attached by hinges to allow free movement, works by redirecting the water past the hull to turn the ship in one direction or another. On smaller sailing ships, they might be attached to a tiller; on larger ones they are manipulated by a steering wheel.

- *Rudder force*
- *Centripetal hull force*

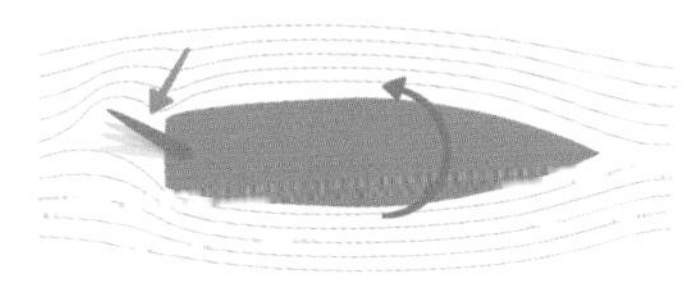

The rudder is turned to port. The force on the rudder makes the ship rotate counterclockwise.

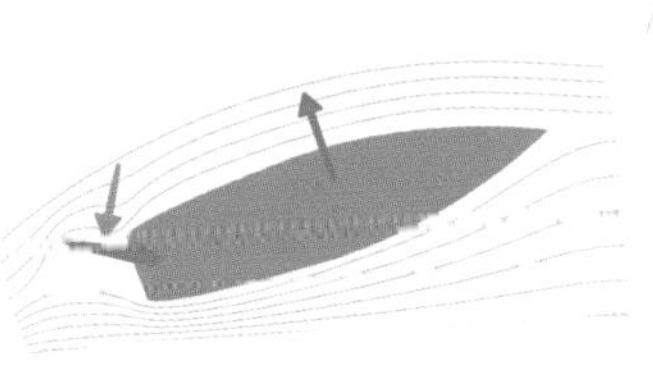

The rotated ship, moving along its original path, acts like a wing undergoing lift. A force is generated on the hull to turn to port.

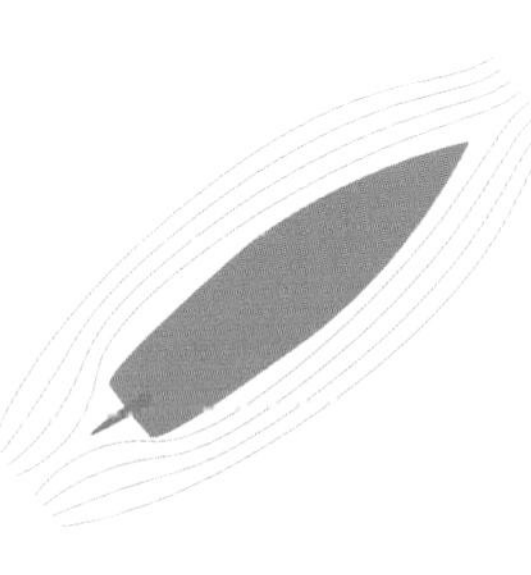

Once on its new course, all turning moments and forces return to equilibrium.

BOW THRUSTERS

While rudders are among the oldest maritime inventions still in use, they are increasingly being supplemented by modern technologies. Foremost among these are "bow thrusters" or "azimuth thrusters"—propulsion devices that can be mounted on a large ship to assist in docking maneuvers.

Until recently, the captain of a large ship has relied on powerful docking tugs to help push the bow and stern of the ship into place as it approaches the berth. Bow thrusters minimize or remove entirely the need for docking assistance as well as reliance on a traditional fixed-propeller and rudder system. Specialized drilling and surveying ships can link their thrusters to satellite navigational systems to maintain a precise fixed position on the earth.

Both fixed and retractable thrusters are in use today. Some rely on mechanical transmission, connected by a gear to an engine; others contain an engine within the pod itself and are called azipods. Typically, a ship with a thruster will carry a sign (a white cross in a red circle) above the waterline, identifying the thruster's location.

FUELING AND BUNKERING

Maritime trade and naval operations depend on regular infusions of fuel and provisions—sometimes in port and sometimes while the ship is at sea. Neither is nearly as straightforward as pulling into a service station for gas and a hamburger.

For commercial ships at anchor, or at a berth in port, fueling, or "bunkering," usually involves pipelines and pumps built into the ship and connected to the fuel tanks below. Once the tanks are emptied (to prevent mixing of different oils), a fuel barge pulls alongside the ship and safety checks are undertaken before the hose is connected to the ship's manifold and the manifold valve opened. As the tank is being filled, the temperature of the oil is monitored, and a sample is drawn to meet various local and international regulations.

But some ships—primarily military ones—must be fueled or provisioned while they are moving at sea. This process, known as connected or underway replenishment, can take a variety of forms. Fuel and provisions can be extended either vertically, from helicopters above, or horizontally, from ships steaming at the same speed alongside. In both cases, a wide variety of materials—fuel, food, ammunition, mail, equipment, and even new personnel—can be transferred to the ship while it is under way.

CONNECTED REFUELING

Once the delivery ship tensions the span wire, the refueling rig is hauled over. The fuel probe has a spring-loaded latching mechanism that holds it in the receiver mounted on the receiving ship.

Once the receiving ship pulls alongside, a messenger line is shot from the delivery ship. The receiving ship hauls in the messenger line and then the span wire that is attached to a pelican hook.

During the whole refueling process, the ships cruise along at 12 to 16 knots within 30 yards (30 m) of each other. Should a maneuvering problem develop, either captain can initiate an emergency breakaway—a rapid but orderly disconnection of the entire rig.

BREAKAWAY MUSIC

Connected refueling, in which ships travel side by side with a fuel line connecting them, is a complex undertaking and requires a great deal of precision: the two ships must match their speeds and course directions over an extended distance. So it's something of an achievement when completed—and often celebrated as such by the playing of what's known as breakaway music.

Naval captains use their discretion in selecting appropriate breakaway music to play over the ship's loudspeakers to mark the completion of a successful transfer of fuel, food, and mail. The choice might reflect the moment (e.g., "On the Road Again" by Willie Nelson) or the ship's name (the U.S.S. *Enterprise* has played the theme from *Star Trek*).

CREW

Keeping a ship moving at sea requires a range of skills and experience. In addition to a captain, ship crews feature a first, or chief, mate, with oversight of ship and crew safety, and a second mate, typically in charge of navigation. A chief engineer carries full responsibility for the engineering operations needed to sustain a voyage and a chief steward oversees the meals, cleaning, and provisioning activities necessary to support life at sea.

Crew sizes can vary greatly. On cargo ships, whose goal is to move as much from place to place quickly and profitably, crew sizes are kept as small as possible. Automation and mechanization have greatly reduced the need for people in navigation, cargo handling, and engineering operations—now leaving crews a fraction of the size they once were. The *Emma Maersk*, one of the world's largest container ships, carries 11,000 twenty-foot equivalent units (TEU) of cargo—but only requires 13 crew.

In contrast, the size of cruise ship crews has grown as ships have gotten larger and traditional ocean liners have given way to floating hotels. But the ratio of crew to passengers has not: *Titanic*'s 1,300 passengers were served by 890 crew. A century later, the largest cruise ship afloat—Royal Caribbean's *Allure of the Seas*, boasts a crew of 2,176 serving up to 5,400 passengers.

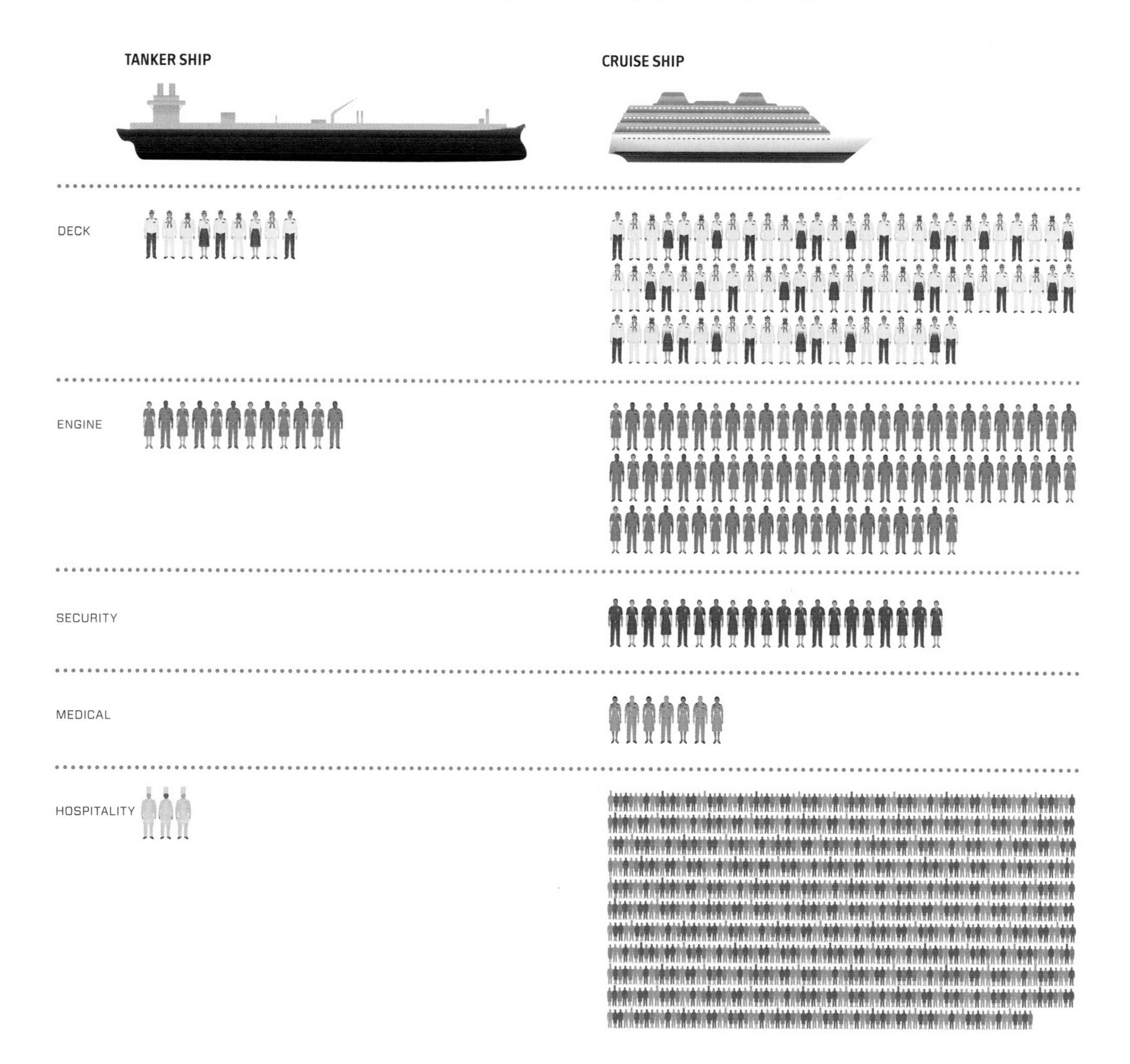

WATER AT SEA

Life at sea has always depended on a steady supply of potable water—historically for cooking and drinking, and today for washing and cooling machines as well. During the Age of Exploration, freshwater for drinking was carried on ships in barrels and topped off with rain caught in tarps during bad weather. Although some primitive forms of desalination occurred on board, an inadequate supply of water would put an end to a journey or require a call at port.

Ocean water contains significant amounts (concentrations) of salt dissolved into it. Saline concentrations are typically measured in parts per million (ppm) of weight, with typical ocean water having a concentration of about 35,00 ppm (or 3.5 percent). Drinking water generally requires a saline concentration of under 1,000 ppm—so a major filtering-out process must be undertaken to make seawater ready to drink. Not surprising, salty water requires more effort to desalinate than that with lower saline concentrations; areas of the Great Salt Lake in Utah, for example, have a saline concentration of up to 250,000 parts per million.

Desalination occurs in nature without prompting: water evaporates from oceans and lakes as it warms, condenses when it hits cooler air, and falls to earth as freshwater. Desalination at sea can likewise use heat to prompt

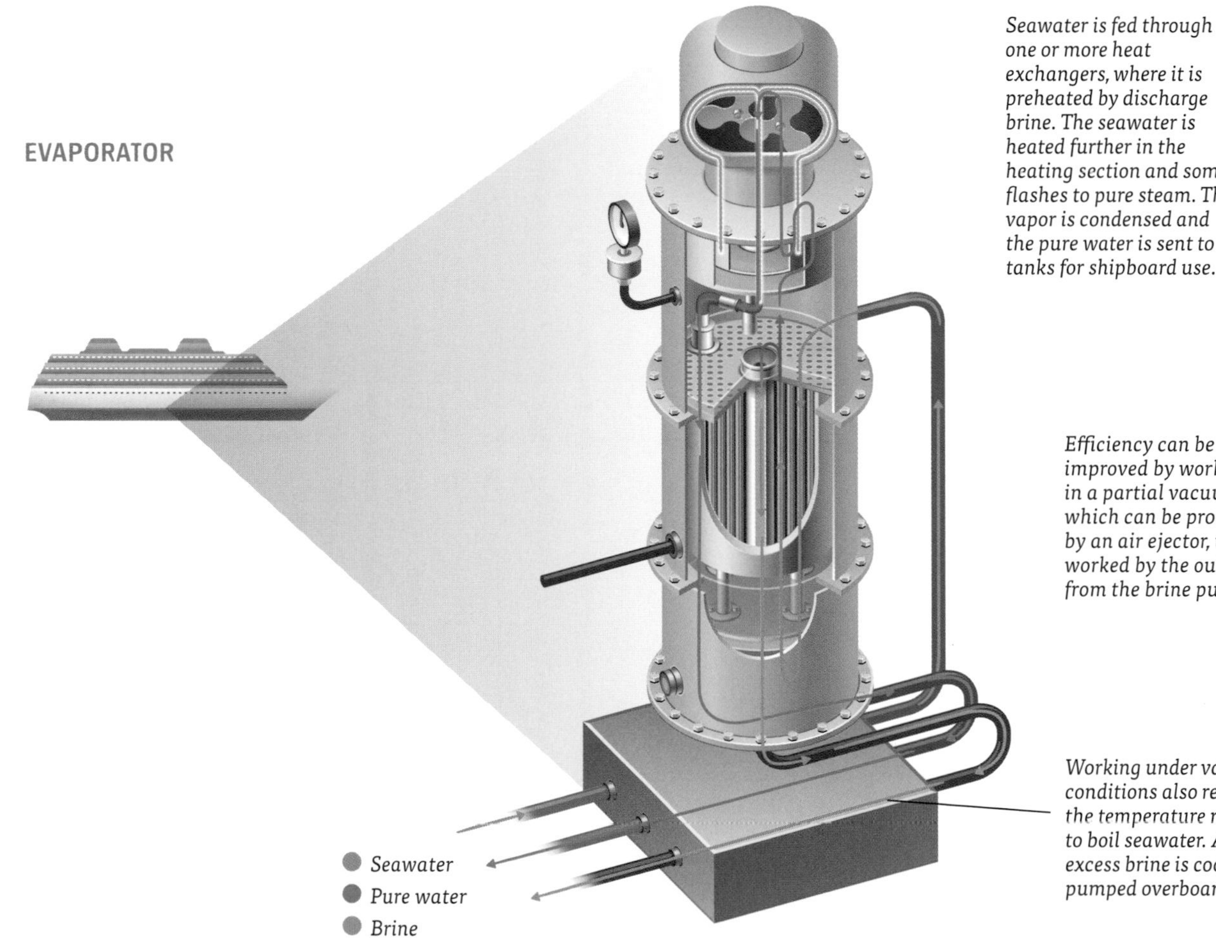

SAFE TO DRINK?

Why can't sailors at sea drink seawater? Because those who do would lose more water in processing the salt than they'd take in from the ocean.

Salty seawater is high in sodium, roughly three times higher than normal body fluids—and hence needs to be removed once it enters the body. To do that, water flows from the blood to the stomach and kidneys, causing excess urination and dehydration. To compensate for fluid loss, the body constricts the blood vessels leading to the major body organs—prompting nausea, weakness, and eventually delirium and brain damage.

vaporization: the condensed vapor given off by boiling seawater is collected in a form pure enough to be drunk while the brine is discharged overboard.

Desalination may also be achieved without the application of heat. By undertaking a process known as reverse osmosis, the salt in seawater can be separated sufficiently to allow the remaining water to be used for human consumption. The process involves putting the seawater under pressure and forcing the water through a filter that purifies it.

Desalination technology is no longer just a maritime technology. An estimated 15,000 land-based desalination plants of both types operate around the globe but produce less than 1 percent of the world's potable water. The technology is relied on most heavily in the Middle East, which boasts roughly 70 percent of the world's desalination capacity, and in arid portions of North Africa—which represents about another 6 percent. Less arid parts of the world, such as California and Florida, also incorporate it into their water supply mix. Even relatively wet places like London feature backup desalination plants in case of major water shortages: the new Beckton Desalination Plant in east London can provide enough drinking water for a million people.

REVERSE OSMOSIS

1. *Seawater is prefiltered and then pumped under high pressure to the saltwater side of the RO module.*
2. *The pressurized seawater counteracts the natural osmotic pressure from the pure-water side and pushes the saltwater through the filter.*
3. *Because of the size of salt molecules, only the smaller water molecules make it to the other side, leaving the salt behind.*
4. *The excess brine is discharged overboard.*

Seawater
Pure water
Brine

SCURVY NO MORE

Gone are the days when sailors at sea got scurvy from a deficiency of vitamin C and had to rely on citrus fruits to replenish their bodies. Today, voyages between ports are much shorter in terms of elapsed time and ways to preserve food—such as freezing and refrigeration—have evolved to make vitamin deficiencies at sea a thing of the past.

Today the biggest sickness at sea is norovirus—a stomach bug that is not life threatening, but is very unpleasant. Found anywhere that people live in close quarters, norovirus can be especially problematic on cruise ships: an outbreak forced Norwegian Cruise Lines' *Crown Princess* to terminate a cruise originating in Florida two days early in February 2012.

DISPOSING OF WASTE

Waste disposal at sea is serious business. One cruise ship alone, over the course of a week's journey, can produce a million gallons of "gray water" (from showers and sinks), 210,000 gallons of sewage, 50 tons of solid waste, and 37,000 gallons of oily bilge water—not to mention 100 gallons of toxic chemicals.

Historically, wastewater and sewage were discharged directly overboard from sailing or motorized craft into the sea. Today, any vessel with a toilet must also have a marine sanitation device to treat wastewater before discharge. The extent to which the water must be treated and where it might be discharged varies by the size of the ship and its location.

Solid waste on larger vessels, with the exception of plastics and hazardous material, is typically incinerated at sea and the ash is dumped overboard. However, ships are not legally required to incinerate: laws in many countries often permit direct discharge of certain types of garbage at specific distances from the coastline.

A number of countries now also regulate ballast water—the water used to stabilize the ship and mitigate vapors from engine exhaust. The primary concern is not oily water (oil is typically removed by a separation process from bilge water before disposal), but rather the transport of invasive species, or nonnative marine organisms, from one place to another in bilge water.

INCINERATION

Waste suitable for incineration is ground into smaller pieces by a rotary shaft as it is loaded into the combustion chamber, helping ensure complete combustion.

Forced-draft fans bring air into the combustion chamber, where it is mixed with fuel that burns the waste at several thousand degrees.

A shaft pushes the incinerator ash to the outside of the chamber, where it is collected and discharged overboard (if allowed) or discharged at the next port.

SEWAGE TREATMENT

In the aeration chamber, raw sewage is mixed with oxygen and bacteria, which decompose it into carbon dioxide, water, and inorganic sewage.

Inorganic sludge and liquid are pumped to a settling tank, where some settles on the bottom and encourages further decomposition of the incoming mixture.

Settling tank liquid is pumped to the chlorination and collection tank, where it's treated to kill bacteria before being pumped overboard (if permitted) or sent to a holding tank.

DISPOSAL DOWN UNDER

After the first two weeks under the sea, nearly every ingredient or foodstuff on a submarine is canned, dried, or frozen. When leaving on a typical patrol, many submarines run out of normal storage for cans and boxes of food—so sailors must walk on layers of cans stacked in passageways and berthing areas until it is consumed.

Much of what gets consumed by naval forces underwater leaves the submarine as human waste—in a not very traditional way. Two-foot-long tin cans, fabricated on board from perforated steel sheets, are anchored with metal weights. After being filled with waste by a special compactor mechanism, the cans are placed inside a trash disposal unit (TDU)—a cylindrical tube extending from the submarine into the ocean. Once cans are placed in the TDU, the tube's valve is opened and the cans are ejected toward the bottom of the ocean, where—one hopes—gravity will keep them there and not compromise the location of the submarine.

FIRE AND FLOOD

Flooding and fire are the two most dreaded casualties to occur on a ship at sea. Many precautions are taken during design and operation to prevent or mitigate these events.

Ships are designed to prevent flooding. Many have double hulls, so that a piercing of the outer skin will not lead to flooding or loss of cargo (in the case of tankers, for example). All but the smallest vessels are divided into horizontal decks and vertical watertight bulkheads, which both give the ship strength and compartmentalize it to isolate and limit damage or flooding. Many ships are designed to a one-, two-, or sometimes even a three-compartment standard, which means that any single watertight compartment, or two or three adjacent watertight compartments, could be completely flooded with seawater and the ship would still remain afloat.

Maritime craft are also designed to minimize the likelihood of fire, with automatic sprinkler systems—similar to hoses present in high-rise and other buildings—that are activated by fire conditions. But in most cases these sprinkler systems are not spraying water on a fire at sea; instead, they rely on a variety of dry chemical or gaseous agents or high-expansion or wet-foam systems. Fire pumps can also be used to ingest seawater and extinguish fires by trained shipboard personnel.

WATERTIGHT DOORS

Watertight doors and hatches are essential to maintain the watertight integrity of the ship and prevent progressive flooding (where flooding water from a damaged compartment is transmitted to adjacent compartments). Below what is known as the horizontal strength deck all movement through watertight bulkheads must be through a watertight door that is usually opened only for passage and then locked, or "dogged," to reestablish the watertight boundary.

Watertight transverse (side to side) bulkheads extend from the collision bulkhead at the bow to the stern's after peak tank and from the keel to the main deck.

Longitudinal watertight bulkheads may form a double hull for further protection.

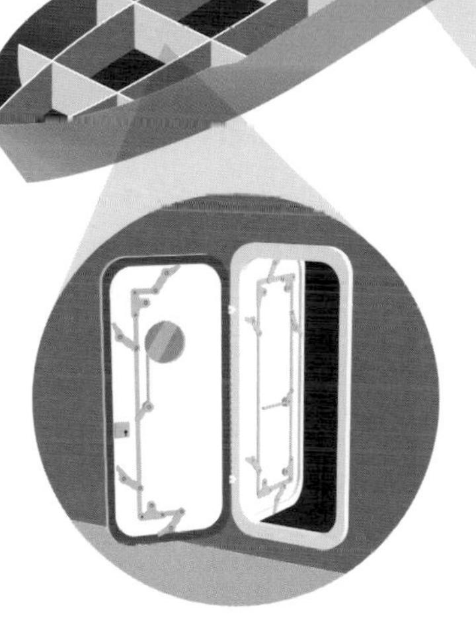

Sliding watertight doors may be remotely actuated through hydraulics or manually via gears or chain drives.

Strength decks feature watertight hatches secured by locking mechanisms known as dogs, to ensure water is not introduced into the hull through weather, waves, or down flooding.

THE S.S. *NORMANDIE*'S END

Although dumping huge quantities of water on a floating ship may seem an inherently bad idea, it certainly has happened—and with great consequence. Few ships have faced such a dramatic end as the S.S. *Normandie*, which sank at Pier 88 on Manhattan's West Side in February 1942.

Sparks from a welder's torch fell on and ignited the ship's life jackets, prompting the New York City Fire Department to respond to the fire. Thousands of gallons of water were sprayed onto the ship to extinguish the fire, destabilizing the enormous vessel at its berth at the end of West 48th Street. Later that day, the *Normandie* responded by rolling over on its port side into the mud of the Hudson River.

ABANDON SHIP

While maritime travel is today an extremely safe way to move from place to place, elaborate preparations are nevertheless required in the event that a ship has to be abandoned at sea. In form, these are similar and analogous to the preparations made by pilots and flight attendants on an airplane and conveyed to passengers at the beginning of each flight (regarding evacuation, oxygen, and use of personal flotation devices); the difference is that marine travelers have a higher chance of escaping from a damaged craft than their counterparts in the air.

Lifeboats and life rafts have been an integral part of marine travel for centuries. Today, lifeboats typically come equipped with oars, flares and mirrors for signaling, and supplies of food and water. More sophisticated ones may include radios or navigation equipment as well as equipment to catch rainwater or fish. Some may include engines or sails, heaters, and even solar water stills. Nearly all are required to contain an emergency position-indicating radio beacon (EPIRB).

Life rafts, in contrast, are a much simpler affair and are not intended for even medium-term survival at sea. They are collapsible and are usually stored in a canister made out of fiberglass. As with some escape equipment on planes, and similar in concept to airbags in cars, they

DAVIT-LAUNCHED LIFEBOAT

Launching personnel ensure the drainage plug is installed and the embarkation ladder is lowered. The lifeboat is lowered to deck level for loading.

Once the boat is embarked, it is slowly lowered by davit to the water. Once in the water, the lines are released and the lowering team boards via a ladder.

FREE-FALL LIFEBOAT

Prelaunch checks of the engine and hull are conducted. The passengers are embarked and restrained in their seats. All hatches and ventilation are shut and secured.

During launching, all passengers must brace for impact with the water.

Once settled, the helmsman restarts the engine to leave the area and initiates distress signals.

LIFE PRESERVERS

The modern life jacket originated in the mid-nineteenth century in the United Kingdom. Originally filled with cork, vests have also been filled with kapok (a tropical tree fiber) and, most recently, foam. Today, the U.S. Coast Guard classifies life jackets I through V for their suitability for offshore and inshore use.

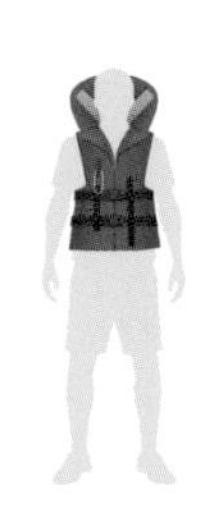

TYPE I

Best for open ocean, rough seas, or remote water, where rescue may be slow in coming. Many are self-righting with a protective collar.

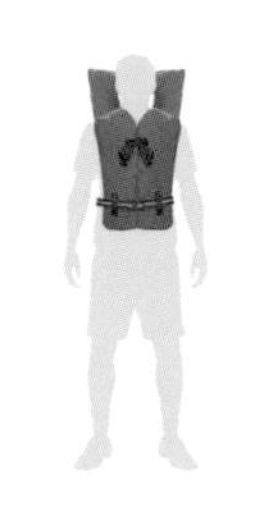

TYPE II

For general boating activities. Good for calm, inland waters or where there is a chance for fast rescue.

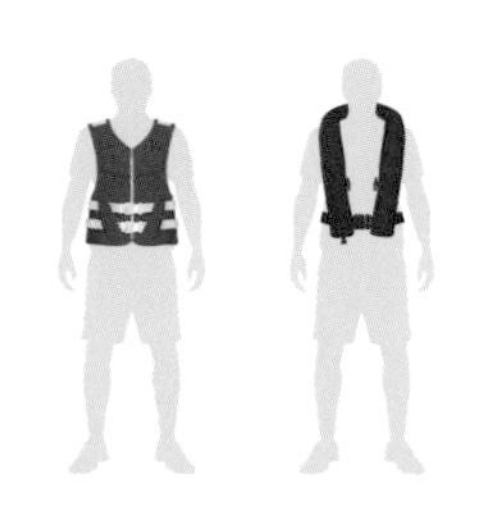

TYPE III

For general boating or specialized activities (e.g., water skiing, canoeing). Good for calm, inland waters or fast rescue.

inflate quickly and automatically via high-pressure or compressed gas, and therefore do not carry anywhere near as much equipment as a lifeboat with a hardened hull.

Regardless of which form of escape mechanism is provided, all passengers are required to have access to a personal flotation device or life jacket. Some are manually deployed: pulling a string pushes a firing pin into a carbon dioxide (CO_2) canister, which inflates the jacket. Others are automatic and rely on a powerful spring held back by a small pellet: when the pellet gets wet, it dissolves and releases the spring—which pushes the pin into the CO_2 gas canister.

LIFE RAFT

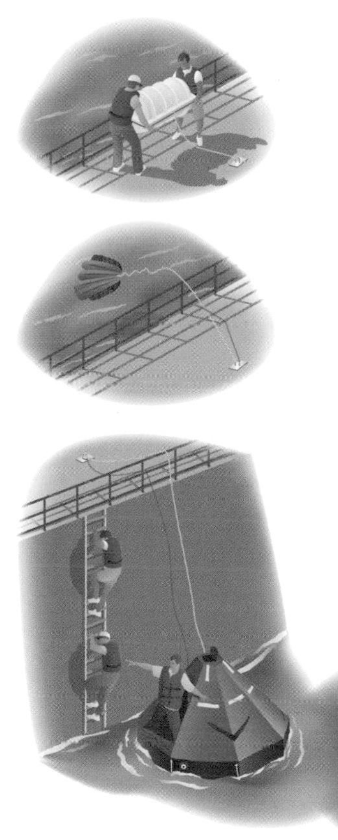

Prior to launching the life raft, the connecting, or painter, line should be secured to the ship and the embarkation ladder lowered.

Depending on the configuration, the raft might be lifted over the side or gravity launched. If the life raft does not self-inflate on impact, it may be manually inflated by pulling a cord.

The raft should be fully boarded before disconnecting the painter line and casting off.

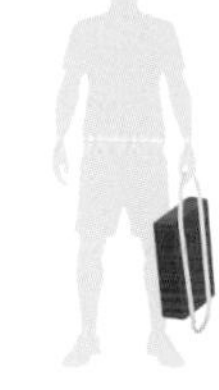

TYPE IV

Throwable devices like ring buoys and boat cushions.

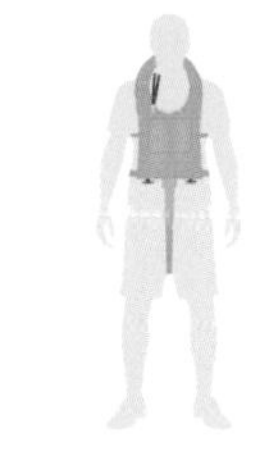

TYPE V

Only for special uses or conditions (e.g., special inflatables used in sailing for man overboard incidents).

OTHERS

Survival immersion suits for open ocean; submarine escape immersion equipment and steinke hoods for submarine escape.

SINKING SHIPS

When the *Titanic* ran into an iceberg in the North Atlantic on the night of April 12, 1912, she was equipped with the technologies necessary to save her passengers, including state-of-the-art lifeboats and a modern ship-to-shore communication system. Unfortunately for her passengers, neither the boats nor radio functioned as they should have. The radio operator on the RMS *Carpathia*, the nearest ship to the *Titanic*, had signed off for the night and never received the *Titanic*'s distress signals. And the lifeboats carried by the ship had a capacity of only 1,200—several thousand shy of accommodating the 3,300 aboard that night.

The sinking of the *Titanic* prompted the convening of the first international convention for the Safety of Life at Sea (SOLAS), held in 1914, and led to a series of new marine safety requirements. Although the number of ships lost at sea declined dramatically over the remainder of the twentieth century, human error has continued to bedevil ship travel right up until the present time. Among the greatest disasters at sea in postwar Europe occurred in the 1990s: the sinking of the ferryboat MS *Estonia* over the course of just one hour in September 1994 took the lives of 852 people.

Even today, size remains no barrier to disaster at sea. The running aground and subsequent sinking of the cruise ship *Costa Concordia* off the Italian coast in January 2012 captivated the world. Though the sinking was slow, and the death toll much lower than that on the *Estonia*, the fact that the sea still retains the ability to claim a ship of such size suggests that a century of new ship safety technologies are only as good as the people who operate them.

NAVIGATION

For as long as ships have existed, people have had to navigate or guide them. In some places, such as along rivers or coastlines, navigation was easy—even if the handling of the boat was harder due to currents. But in other places, particularly for travel between islands or continents and away from land, navigation proved as much if not more of a challenge than the handling of the vessel itself.

Navigation as a science is said to have originated on the Indus River, in India, some 5,000 years ago. Trade based in the Arab world contributed greatly to its development, giving rise to a number of tools for celestial navigation that allowed the crossing of oceans and seas. But trade between nations did not prompt extensive oceanic exploration until roughly the fifteenth century. Once trade between Europe and Asia became a high priority, new sorts of sailing ships were manufactured to ply these routes—and new forms of navigation developed to support them.

Evidence suggests that explorers like Columbus relied in some measure on dead reckoning, that is, estimating a current position based upon a previously established "fix," or position, modified to take account of the ship's estimated speed and direction of travel since that point. (Speed was measured by a chip log that included a weighted rope fed out over the back of the boat; knots at

TOOLS OF TRADE

As early as the sixth century B.C., the ancient Greeks pioneered celestial navigation and were able to expand their reach and empire across the Mediterranean.

Following the rise of Islam in the seventh and eighth centuries, Arab spice traders controlled the routes from the Mediterranean to India and refined celestial tools, such as the astrolabe.

regular intervals allowed a reliable measure of counting how fast a boat was going during a given period of time.) Unfortunately, dead reckoning couldn't take into account winds or currents—which meant that many mariners ended up reaching lands they didn't intend to visit.

Most of the early navigational aids took the form of celestial navigation—determining one's position based upon the relationship of a celestial body to the horizon. Some of them, like astrolabes, predate the Age of Discovery; others followed it. Marine sextants and cross-staffs helped calculate latitude by measuring the angle between the horizon and the sun, moon, or stars. Longitude was calculated by measuring the difference in the time of day between two locations; the seagoing chronometer, invented in the second half of the eighteenth century, revolutionized mapmaking by allowing explorers like Captain Cook to circumnavigate the globe and record with great accuracy what they saw.

Marine compasses would later help determine direction relative to the magnetic north pole; the first gyroscopic compass—which always pointed to true north—was invented in 1907. Radar and radio waves would come two decades later as a way to navigate in fog and to identify targets at sea and avoid collisions. Like many maritime technologies, they emerged from the military: the U.S. Navy used the term "radar" as an abbreviation for "radio detection and ranging" starting in 1940.

Radar technology allowed mariners to identify both the direction and the distance between ships at sea. But it would not be until satellite technology debuted, later in the twentieth century, that marine navigation would become a precise science. Today, the early satellite-based systems, such as NAVSAT, that relied on tracking stations on earth have largely been replaced by more sophisticated global positioning systems (GPS), which now dominate navigation on land and air as well as on the seas.

By the twelfth century, trade between Europe and Asia had expanded across the vast silk and spice roads and the Chinese-invented magnetic compass spread across the continents.

Starting in the fifteenth century, European explorers searched the open oceans for new trade routes to Asia and invented tools, such as the cross-staff and chronometer, for improved celestial navigation.

By the eighteenth century, trade reached around the globe. The marine sextant and improved marine chronometers allowed sailors to accurately use celestial bodies to fix their positions anywhere on earth.

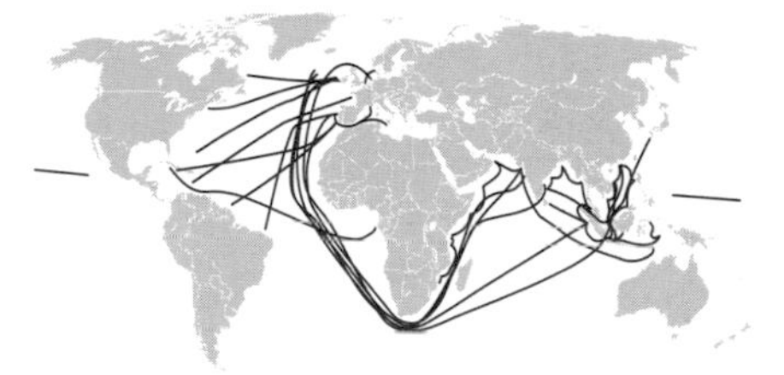

ELECTRONIC NAVIGATION

Maritime navigation has come a long way since the days of consulting the stars and the horizon or the sun and its shadows. Much of that progress has occurred in the last half century, thanks to a revolution in satellite technology that has benefited not only ships but also airplanes, trains, and cars.

Until recently, maritime navigators within 2,000 miles of American or Canadian shores relied on a land-based long-range navigation system (LORAN) to identify their locations at sea. A master transmitting station and at least two secondary stations sent radio pulses at precise intervals. By identifying the difference in time it took to receive the signal and converting that time into latitude and longitude, a ship's crew could reliably calculate its location.

The first operational satellite navigation system was known as the navy navigation satellite system (NAVSAT) and involved a system of orbiting satellites and a network of tracking stations that would monitor their positions and use those to determine points on earth. Originally developed by the U.S. Navy to provide accurate position data to nuclear ballistic missiles, it moved into civilian usage and was relied on heavily through the early 1990s.

Both LORAN and NAVSAT have largely been replaced by a space-based radio-positioning technology known as the global positioning system (GPS). Created by the U.S. military and then opened up for commercial use, GPS measures the distance between a small number of orbiting satellites and a receiver on the earth and computes the receiver's position from those measurements.

GPS SATELLITES

GPS, or global positioning system, was developed by the U.S. government for military purposes. It relies on a network of two dozen satellites that circle the earth at an altitude of 12,000 miles above it. Each satellite weighs 3,000–4,000 pounds and rotates around the earth twice each day, powered solely by solar energy.

At anytime during the day, from any point on earth, four satellites are visible. Each satellite transmits a radio signal that contains its location in space and the current time (based on an automated clock); the signals are synchronized and sent at the same time. A GPS receiver on the ground finds these four satellites, calculates the distance to each from its point on the ground, and uses those distances to determine its precise location.

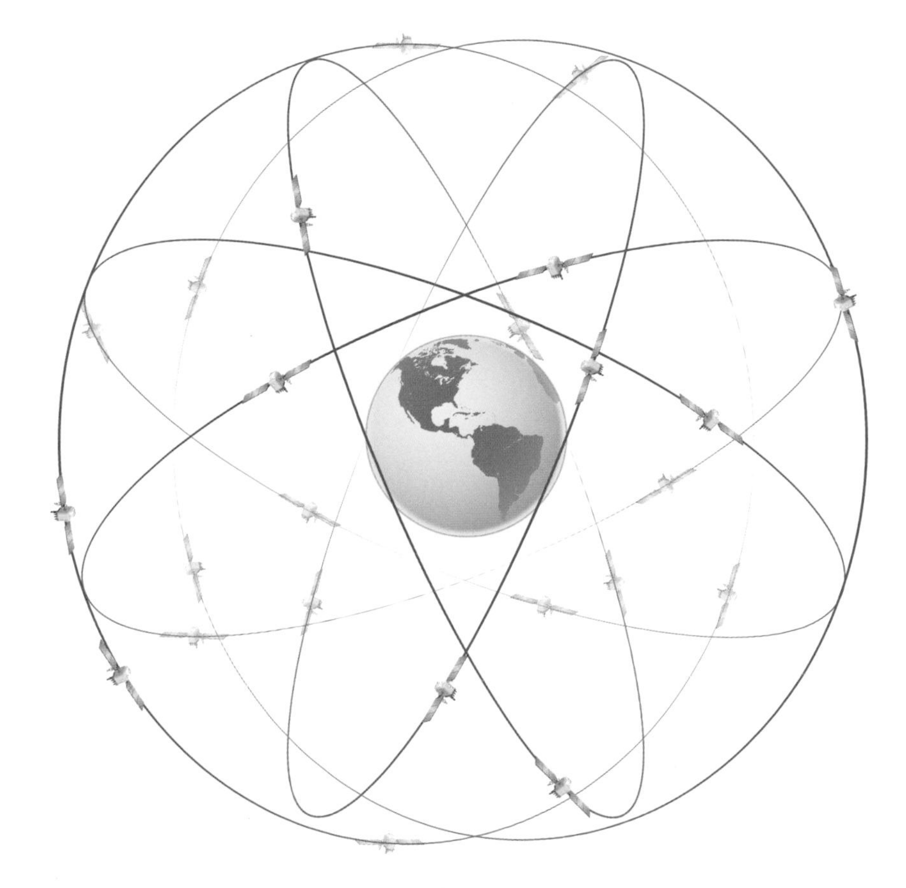

UNDERWATER NAVIGATION

Modern military submarines covertly travel underwater and do not typically use their active sonar systems to detect underwater hazards. Surfacing or coming to periscope depth to fix their position with GPS risks detection by ships and aircraft. Instead, submarines rely on inertial navigation systems (INS) to accurately estimate their underwater position. These systems are a modern update of the dead reckoning methods of Columbus' era that use motion sensors (accelerometers) and rotation sensors (gyroscopes) to continually calculate the position of the submarine without external fixes. While the fix may be accurate, the submarine must also rely on charts that may not show all the details of the underwater terrain—demonstrated most notably in 2005 when the U.S.S. *San Francisco* ran into an underwater mountain at top speed.

HOW GPS WORKS

❶ *A ground-based receiver detects the presence of a satellite in orbit and receives the satellite location and time data. Using these data and the speed of the radio signal, the receiver is able to calculate its distance from the known location of that satellite.*

❷ *At this point, the receiver location is somewhere on a sphere centered around the known satellite location with a radius equal to the estimated distance.*

❸ *The signal and range to a second satellite defines a second sphere with a different center and radius. The receiver location is further refined to where these two spheres cross (a two-dimensional circle).*

❹ *Using a third (or if available fourth) satellite, the receiver location is pinpointed at the intersection of the imaginary spheres (the earth is used as an additional sphere to eliminate intersection points in space). Because GPS receivers do not have an atomic clock like the satellites (it would be too costly), the receiver must use some tricks to self-correct for timing errors in estimating the spheres. This yields position accuracies up to approximately 30 feet (10 meters). With differential GPS, a signal from an additional land-based station can further refine the accuracy down to three feet (1 m).*

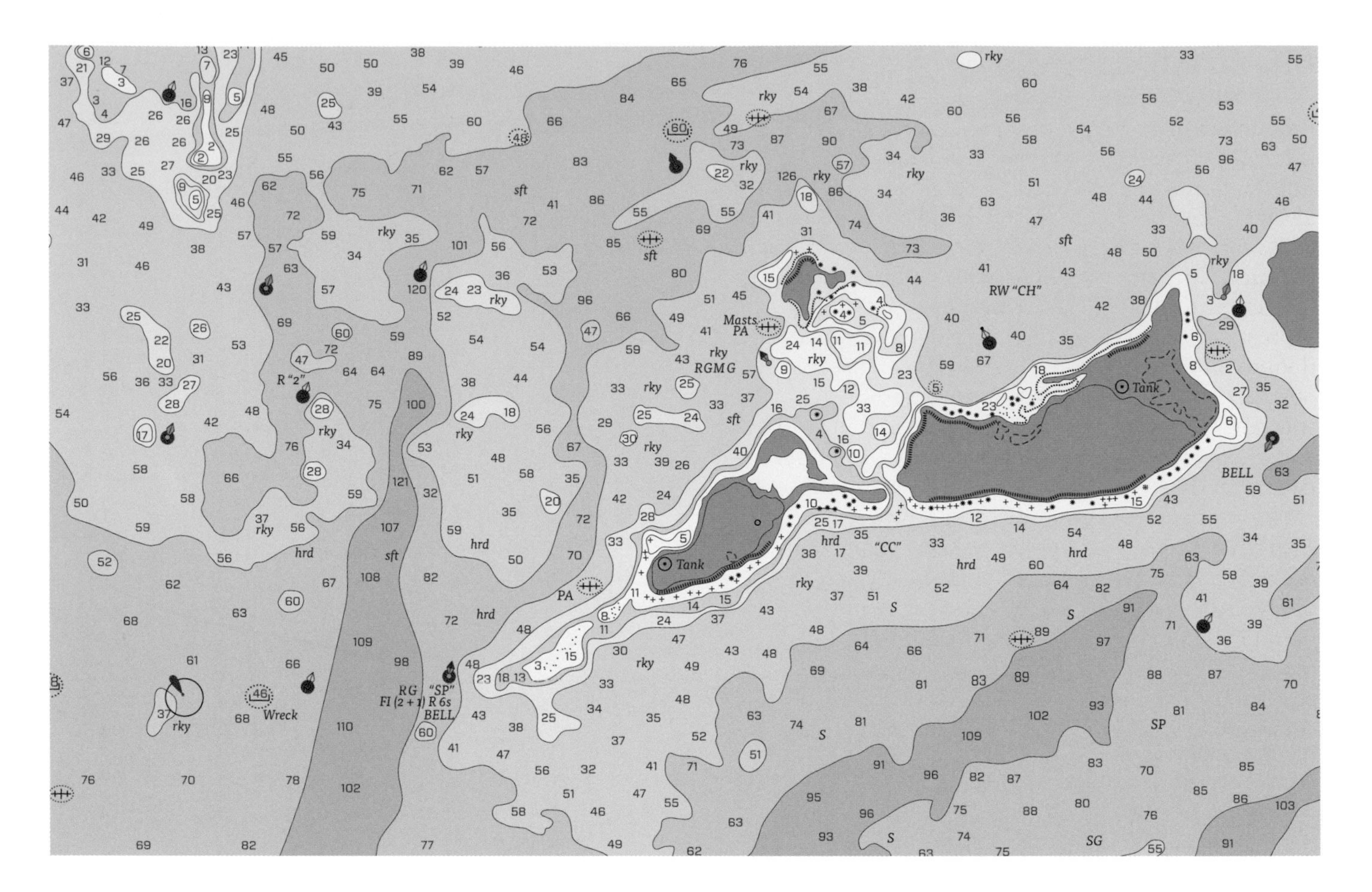

CHARTING A COURSE

Navigation today would not be possible without a vast array of information about the depth of water, and in particular about any obstructions that lie beneath it. This information is typically presented on a nautical chart—a sort of road map of the sea, its topography, and its adjacent coastline.

Nautical charts vary greatly. Sailing charts cover great blocks of ocean, at a ratio of greater than 1:600,000. Those used for coastwise navigation are more granular—usually from 1:50,000 to 1:150,000. Charts covering bays and harbors, where depths and obstructions become more critical, are more detailed, typically at a 1:50,000 ratio or larger.

Features on a nautical chart range from the depth of the water to the identification of beacons to the location of shipwrecks. Similar to the way a topographical chart used by a mountaineer will show the location of ridges along a trail, a nautical chart will feature depth contours—lines connecting points of equal depth.

Today's charts are produced both electronically and on paper. In addition to hazards and depths, they show marine sanctuaries or zones where speeds may be controlled due to risk of collision with protected sea mammals. They may also offer up-to-the-minute updates on ongoing navigational projects affecting commonly used sea routes.

READING A CHART

Charts show limited topographic and natural land features (e.g., sandy shoreline) along the coast.

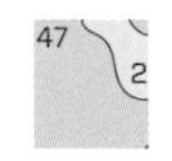

Water depths are prominently featured with sounding data and shoal water (lighter blue areas).

The nature of the seabed is also featured to aid in anchorage selection (rky—rocky, sft—soft, etc).

Underwater wrecks and obstructions should be avoided as they may extend well above the charted sea depth.

Aids and services such as fixed lights, buoys, beacons, fog signals, and radar position systems (RACONs) are shown.

Larger radio towers, water tanks, and bridges might be shown to aid in visual navigation.

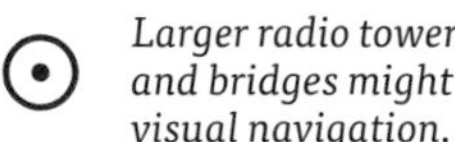

MEASURING DEPTH

Production of a nautical chart relies on a wealth of information about water depths. Although this information is relatively easy to obtain nowadays, it wasn't always. Well prior to sonar technologies, a rope with a lead sinker would be tossed overboard and measured once the sinker hit bottom and the rope could be pulled taut.

Over time, these survey vessels evolved into highly sophisticated hydrographic survey ships that scan the sea floor to produce information about both depths and obstructions. Modern hydrographic vessels use specialized sonar (primarily side-scan and multibeam sonar) to produce nautical charts. By analyzing echoes from sounds transmitted to the sea floor, the sonar creates a virtual picture of the ocean floor—including such details as the size and shape of obstructions, the texture and materials of the sea floor, and the location of underwater pipelines and cables.

Sonar technology has, like many other maritime tools, become more sophisticated with the advent of computers. Initially the purview of the military, commercial sonar came to market in the second half of the twentieth century. But until 1990 or so, it was very much a paper-based affair. Today, digital scan converters are used to compute the precise parameters of the sea floor and transmit the information in real time to boat captains. Modern sonar technology is also prized by the commercial fishing industry as even relatively inexpensive systems can highlight fairly precisely the whereabouts of schools of fish.

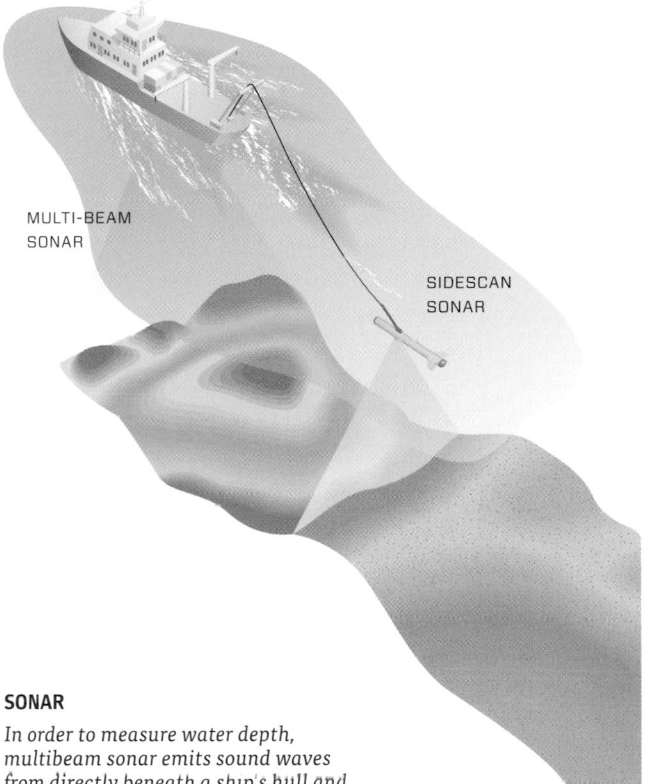

SONAR

In order to measure water depth, multibeam sonar emits sound waves from directly beneath a ship's hull and measures and records the time for the acoustic signal to travel to the sea floor (or object) and back to the receiver.

In side-scan sonar, the transmitted energy sweeps the sea floor from directly under the sonar array. The strength of the return echo is continually recorded, creating a "picture" of the ocean bottom.

THE WORLD IS FLAT

It's not easy to represent the three-dimensional world in two dimensions. One way of doing so is by a gnomic projection, which represents great circle routes as straight lines and reflects the way radio signals travel. Since the shortest distance between two points is shown as a straight line, these charts are often used for open ocean voyage planning.

Mercator projections are often used for navigation because any straight line on the map (known as a rhumb line) is a line of constant direction, but not necessarily the shortest distance between points. The sizes and shapes of large areas are distorted the farther away an object is from the equator (Greenland looks the same size as Africa, when it is much smaller).

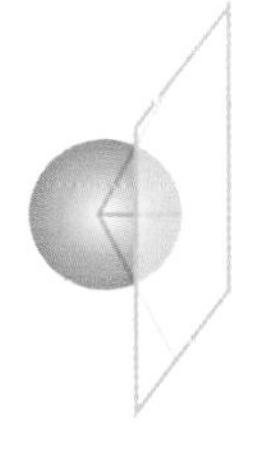

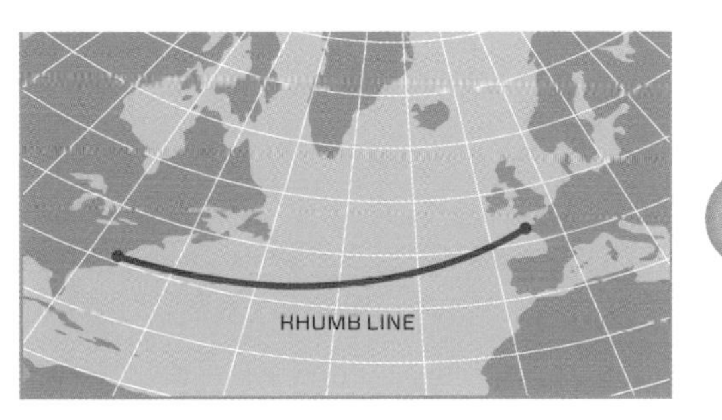

Gnomic projection

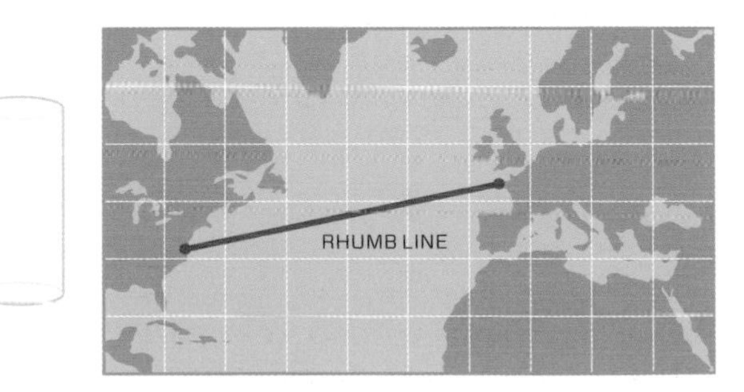

Mercator projection

LIGHTING THE WAY

Over time, a network of navigational aids has evolved to aid the mariner in pursuing a charted course without incident. These aids range from simple buoys marking channels to elaborate lighthouses.

In the nineteenth century, much of the work of guiding mariners along the shore fell to "lightships," or "light vessels," rather than lighthouses. Lightships were stationed off critical coastlines and harbors to guide mariners in dark and stormy conditions. Early lightship crews hoisted oil lamps to the top of the ship's mast when light was needed for navigation, typically from dusk to dawn, as well as in foggy conditions; later lightships sported permanently fixed lanterns. Today's few remaining operating lightships are automated, and many rely on solar power.

Lighthouses were developed to replace lightships as a cheaper and more sensible solution to problems of coastal navigation. Some were constructed in the water and affixed to the sea bed by steel caissons and screw piles. More often they were put on top of a cliff, or just below one, to be visible in fog. Their height determined the distance from which they could be seen: the square root of the height of the lighthouse (in feet) multiplied by 1.17 determined the distance to the horizon in nautical miles.

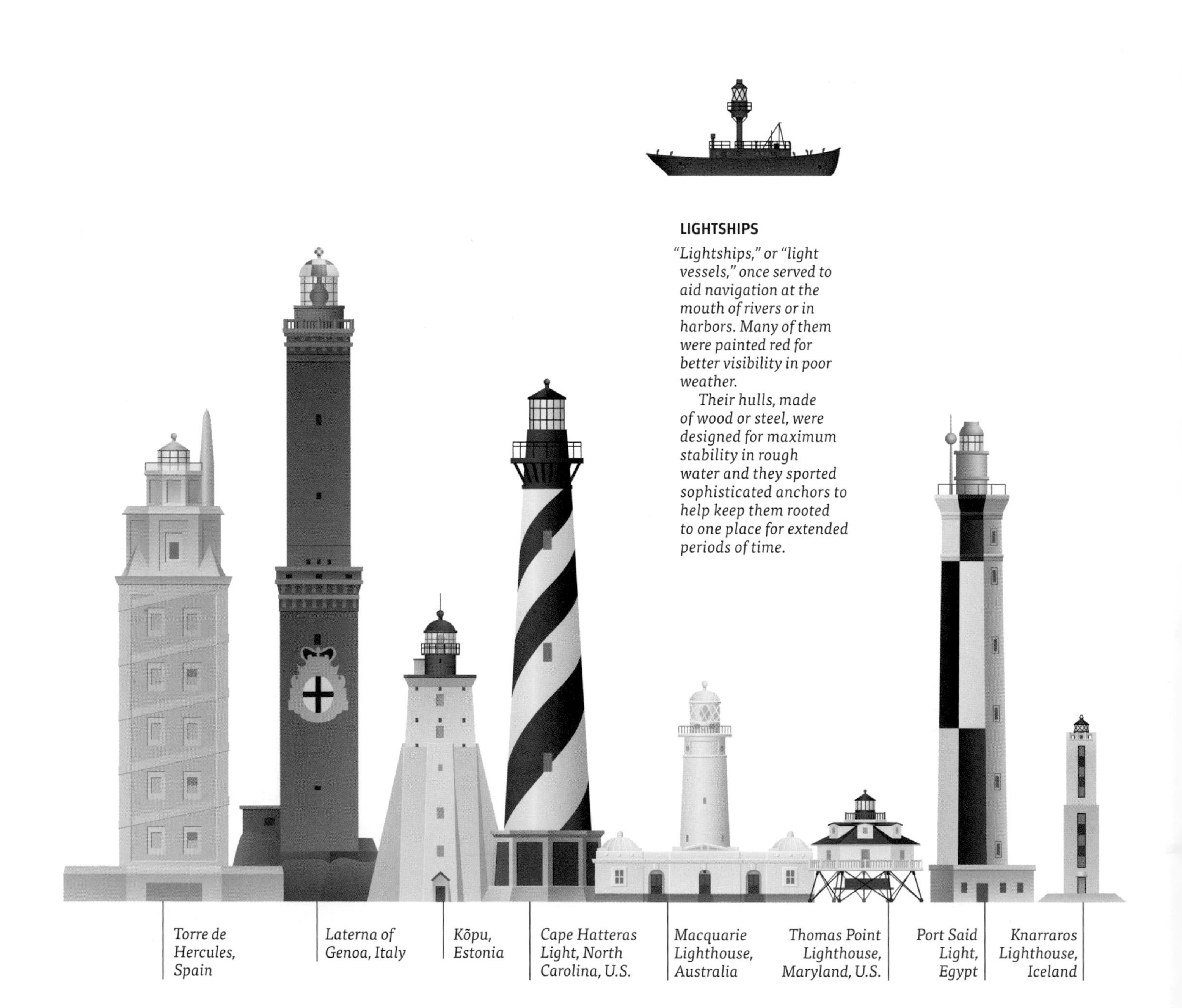

Today, many of the world's lighthouses are retired, having been replaced by modern positioning electronics. Those still operating are largely automated and are typically powered by solar cells or batteries—a far cry from the whale oil and wick lamps, and even the kerosene and gas, that once provided the source of their light. Most still depend on a form of Fresnel lens, a special form of parabolic reflector or prism that refracts and concentrates light rays into a powerful directed beam that can be seen as far as 20 miles away. Often, the high-intensity light beam coming off the lens will rotate or flash in an omnidirectional fashion.

Cabo Branco Lighthouse, Brazil

Yokohama Marine Tower, Japan

Enoshima Lighthouse, Japan

THE FRESNEL LENS

The Fresnel lens was invented in the UK in 1822. The opposite of telescope glass, which refracts light to seem nearer, Fresnel lenses concentrate light into a powerful beam that can be seen at a distance. Their stepped surface bends light as much as heavier and thicker glass lens would do, making them a good choice not only for lighthouses but also for car headlights—which rely on a form of Fresnel lens molded from plastic.

THE LITTLE RED LIGHTHOUSE

Perhaps no lighthouse in the western world is as celebrated as the Little Red Lighthouse, which sits just below the George Washington Bridge on the eastern shore of the Hudson River in northern Manhattan. Memorialized by Hildegarde Swift in a treasured 1942 children's book, *The Little Red Lighthouse and the Great Gray Bridge*, its fate was typical of lighthouses of its era.

Constructed in 1921 on the site of a previous lighthouse, Jeffrey's Hook Light served to warn mariners of a rocky outcropping in the Hudson River. Less than a decade later, the monumental George Washington Bridge rose at the site—its eastern pier built almost directly over the lighthouse and crowned by a powerful beam designed to warn mariners and aviators of the bridge's presence.

While the book depicts a happy coexistence between the small lighthouse and the large bridge, the partnership would not endure long. The lighthouse ceased operation in 1947, and today is maintained by the New York City Department of Parks and Recreation as a much-loved historic landmark.

In Europe and most of the world, green buoys mark the port, or left, side of the channel when sailing toward land. Green buoys have odd numbers and green lights.

In the Americas, Japan, Philippines, and Korea, red buoys mark the starboard side of the channel when sailing toward land ("red right returning"). Red buoys have even numbers and red lights.

The four cardinal buoys are differentiated by the directional signals atop them, which indicate the safe side of a danger. (In this illustration, the signal points down, indicating safe water to the south of the buoy.)

BUOY BASICS

Navigation close to shore would be almost impossible without buoys. Buoys perform a number of functions for the mariner: they mark navigable channels, identify locations and obstacles in harbors and sea lanes, and help direct ships to their berths in port.

Most familiar to the casual seafarer are what's known as lateral buoys, the traffic signals that guide ship crews safely along a channel or waterway. These buoys rely on a combination of colors and numbers to indicate a traffic lane and direction: "red, right, returning" means keep the red buoys (typically even numbered) on the right, or starboard, side of a ship when sailing toward land. (In contrast, the odd-numbered green buoys will be on the right when sailing out to sea.) Buoy numbers start low and get higher as a ship approaches land.

Buoys do more than provide directional assistance. Cardinal buoys might be placed on the seaward side of a shipwreck or obstruction and signal the direction of a hazard or safe water; large navigational buoys (up to 40 feet in diameter, with lights 36 feet high) often mark the entrance to a harbor at major coastal ports. Not all are red or green: red-and-white–striped buoys typically mark the end of a channel while yellow buoys often mark things like anchorage, fishing, or dredging areas.

Buoys vary widely in their appearance. They take a variety of shapes—can, cone, pillar, or sphere among others. Some are designed and constructed around sound signals—whistles, horns, gongs, or bells. Most of these sound buoys have an underwater shape that allows and encourages them to move actively (and thus noisily) in the sea.

The average buoy consists of an above-water structure, a long length of chain (typically three times the water depth), and a sinker made of concrete or cast iron. The above-water structure may contain a sound or light unit—the latter including a battery pack, flasher, lamp changer (which automatically replaces burned-out bulbs), a lens, and casing. In good visibility, a buoy's lights can be seen as far away as four to six miles.

BEACONS IN THE FOG

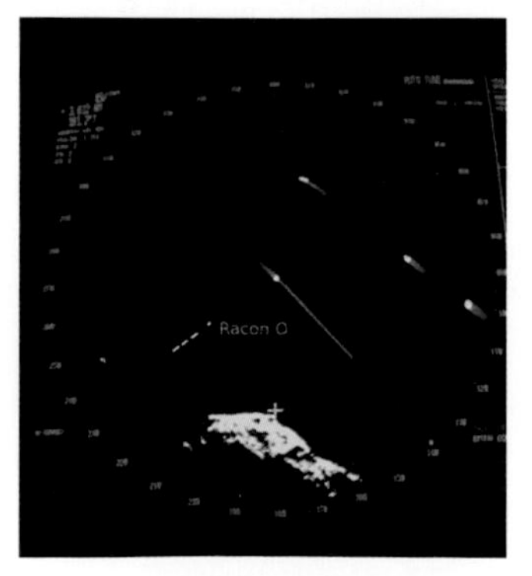

Devised from the words "radar" and "beacon," racons are radar transponder beacons that receive and retransmit signals to allow themselves to be clearly identified on a ship's radar display. The transponder might be mounted on buoys or alternatively on obstacles near the ship's route—such as bridge spans or even coral reefs. They allow the vessel's crew to identify its own location but also, and more important, identify the location of obstacles that the ship needs to avoid.

Racons typically are prompted by receiving a radar pulse from a ship. They respond to that pulse with a signal on the same frequency, and their response shows up on the ship's radar screen as a series of Morse code dots and dashes that correlate to a code on the chart. In order to prevent a racon signal from masking other important radar information, the transponders operate roughly half of the time instead of on a continuous basis.

BUOY TENDERS

Buoys contain lamp changers—multiple sockets fitted around a central hub. When one bulb, or lamp, burns out, another lamp rotates into position to replace it.

Notwithstanding this automated feature, buoys need to be serviced regularly—both for bulb replenishment and for more general maintenance. In the United States and elsewhere, this work is carried out by a buoy tender—a Coast Guard vessel customized to lift a buoy out of the water and undertake buoy maintenance work on its deck.

The unique attribute of the buoy tender is its ability to stay in one place while servicing a buoy. A customized thruster propulsion system has the ability to rotate 360 degrees and allow the tender to maintain almost static positioning and avoid twisting buoy chains during buoy maintenance. These ships utilize onboard cranes to lift and swing buoys and their sinkers onto their decks.

A typical buoy is made of three parts—a top mark, a buoy body, and a tail tube to counterbalance the top mark. A fully configured buoy can weigh approximately 10 tons (excluding moorings).

TENDER

Buoy tenders are purpose-built ships designed to lift up to 20 tons of buoy and sinker out of the water for inspection, maintenance, and repair.

Depending on the depth of the bottom, size, and purpose of the buoy, tethers may employ anchoring chains, nylon lines, and/or floats (typically glass).

Bottom sinkers are typically concrete or cast iron and can weigh up to 10 tons.

REPAIR

Buoys are typically inspected every year. A true dirty job, the entire buoy system is cleaned of mud and mollusk growth, damaged chain is replaced, and lights and sound signals are inspected or repaired.

TERRITORIAL WATERS

The belt of coastal water that reaches out from a coastline to somewhere around 12 nautical miles from shore (a nautical mile is slightly longer than a land-based mile) is often referred to as territorial waters. Though in legal terms these waters are considered the property of the adjacent coastal country, foreign ships are guaranteed what's known as innocent passage—that is, the right to travel safely through the waters of another nation.

Beyond the first 12 or so nautical miles (13.8 mi; 22 km), an international agreement governs the definition and rules associated with a variety of other zones. Up to about 24 nautical miles (27.6 mi; 44 km) from shore (and including the territorial waters), a "contiguous zone" gives the adjacent country limited rights to enforce customs and immigration laws. Overlapping that zone, from about 12 to 200 nautical miles (13.8-230 mi; 22-370 km), is what's known as the exclusive economic zone. Here a coastal nation has rights over the economic resources that lie under the sea—be they oil, products of other forms of mining, or fishing.

Typically, the exclusive economic zone will include part of the continental shelf, which projects out under the water from individual coastlines. But where the continental shelf protrudes a great distance, states have rights to its seabed resources beyond the 200-mile exclusive economic zone and up to a 350-mile limit. Determining where to draw these various boundaries can be something of a controversial exercise, particularly when countries are close together or border one another.

Territorial sea extends up to 12 nautical miles from the coast.

The contiguous zone extends up to 24 nautical miles (28 mi; 44 km) from the coast.

An exclusive economic zone extends to a maximum of 200 nautical miles (230 mi; 370.4 km) from the coast. A nation has control of all economic resources within its exclusive economic zone (e.g., fishing, mining, oil exploration, etc.).

NAUTICAL MILES FROM SHORE

0

50

100

150

200

Policing is relied on to prevent or punish infringement of customs, immigration, or discharge laws and regulations.

UNDERWATER NAVIGATION

Underwater navigation offers some unusual challenges to the mariner. How do submarines prevent collisions with other friendly submarines while operating covertly with limited visibility and in the absence of any form of undersea water traffic control? The answer dates back more than half a century.

During the cold war, NATO adopted a policy of assigning submarines either a moving or stationary three-dimensional box—where they were expected to remain unless they surfaced. This water space management continues as a way to coordinate naval movements among allied nations and prevent potentially catastrophic underwater collisions.

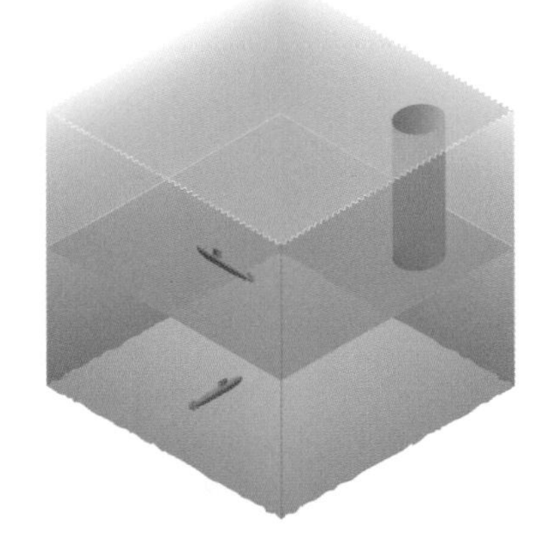

Within stationary boxes, submarine operating areas may overlap and be separated by depth. "Stovepipes" are provided to allow the deeper boat to get to the surface in an emergency or to communicate.

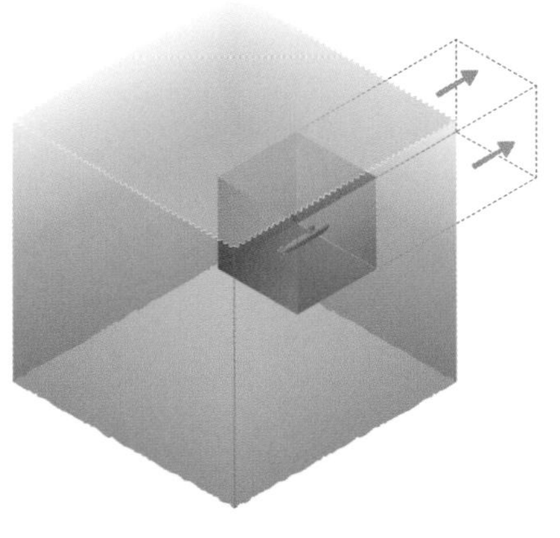

Rectangular "moving havens" are typically several hundred square nautical miles in area and follow a predetermined track. They typically move at a constant speed until they reach a stationary operating area.

TRAFFIC AT SEA

Numerous technologies are used to minimize the number of vessel collisions along the world's waterways. Ships above a certain size and those carrying passengers are required to participate in what's known as an automated identification system (AIS), a tracking system that permits a ship at sea to communicate to other ships its location and its identity in real time. It can also provide a range of other information, such as its compass heading, its speed, and its rate of turn—all of which serve as key data in accident investigations.

More complicated than sailing in open waters is navigating busy channels in port. To manage the movements of vast numbers of ships, the world's busiest harbors often operate what are known as traffic separation schemes—a sort of one-way system for marine traffic. These schemes establish lanes for outbound and inbound traffic as well as no-go areas between them.

Monitoring of vessel traffic in busy waterways is sometimes done through one or another form of vessel traffic system (VTS). These systems rely on a tracking and tagging system, similar to air traffic control, which identifies the position of vessels and directs them on a particular course. Typically, such systems are run by harbor or port authorities and rely on AIS as well as radar, cameras linked to closed-circuit television, or VHF radio communications.

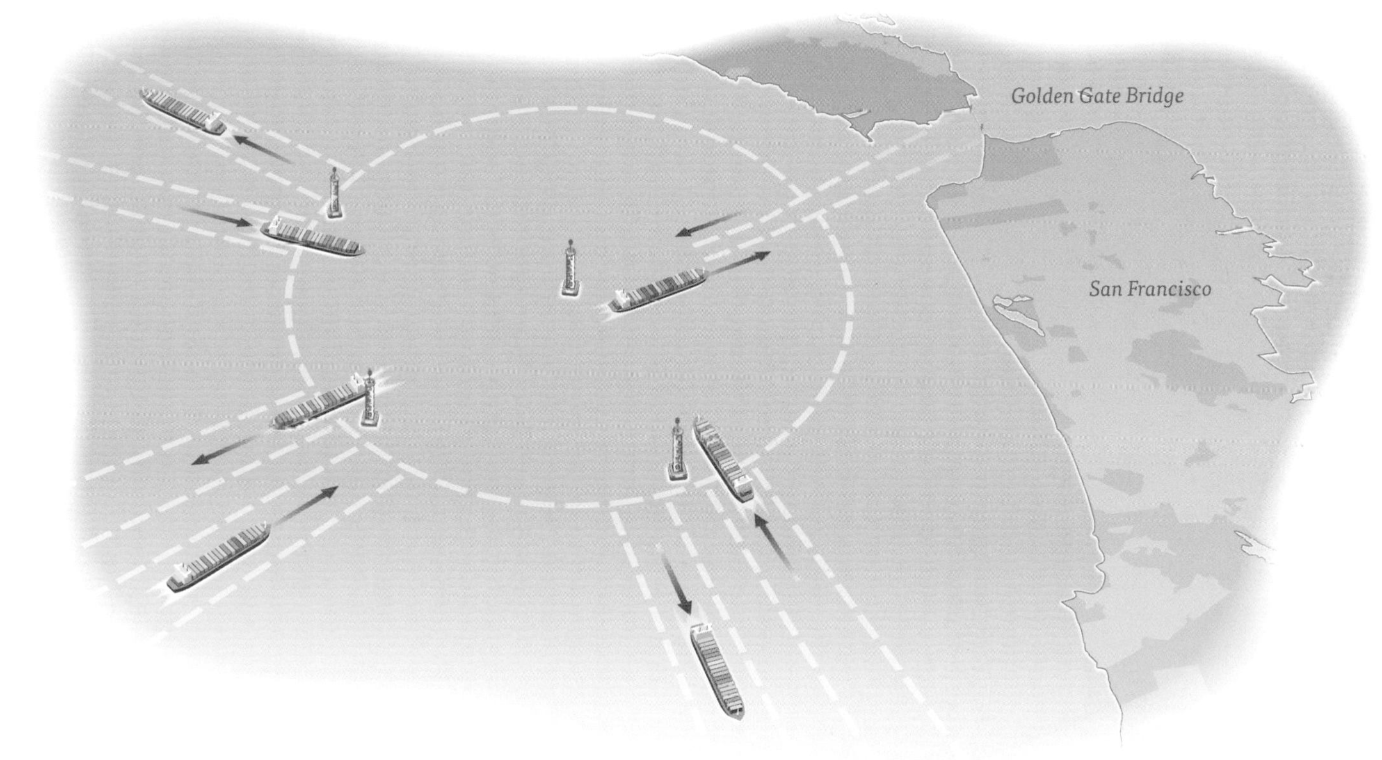

Like the median of a highway, the body of water between two opposite lanes is restricted, so the risk of head-on collisions is reduced.

Once a ship enters a traffic lane it should sail in the general direction of that lane.

Vessels crossing lanes should do so at 90 degree angles to minimize time in the lane.

TRAFFIC IN PORT

Vessel traffic systems (VTS), designed to manage marine traffic moving in and out of busy ports, came late to the United States. VTS appeared first in Europe, in the Port of Liverpool (1949) and then in Rotterdam (1956). A collision under the Golden Gate Bridge in 1971 hastened its introduction in San Francisco Bay and in Puget Sound in Washington, but these systems remained voluntary in much of the United States until another big accident—the *Exxon Valdez* oil spill in 1989—made participation in VTS schemes mandatory in certain U.S. ports. Today the U.S. Coast Guard operates 12 VTS systems across the country.

SHIP LIGHTS

The rules of the road for vessels at sea prescribe priorities and rights-of-way for certain vessels over others. For two vessels within sight of each other, the nature of the vessel may be easy to ascertain (for example, a vessel under sail), but most may not be as obvious (for instance, a vessel that has run aground). During the day, the rules call for certain combinations of black spherical and conical day shapes to be shown. At night, lights (classified as masthead, stern, and side, or running, lights) are required so that vessels can see one another and determine their direction (from the side lights) as well as determine rights-of-way.

A vessel engaged in trawling (dragging nets behind it).

A vessel restricted in its ability to maneuver, but under way (e.g., a buoy tender lifting buoys over the side).

A vessel at anchor (more than 100 m in length).

A power-driven vessel under way (more than 50 m in length).

A towing vessel with the tow following more than 200 m behind (the tow would also be lighted).

RULES OF THE ROAD

Maritime travel is governed by a wide-ranging set of rules of the road. These vary from the general and intuitive (travel at a safe speed, try to avoid collision, cross traffic lanes at right angles) to the more prescriptive, such as those determining right-of-way—for example, "if to starboard red appears, 'tis your duty to keep clear."

These rules also determine a sort of cascade of maritime etiquette, based on the level of difficulty one ship has in maneuvering around another: powerboats are expected to give way to ships under sail, sailboats must give way to fishing boats, and fishing boats must give way to seemingly captainless ships or to those whose ability to maneuver is otherwise constrained.

Maritime regulations also determine sounds and lights that must be used for certain types of movement—for example, for pushing or pulling barges or other forms of vessel towing. Coastal waterways will typically be subject to international navigation rules set by the International Maritime Organization (an arm of the United Nations); inland waterways will be governed by a set of rules determined by adjacent countries or, in the case of Europe, by the European Community.

Stand-on vehicle
Give-way vehicle

Like a car passing another on the highway, the passing, or "give-way," vessel is required to stay clear of the vessel being overtaken until it is well in front of the vessel being passed (the "stand-on" vessel).

Likewise, when vessels meet in a crossing situation (think of a four-way stop sign), the vessel to the right, or starboard, is the stand-on vessel. The give-way vessel should slow down or alter course to starboard (the right) and pass astern of the vessel. In general, a ship should never alter course to port (the left) in a crossing or meeting situation to avoid colliding with the stand-on vessel.

SEA LANGUAGE

Long before radio communication was invented, merchant and naval ships at sea relied on aural or visual signals—such as foghorns, bells, and flags—to send each other messages or warnings. The vocabulary was not as limited as one might think: communication by flag alone contained 600 basic signals.

Flag semaphore was widely used at sea during the nineteenth century, but it did not originate there: it originated in France in the late eighteenth century as a way to transmit information during battle on land. The utility of the system was not lost on mariners, and signaling by flag became a routine part of life at sea until radio communication became mandatory in the early twentieth century.

Semaphore relies on a pair of handheld flags being placed in a variety of positions to convey numbers and letters. At sea, the colors of the flags are typically red and yellow; on land they are blue and white. Although radio communication has all but obviated the need for them, flag signals are still used in the military during replenishment at sea and on other occasions requiring radio silence.

SEMAPHORE

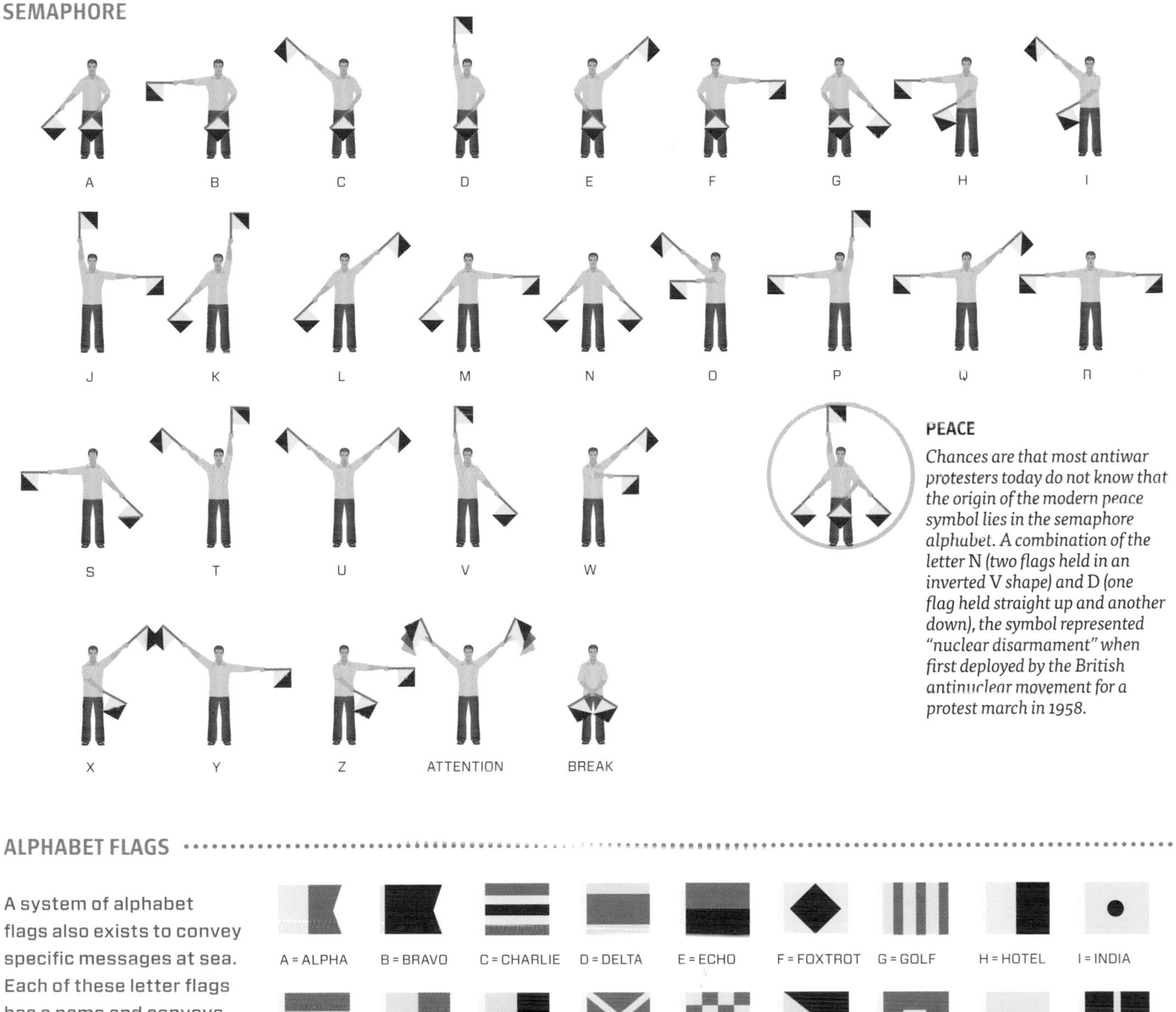

PEACE

Chances are that most antiwar protesters today do not know that the origin of the modern peace symbol lies in the semaphore alphabet. A combination of the letter N (two flags held in an inverted V shape) and D (one flag held straight up and another down), the symbol represented "nuclear disarmament" when first deployed by the British antinuclear movement for a protest march in 1958.

ALPHABET FLAGS

A system of alphabet flags also exists to convey specific messages at sea. Each of these letter flags has a name and conveys a specific message: for example the "a," or "alpha," flag, which is blue and white, conveys "I have a diver down under: keep well clear."

A = ALPHA, B = BRAVO, C = CHARLIE, D = DELTA, E = ECHO, F = FOXTROT, G = GOLF, H = HOTEL, I = INDIA

J = JULIETTE, K = KILO, L = LIMA, M = MIKE, N = NOVEMBER, O = OSCAR, P = PAPA, Q = QUEBEC, R = ROMEO

S = SIERRA, T = TANGO, U = UNIFORM, V = VICTOR, W = WHISKEY, X = XRAY, Y = YANKEE, Z = ZULU

COMMUNICATIONS AFLOAT

In 1899, just three years after Guglielmo Marconi received a British patent for a form of wireless radio using high-frequency electric oscillations, the first ship-to-shore message was transmitted from lightship #70 to the coastal receiving station at the Cliff House in San Francisco. "Radio telegraph men" soon began appearing on ships calling on coastal ports, and within a decade round-the-clock radio coverage was required on vessels of a certain size. (Ironically, both the *Titanic* and its rescue ship, *Carpathia*, were staffed by Marconi wireless operators—though the lone *Carpathia* operator had retired for the night and never received the distress call from the sinking ship.)

Today, a British satellite telecommunications company named Inmarsat provides communications for most seagoing vessels. The company maintains a set of roughly a dozen geostationary satellites that provide coverage of virtually all of the world's navigable waters. Reception between ship and satellite is unaffected by weather, location, or time of day.

Maritime communication requirements in the modern age, particularly for merchant ships, are very specific. The global maritime distress and safety system (GMDSS) prescribes minimum technological specifications for ships based on distance from shore and location on the water. These primarily address radio transmission requirements but also include search-and-rescue transponders and participation in a satellite-based emergency-position system.

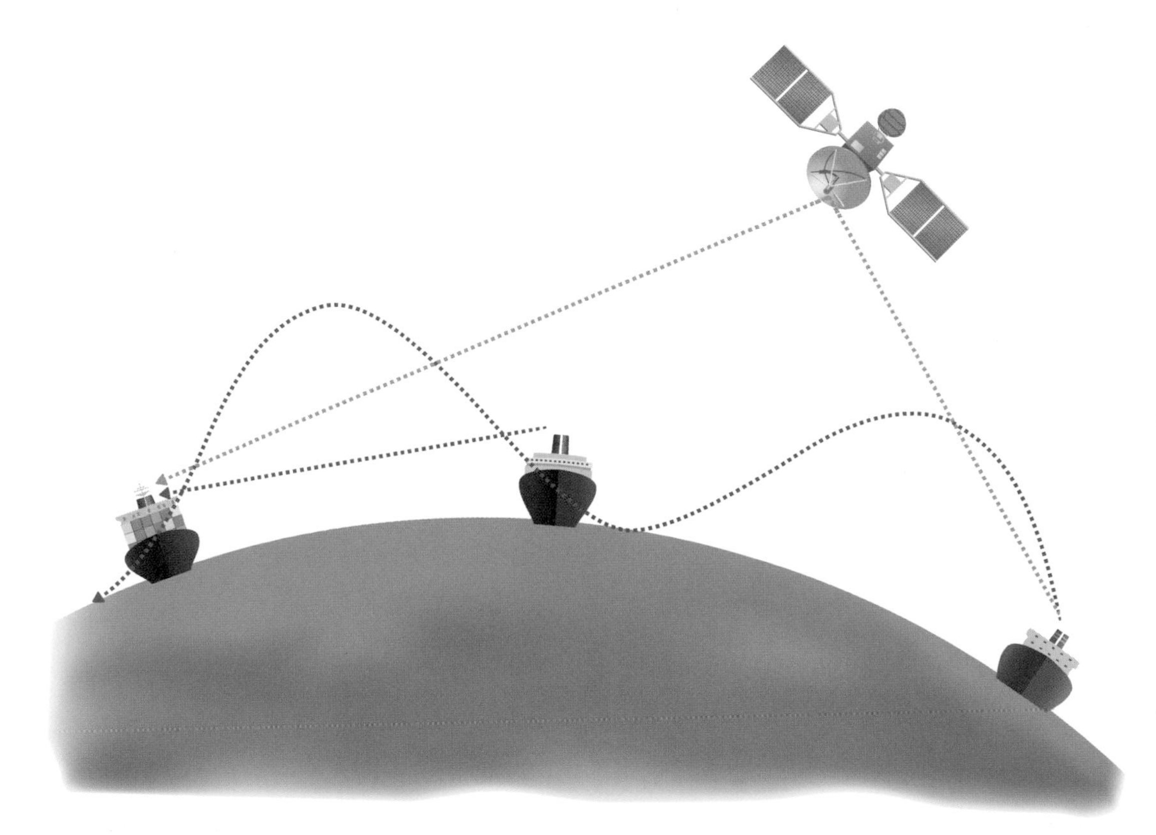

- *Marine VHF radio (also known as bridge-to-bridge) is installed on all large ships and most seagoing small craft. Used for a wide variety of purposes, including summoning rescue services and communicating ashore, it works over a range of only 10 to 20 nautical miles.*
- *The waves of HF radio can bounce off the atmosphere and ocean surface and can travel up to 10,000 nautical miles. It can be used for voice communications and data (although transmission rates are slow).*
- *Satellite communication is available but equipment and access are expensive. It allows access to voice and data almost anywhere on the ocean.*

PIRATE RADIO

"Pirate radio" isn't something dreamed up for a movie. Amateur radio operators were desperate to avoid regulation as far back as 1910—when they sent false messages and distress calls to launch naval boats on misguided missions.

Pirate radio made news again in the early 1960s in Britain in response to the unmet demand for pop and rock music. Broadcasts from ships in international waters technically did not run afoul of the BBC's monopoly on land-based radio—until the Marine Broadcasting Offences Act in 1967 officially outlawed offshore broadcasting and the pirates moved back onto land. Today there are an estimated 150 unlicensed broadcasters in the UK, including dozens in London.

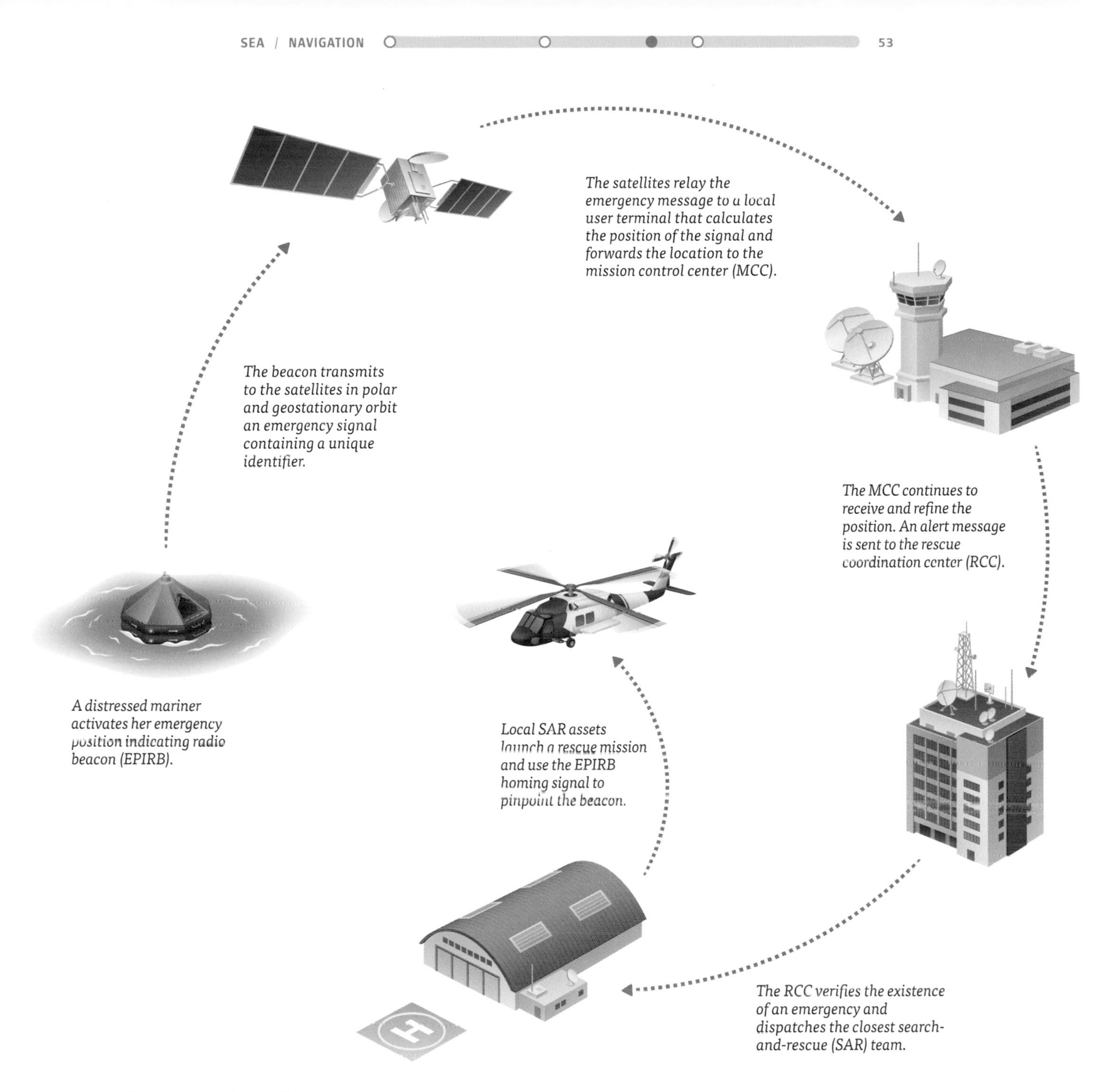

IN AN EMERGENCY

Conventions for emergency communications at sea date back to 1906, when the Berlin Radio Telegraphic Convention adopted the "SOS" code as the official international distress call: three dots, followed by three dashes, followed by three dots. Today, a particular frequency—Channel 16—is the international calling and distress channel.

Among the most significant advances in emergency communications at sea are what are known as emergency position indicating radio beacons (EPIRBs). These beacons are designed to transmit an alert via satellite connection in an emergency and can be either automatic or manual. Typically, they are mounted on a vessel in a way that allows them to float free in the event of an accident or sinking.

In addition to the beacons themselves, an international satellite-based search-and-rescue system, Cospas/Sarsat, based in Montreal, locates the signal when an EPIRB radio beacon is activated. Initially limited to four countries, the system today operates more broadly in cooperation with bodies such as the Civil Aviation Authority, International Maritime Organization, and other international entities.

Separately, a digital selective calling (DSC) system allows boat operators in distress to push one button to alert other ships and coastal monitoring systems to its identity and emergency. The system includes a built-in GPS connection so that the call also includes the vessel's precise location.

ANATOMY OF A WAVE

Everything from earthquakes to passing ships to skipping stones can cause waves, but the wind acting on the surface of the water generates most ocean waves. In deep water, a wave is really just a forward motion of energy, not water: the individual water molecules do not even move forward with a passing wave (they actually rotate in one place). The friction between the passing air and the water molecules forms wind-generated waves, causing ripples to develop on the water surface (think of lightly running your hand across a freshly made bed and the ripples that form on the sheets).

Numerous factors influence how big those initial ripples will become. Principal among them is wind speed: the faster the wind, the larger the waves. Likewise, the area over which the wind blows without obstructions from land (also known as fetch) influences the size and formation of waves. Wave generation is also affected by the amount of time (duration) that the wind has been blowing. When a weather front first enters an area, it can take hours for the wind to transfer its energy and the seas to develop. Once the front leaves, it can take hours more for the waves to subside under the force of gravity.

The depth of the water will also influence wave formation. In deep, open ocean, waves will generally build in steepness until their distance from crest (top) to trough (bottom) reaches one sixth of their length (distance from crest to crest). Above that height, the waves will break. Closer to shore, however, waves are influenced by the ocean bottom and may break when their height approaches the bottom depth.

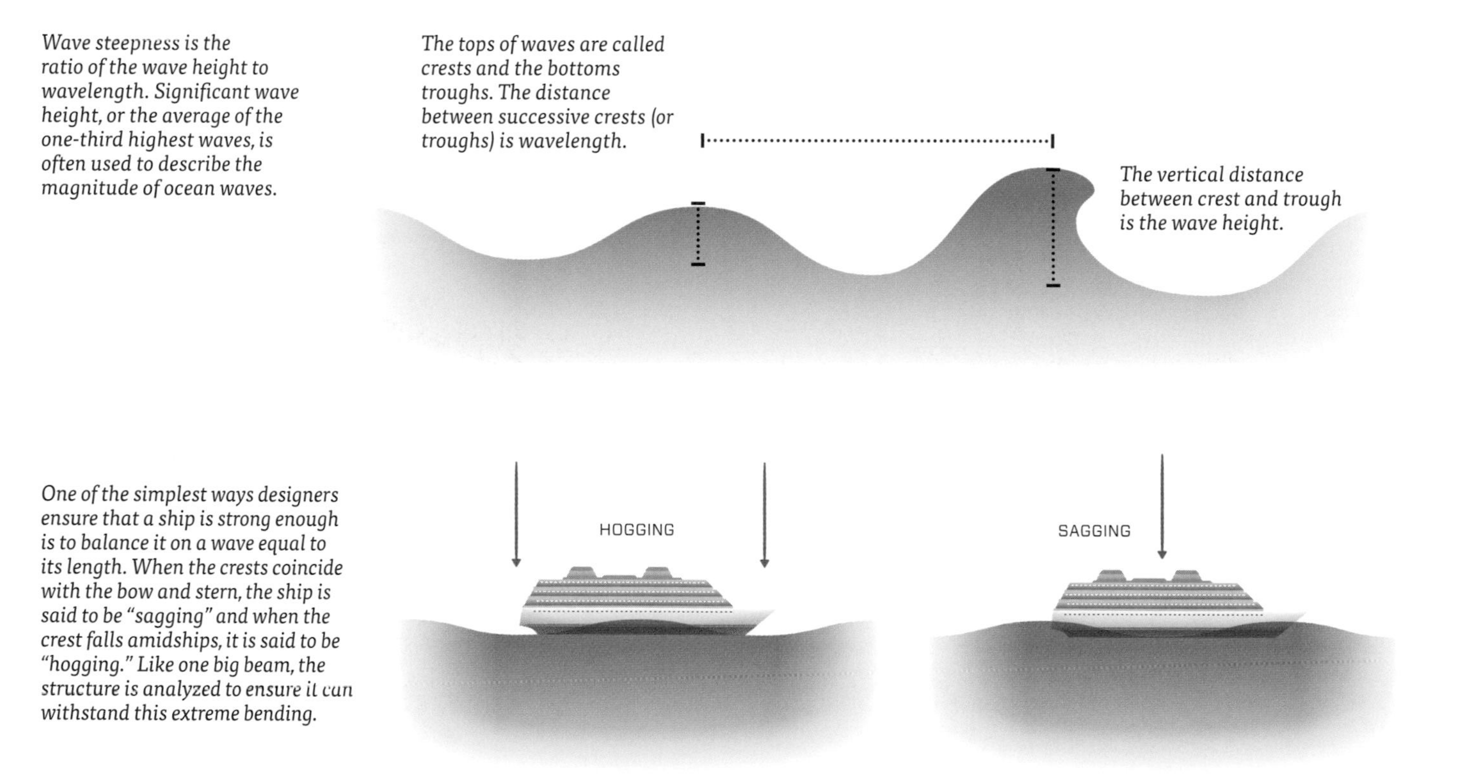

Wave steepness is the ratio of the wave height to wavelength. Significant wave height, or the average of the one-third highest waves, is often used to describe the magnitude of ocean waves.

The tops of waves are called crests and the bottoms troughs. The distance between successive crests (or troughs) is wavelength.

The vertical distance between crest and trough is the wave height.

One of the simplest ways designers ensure that a ship is strong enough is to balance it on a wave equal to its length. When the crests coincide with the bow and stern, the ship is said to be "sagging" and when the crest falls amidships, it is said to be "hogging." Like one big beam, the structure is analyzed to ensure it can withstand this extreme bending.

TSUNAMIS

Tsunamis (Japanese for harbor waves), often mistakenly referred to as tidal waves, are precipitated by a cataclysmic event in the open ocean—an earthquake, for instance. The resulting high-energy waves are extremely long (often hundreds of miles long) and move very fast (in excess of 500 mph/800 km/hr). To a vessel in the open sea, a passing tsunami wave might not even be noticeable with a height of about three feet (one m). However, as the ocean becomes shallower, the high-energy, fast-moving waves build into steep, high-amplitude leviathans that can cause extensive destruction to coastal communities.

WHEN WATER ATTACKS

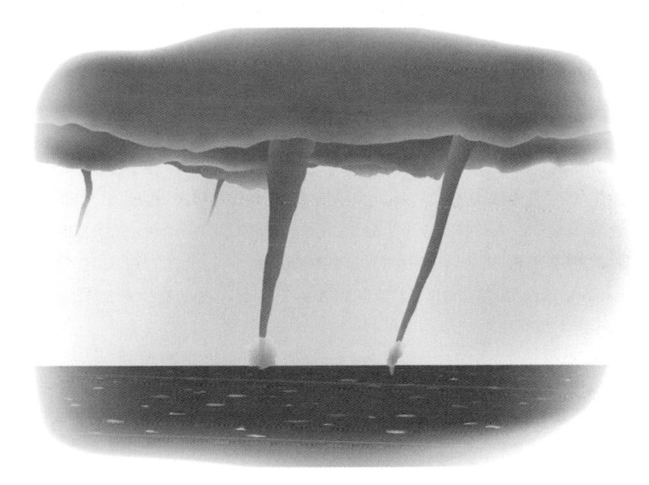

MAELSTROMS

Maelstroms are large whirlpools, typically found in places with extreme tides and currents. The word "maelstrom" is derived from the Nordic words "grind" and "stream," which is fitting: some of the largest permanent maelstroms in the world can be found off the coast of Norway. Homer famously wrote about Charybdis, a maelstromlike sea monster that threatened to destroy Odysseus' boat in The Odyssey.

WATER SPOUTS

Water spouts, a not uncommon seasonal occurrence in the Florida keys and Great Lakes among other places, are caused by a tornado's occurring over water. There are two types of spouts. A tornadic water spout refers to a spout that has drifted over water from its origins on land. Fair-weather spouts are caused by warm air and high levels of humidity in the lower atmosphere and are found most frequently in tropical and subtropical climates.

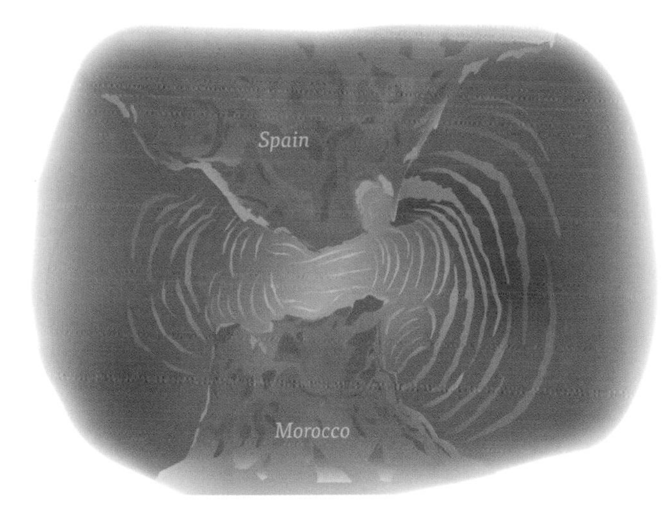

INTERNAL WAVES

Strange as it may sound, waves are not always found on the surface of the ocean. When bodies of water of different density (typically due to temperature or salinity differences) meet, such as at the mouths of rivers or major straits, internal waves can be created. Though not visible at the surface, the wave energy produced by these internal waves can lead to what is sometimes referred to as dead water—where ships in apparently calm conditions find it hard to move forward.

ROGUE WAVES

Rogue waves, often referred to as freak, monster, killer, or extreme waves, are among the most feared of sea dangers. Typically, they occur in the open sea, where the water is deep. Often they are explained as the result of currents converging, or wind pushing on the water surface, or even the shape of the coastline—but nobody really knows what causes them. The infamous sinking of the Edmund Fitzgerald *in the Great Lakes in 1975 has been attributed to a rogue wave.*

AVOIDING THE ELEMENTS

Weather-related and seasonal dangers can offer significant challenges to navigation on the seas. Some of these, such as ice or hurricane activity, are predictable based on time of year; others, such as random storms at sea, are not.

A number of tools are used by mariners to avoid these dangers. In addition to the meteorological forecasts provided by national weather forecasting bodies, a variety of commercial providers offer customized routing services and forecasts. Just as air traffic control will route an airplane pilot around a storm, shipping companies rely on private weather services to recommend the safest and cheapest route (in terms of fuel consumption) based on meteorological models, satellite data, and observations.

Ice offers some of the most significant challenges to navigation. This comes in the form of ice sheets and ice shelves (extensions of ice sheets into the sea), ice tongues (floating extensions of a glacier), and icebergs. The last are perhaps the most dangerous as only a small section of an iceberg may be visible to a ship at sea: on average, about one eighth of an iceberg is visible above water.

ICEBERGS

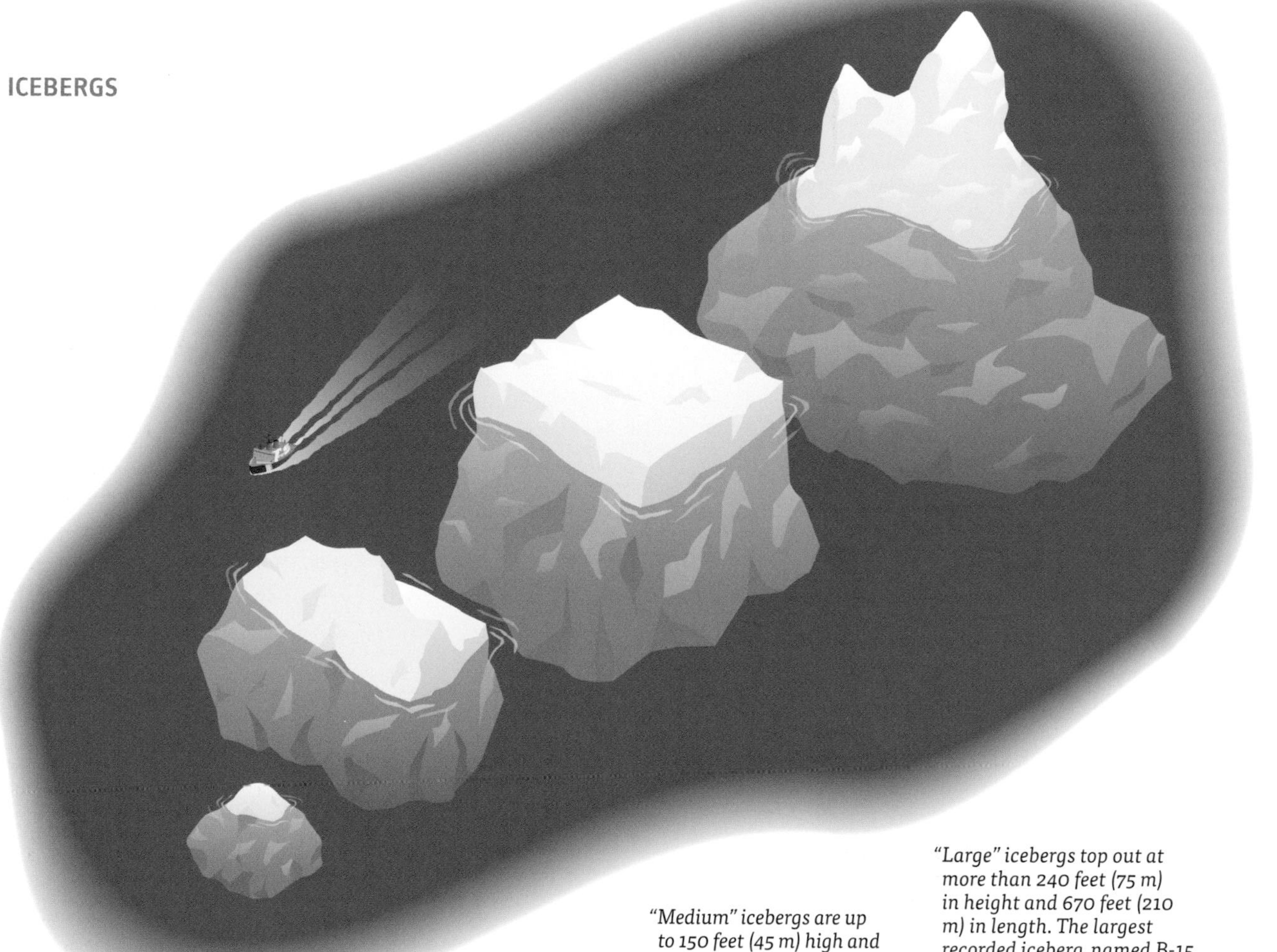

The smallest icebergs are known as growlers (derived from the noise they make as air escapes) and usually are smaller than a car. Icebergs slightly larger are known as bergy bits and are about the size of a small house.

"Small" icebergs can reach 50 feet (15 m) high and 200 feet (60 m) in length. These are some of the more common sizes found in the North Atlantic.

"Medium" icebergs are up to 150 feet (45 m) high and 400 feet (120 m) in length. Besides size, icebergs are also classified by shape. Tabular icebergs are flat topped. Nontabular shapes include domes, pinnacles, wedges, and blocks.

"Large" icebergs top out at more than 240 feet (75 m) in height and 670 feet (210 m) in length. The largest recorded iceberg, named B-15, broke off from Antarctica in 2000 and was larger than the island of Jamaica.

Icebergs are basically floating hunks of freshwater ice that have broken off from ancient glaciers. Though they are found at both North and South poles, those in the south tend to hug the coast of Antarctica and thus prove far less troubling to mariners. In contrast, an estimated 10,000 icebergs break off, or "calve," each year from the ancient glaciers of western Greenland, where the Labrador current pushes them southward. Roughly 200 to 300 of them make it into North Atlantic shipping lanes each year, where they can take anywhere from a period of months to a period of years to melt.

HURRICANES

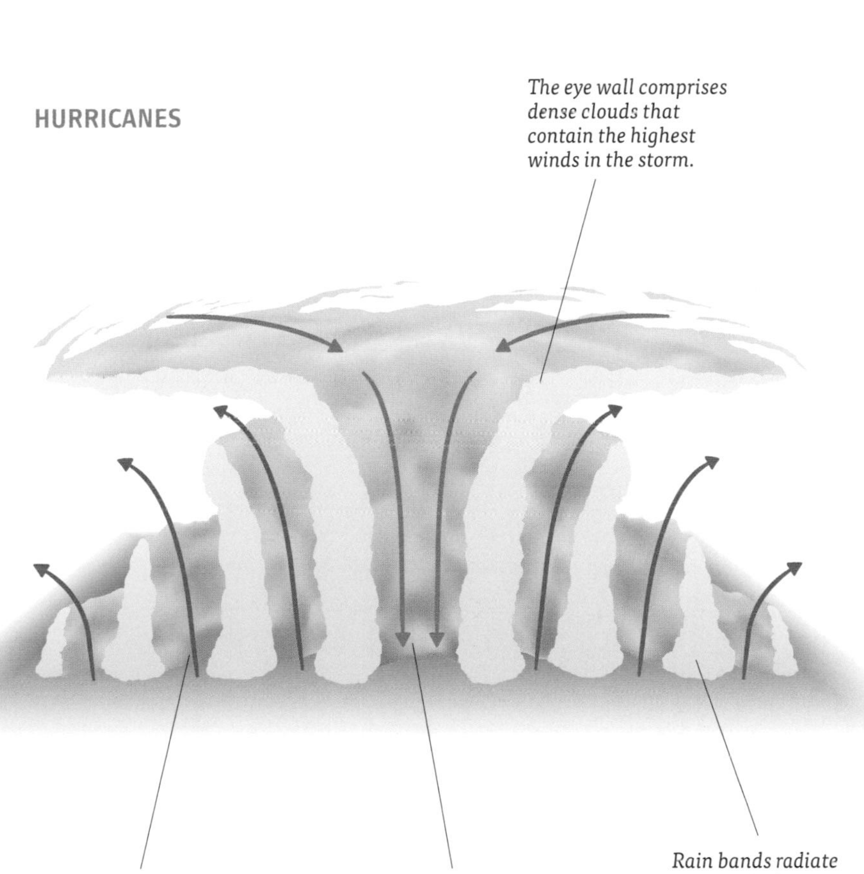

MAPPING ICE

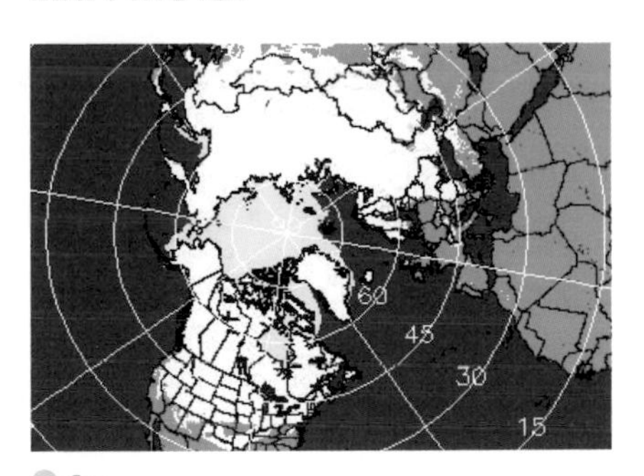

Ice
Snow
Land
Water

While icebergs are a hazard to ships, of equal concern to ships plying scheduled commercial routes in the polar regions is the seasonal formation of ice itself. Many ship channels, particularly in or near the Arctic, freeze over on an annual basis and become impassable at a certain point during the winter. Monitoring the ice floes and selecting a route around this ice before and just after these closures is critical to maintaining ship schedules.

In North America, the Canadian Ice Service (CIS) provides accurate information about ice in the country's navigable waters—rivers and lakes as well as coastal. Together with its U.S. counterpart, the U.S. National/Naval Ice Center, it produces maps of areas of particular concern, including the Great Lakes and Bering Strait, and identifies the formation and location of specific icebergs—such as those in Newfoundland and Labrador—likely to drift into major shipping lanes.

The formation of ice is not simply monitored passively in certain parts of the world. Particularly in the Baltic Sea, where many ports depend on year-round access, icebreakers are critical to commerce. Given the geography, it is perhaps not surprising that the vast majority of the world's icebreakers are Russian.

The United States has two active icebreakers—a "heavy" ship and a "medium" one (primarily used for scientific research). Requests for the funding of an additional heavy icebreaker have been submitted to Congress.

CLOSER TO SHORE

As the oldest form of movement of people and goods, maritime transportation has had an outsized impact on the evolution of the world's great cities. With few exceptions, the most powerful economies over time have been ports—for example, Athens, Alexandria, Istanbul, London, New York, and Hong Kong. Their harbors and waterfronts were in many ways their lifeblood, and commerce and trade the backbone of their economies.

The physical configuration of these cities reflected the dominant role of trade: wharves and docks were central to the economy and the towns more or less grew up around them. Countinghouses, transport, banks, and other support services located nearby sprang to life whenever ships were in port. Over time, some of these services would move away from the port, yet remained tied to it economically.

Shipping affected the social life of a city as well. Loading and unloading ships was hard work, but it was fairly regular and proved a magnet to those without the skills or connections to move beyond manual labor. In modern times, waterfronts became home to communities of immigrants and other outsiders—most of whom arrived there on ships and settled not far from the piers on which they landed.

While maritime trade retained a very tangible presence at the heart of port cities well into the twentieth century, waterfronts around the world would be transformed as a result of the invention of the container in the late 1950s. Container technology demanded vast upland area for storage and loading of containers and easy access to highways and railroads. Many ports found suitable land to handle containers on the outskirts of the city, and within a relatively short period of time a majority of their downtown piers fell silent.

Today, many of the world's waterfronts bear only a faint resemblance to what they looked like fifty years ago. The working waterfront is now often remote from city centers and populated more by machines than it is people. Many downtown piers and wharves, after decades of disuse, have become places of recreation and gentrification—sporting venues, parks, concert halls, and gleaming new commercial and residential buildings. The cries and shouts of longshoremen have, in places, been replaced by those of children in playgrounds.

Yet look beyond the waterfront esplanades and you will still see silhouettes in steel making for open water from berths and wharves located in remote industrial backwaters. If anything, the ships those silhouettes represent are bigger and faster than ever—carrying more diverse cargoes to more destinations around the world.

TIGER PORTS

Since the earliest days of sea travel, the world's biggest ports have reflected the world's most powerful trading nations. From Egypt and China through to European colonization of the New World, trade and commerce were the force behind economic and political development. Today that connection remains: many of the world's biggest ports are in Asia, reflecting the huge population growth and dynamic economies of that region.

SHANGHAI

The city of Shanghai has long been a center for international trade. Comprised of both river ports and a deep-sea facility, Shanghai is now the largest port in China and the world's busiest container port.

LONG BEACH

Long Beach and it sister port, the Port of Los Angeles, comprise the largest port complex in the United States. Powered by Asian trade, the port moves large volumes of incoming traffic east on double-stack trains serving the East Coast.

While their berths may be farther from city centers than they once were, today's cargo ships still demand a variety of critical services from the harbors and ports they serve. Above all they require people—local pilots, to navigate tricky channels or rivers, and Coast Guard authorities, to manage traffic in the harbor and ensure a place to anchor if needed. And they require increasingly expensive physical facilities—deep, dredged channels and long, safe berths with machines able to move vast amounts of cargo quickly and efficiently.

Cargo ships are not alone in needing assistance within the port. Recreational vessels and ferries likewise rely on a range of marine services for navigation and safety purposes. Cities and towns need to keep their municipal docks dredged and their waters free of debris, and must respond to fires and other emergencies on the water. And all boats, whatever their purpose, need regular repair and maintenance.

Various craft and equipment not usually seen on the ocean exist to meet these local maritime needs. Icebreakers, lightships, garbage skimmers, drydocks, tugs, and dredgers are just a few of the invisible workhorses of the world's ports, operating around the clock—and closer to shore—to meet the needs of waterborne traffic.

ON THE WATERFRONT

Though merchant mariners believe that nothing compares with the lure of the open sea, work along the shore—an expression that gave rise to the word "longshoreman"—has had an unusually colorful history in many countries, including the United States. Waves of immigration in the late nineteenth and early twentieth centuries brought skilled men to cities along America's eastern seaboard, but language handicaps, prejudice, and a lack of open land forced them to seek manual labor on the docks.

Dock gangs thus strongly reflected old-country ties: in Baltimore, nearly 80 percent of the port's longshoremen in the 1930s were Polish. In New York, the Irish and Germans controlled docks on Manhattan's West Side while the Italians had outposts both in Manhattan and Brooklyn.

But having the right ethnic background guaranteed no one a job. Men had to "shape up" on the docks to get work when a ship entered port and pay dearly for it: a toothpick behind the ear might signal willingness to pay off the dock boss for a day's work. Bucking the system, as fans of the movie *On the Waterfront* know, was not an option.

Today, dock hiring is still geographically based in places like New York: each pier has a local branch of the union from which much of its gang is drawn. But modern longshoremen are registered, seniority and skill are respected, and work is no longer really casual. Ever since a deal to allow automation provided qualified longshoremen with a guaranteed annual income whether they worked ships or not, longshore incomes have been more regularized.

SINGAPORE

The tiny island state of Singapore boasts one of the largest ports in the world. The bulk of the cargo moving through Singapore is not staying there: it is being transshipped from one location to another.

HONG KONG

Hong Kong's strategic location has made it a center of maritime activity. Compared with other ports, it is land starved—but it has pioneered technology that allows it to handle more cargo per square foot than anywhere else in the world.

ROTTERDAM

For decades the world's largest port, Rotterdam remains Europe's biggest—moving vast amounts of cargo into the continent by barge, rail, and road. It is also among the world's most automated, pioneering the use of robotic machines to move containers within its terminals.

HOME IN THE HARBOR

Busy harbors play host to thousands of vessels each year. New York is visited by about 5,000; Singapore welcomes several times that. On top of these registered callers are many more boats—including ferries, recreational craft, and workboats—whose journeys go unrecorded and are largely confined to the harbor itself.

Together, these denizens of the harbor represent a wide range of ship purposes and types. Some vessels are very big—cruise ships and oil tankers, for example, are among the largest. Some, such as tugboats and local ferryboats, are much smaller. Most travel under their own power, but a variety of engineless working boats—like dredges and barges—may rely on the power of a tug to move them safely through harbor waters.

Not all harbor vessels are used for transport. Police boats and Coast Guard vessels can be found patrolling the waters of the world's major harbors, enforcing safety and speed regulations and providing assistance to mariners in trouble. Other boats are working to keep the harbor open for business, including trash skimmers that collect floating debris and dredgers that provide the necessary depths of water in navigation channels.

PILOT BOAT

Often painted with a large "PILOT" across the side, they also fly a "H"otel signal flag and white over red all-around masthead lights.

DIVE BOAT

When there are divers over the side, these boats will display an "A"lpha signal flag, the commercial "diver down" flag, or both.

TRASH SKIMMER

These versatile boats can pick up solid and oil waste from harbors, protecting other vessels and the environment.

FIREBOAT

With extremely large pumps, these fast response boats can spray over 50,000 gallons of sea water per minute on a vessel or port fire.

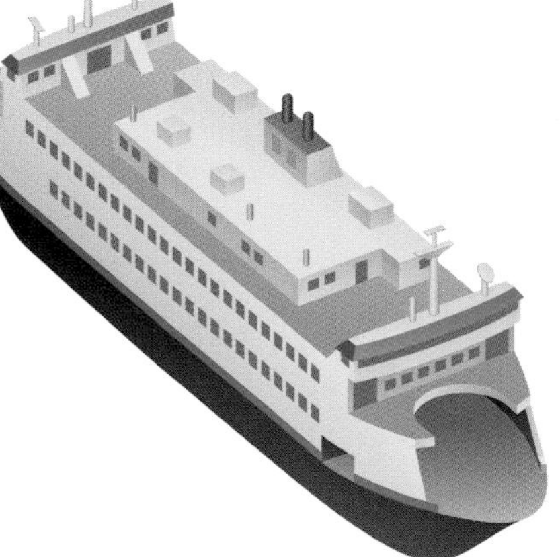

CAR FERRY

The largest car ferry operating between Ireland and Wales can carry 1,342 cars, 240 trucks, and over 2,000 passengers and crew.

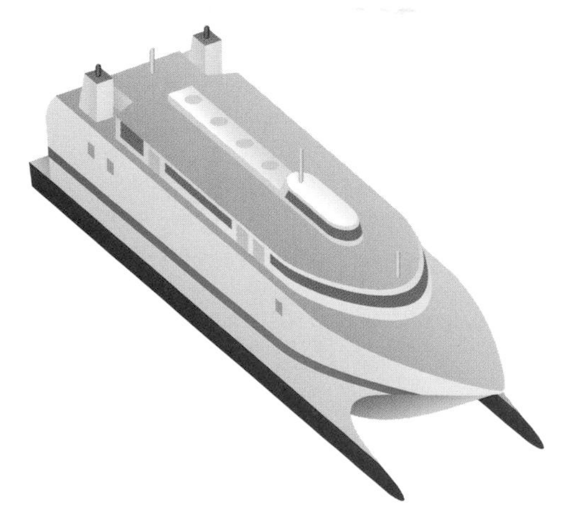

FAST FERRY

Fast ferries typically utilize catamaran, trimaran, or hydrofoil hull forms. The fastest in the world is said to travel 52 knots (60 mph; 96 km/h).

MOTOR LIFEBOAT

Designed never to capsize, patrol motor lifeboats perform rescues in dangerous surf zones.

MARK V

Heavily armed and armored, these shallow-water special operations boats can travel in excess of 65 knots (75 mph; 120 km/h).

SEAL DELIVERY VEHICLE

Designed to be launched by a submarine or ship, these minisubmarines can covertly deliver a small team of commandos.

TUGBOATS

Tugboats are the invisible workhorses of the harbor. Even ships that move under their own power often rely on tugs to help maneuver into slips and berths. They are enormously powerful—strong enough to pull or push heavily laden barges or bulk carriers many, many times their size. Many tugboats feature Voith Schneider propellers, a series of spinning, vertical plates that allow the ship to move instantaneously in almost any direction.

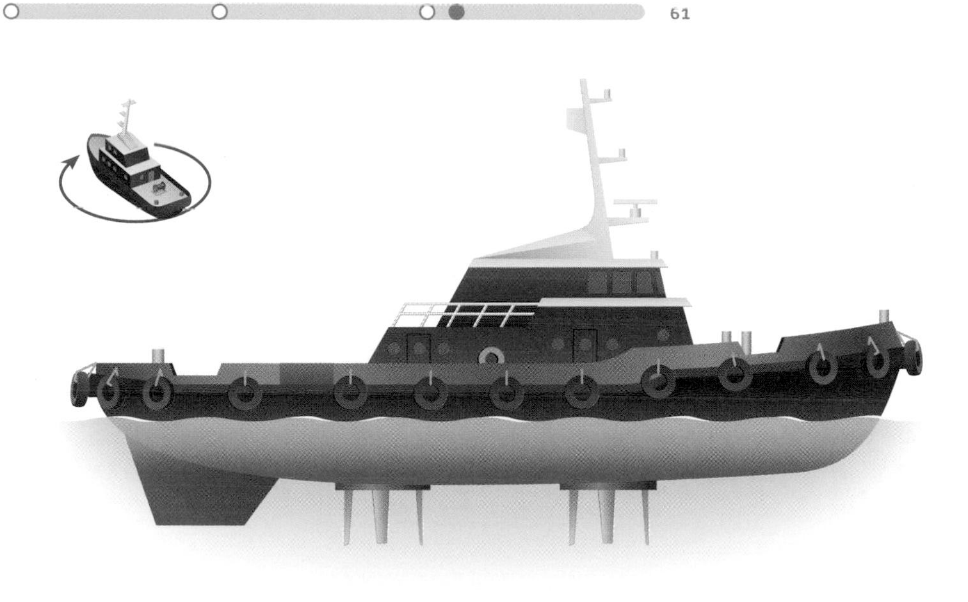

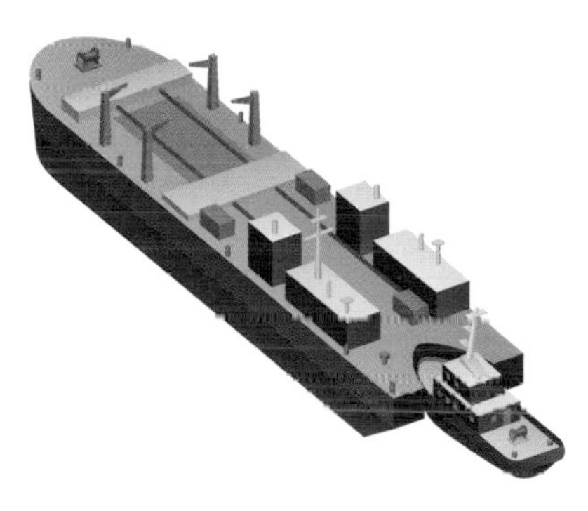

AT/B

An articulated tug barge is actually two vessels—a tugboat and the specially designed barge that connects to it.

FLOATING CRANE

Often used for offshore construction or maintenance, some floating cranes are semisubmersible.

HOPPER BARGE

A hopper barge is designed to carry materials, like rocks, sand, soil, and rubbish.

LCS

Designed to operate in the littorals, the littoral combat ship is a trimaran warship that can travel 44 knots (51 mph; 81 km/h).

STILETTO

An experimental warship utilizing a cathedral-shaped hull, it can travel up to 51 knots (59 mph; 94 km/h).

LCAC

The landing craft air-cushioned vehicle is used to ferry troops and material from offshore supply ships onto a beach.

The ship's hull plating is 2 to 3 inches (5 to 7.5 cm) thick at the bow, stern, and waterline. If ice penetrates the hull, a double hull and extra watertight compartments prevent flooding.

Some icebreakers utilize a bubbler system, in which compressed air is forced through rows of holes across the bow to form a cushion between ship and ice. Others rely on low-friction coatings to resist ice accumulation.

The bow of an icebreaker is U-shaped to maximize its ice-breaking capability. The tradeoff is that these ships aren't very seaworthy in the open ocean.

BREAKING ICE

In some of the most northerly parts of the world, icebreaking ships are critical to the year-round existence of local communities. Nowhere is this truer than in Russia, where remote coastal towns depend on these ships to maintain their connections with the outside world during the long winter. As a result, Russia has been and remains a driving force in icebreaking technology.

Modern icebreakers usually have thick, rounded keels, with shielded propellers at the bow and stern of the boat. They feature very powerful engines and searchlights as well as pumps that move the weight of ballast water from side to side as a tool to break up ice. Most are also equipped with blowers that produce air bubbles just under the waterline to assist in loosening the ice. Some also carry helicopters—which offer the crew a unique vantage point from which to see patterns of ice and locate open water.

Icebreakers work unlike any other boat: they literally beach themselves on top of an ice floe or shelf and use their weight to loosen part of it. In most cases, it is their reinforced, rounded bow that undertakes most of the work. Recently, however, experimental double-acting icebreakers have been built. Their pointed bow allows more stable and efficient movement through open water than a traditional rounded-hull icebreaker, but they can spin on their own axis once they encounter ice—so that their rounded stern can take on icebreaking duty.

RUSSIAN NUKES

Russia has some of the coldest weather in the world. It also has the world's largest fleet of icebreakers—serving both coastal ports and polar sea routes. Six powerful icebreakers with civilian crews regularly ply the waters of the Arctic north of Siberia, serving other locations as needed. But it is not their fleet size or the strength of their hulls that is most notable: it is their engines.

Since the debut of the *Lenin* in 1959, the Russian fleet has relied on nuclear fuel to provide the immense power each vessel needs to move through thick blocks of ice. Today, the fleet includes the NS 50 *Let Pobedy*, the largest icebreaker in the world. Roughly 522 feet (159 m) in length, it houses two atomic power plants and is served by support ships that provide new (or remove spent) fuel, test for radiation, and monitor workers' health.

KEEPING CHANNELS DEEP

Few ports in the world are naturally deep enough to permit passage of the largest of the world's passenger and cargo vessels. Most are fed by rivers or watersheds that carry a steady stream of silt, sand, and other debris into their waters. These sediments deposit themselves as shoals within the harbor and must be removed regularly to provide the deep channels needed by today's ships.

Most dredging is classified as either maintenance dredging, needed to maintain preexisting harbor depths, or as deepening, involving the removal of material from the bottom of a channel or berth. As ships have gotten larger and draw more water, ports have competed to keep their business by deepening navigational channels to accommodate them. Today's largest containerships demand 55 feet of water or more—a tall order for ports like New York, where the natural depth of certain key channels is under 20 feet.

The equipment used to dredge the world's harbors is big, expensive, and varied—representing a number of different dredging technologies: mechanical, hydraulic, and hydrostatic. To a large extent, the choice of equipment is a function of its suitability in grabbing or sucking a particular type of sediment or rock—ranging from fine sand to heavy boulders—from the harbor bottom.

Material from maintenance dredging or deepening work can often be repurposed to the benefit of the port. In some places, dredged material can be used to create more land along the waterfront for commercial or recreational purposes. In others, it might be used to create fish habitat (typically in the form of reefs) or bird habitat (often in the form of new islands). Some of the cleanest, sandiest dredged material can be used to replenish beaches victimized by storms or subject to more natural forms of erosion.

But dredged material is not always clean. Today, in many developed parts of the world, dredged material must be tested for contaminants and disposed of according to a rigorous protocol, with only the cleanest spoils suitable to be dumped back in the ocean or placed on nearby beaches.

In the United States and other countries, contaminated sediment must be remediated, or cleaned, before it can be dumped. Remediation often involves mixing the dredged material with Portland cement or other material that will bind it—making it safe enough to be used as fill to construct golf courses or other lightly used forms of topography.

MECHANICAL DREDGE

Mechanical dredges boast long arms that reach down to scoop material on the sea floor and deposit it in a barge. A variety of types of buckets might be attached to the arm depending on the material to be dredged. Some barges are known as dump scows: their bottoms open up for ocean disposal.

HYDRAULIC DREDGE

Hydraulic dredges suck the material and surrounding water into either a pipeline, which connects it to a receiving site, or into a storage receptacle, where it is drained of water and then discharged at a disposal site. The latter form is known as a hopper dredge and is best suited to heavier material.

HOPPER DREDGE

Hopper dredges operate much like giant, self-propelled vacuum cleaners. To dredge a channel, the large drag arms are lowered over the side to the channel bottom. While moving slowly forward at two to three knots (up to four mph or seven km/h), the drag arms suck a water and sand mixture, known as slurry, from the channel bottom through pipelines and into the hopper.

BRIDGING LAND

The earliest man-made waterways were not for transportation; they were irrigation canals serving agrarian societies in the Middle East, India, and Egypt. Canals for transportation appeared around 600 A.D.: the 98-foot-wide Grand Canal of China was built to carry the Chinese emperor between Beijing and Hangzhou—a distance of just over 1,100 miles (almost 1,800 kilometers).

The early canals were basically trenches filled with water and lined with a watertight material. They became popular during the Middle Ages as the cheapest and most efficient way to travel through countryside that featured primitive and often unsafe roads. The Industrial Revolution and its need to move raw and finished materials cheaply greatly contributed to the use and popularity of these waterways.

The most ambitious canals, with respect to transport, were those that connected bodies of water across land. In the United States, the Erie Canal was completed in 1825 to connect New York City and the Mid-Atlantic region to the Great Lakes. More than forty years later, the Suez Canal would provide passage between the Mediterranean and Red seas. And in 1914 the Panama Canal opened, linking

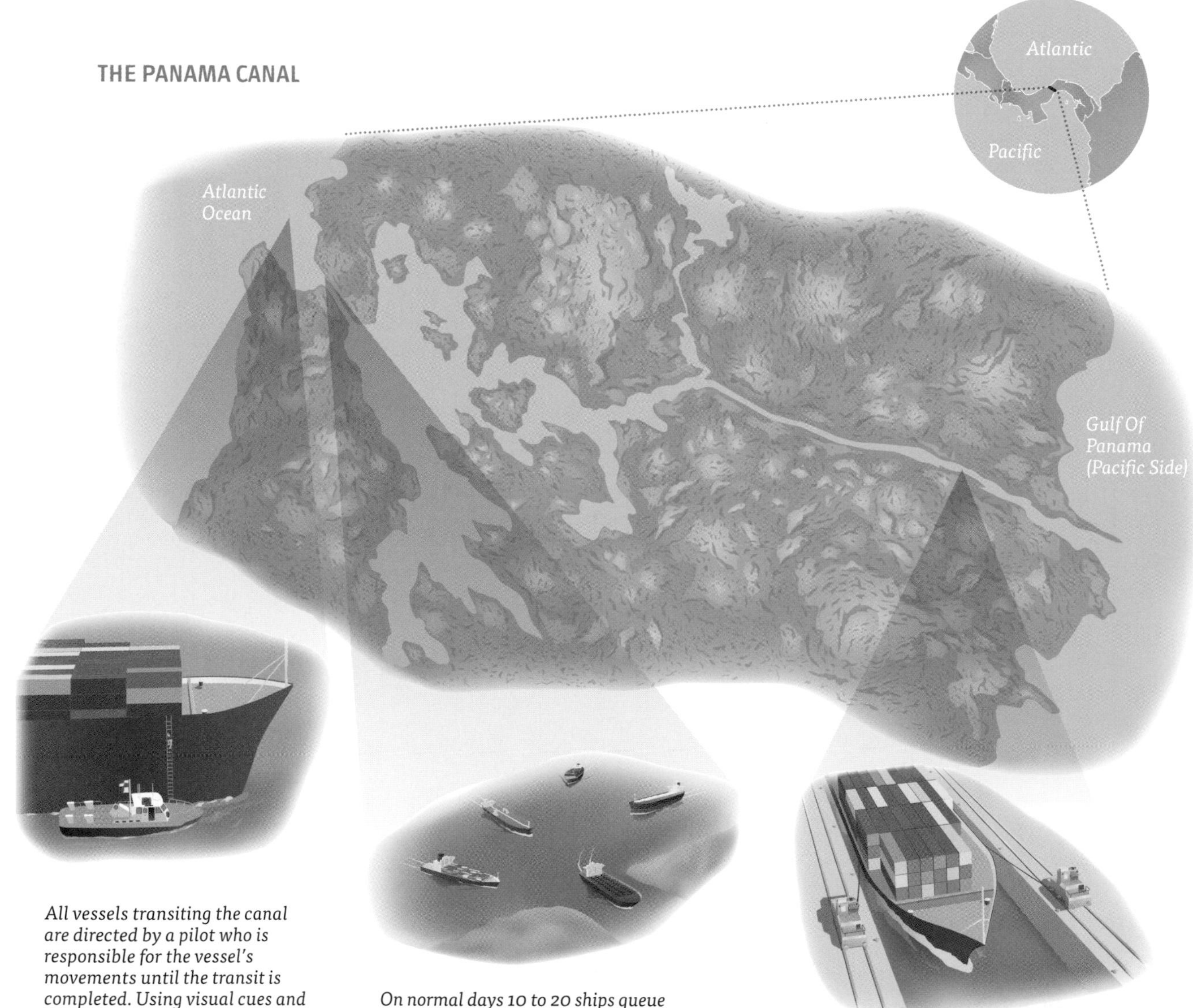

All vessels transiting the canal are directed by a pilot who is responsible for the vessel's movements until the transit is completed. Using visual cues and years of experience, the pilot aligns the ship within the canal.

On normal days 10 to 20 ships queue to transit the canal but demand can approach 100 ships per day. Some shipowners will pay huge sums to jump the queue—often hundreds of thousands of dollars.

While most ships transit the canal under their own power, larger ships might be tied to electric locomotives running on tracks. Known as mulas or mules (they were never actually animals), they mostly provide emergency stopping power.

the Pacific and Atlantic with a hundred-mile journey across the Panamanian isthmus.

Over the course of the nineteenth century, many canals (including the Erie) ceded their traffic to the expanding railways. But critically situated canals like Panama and Suez continued to play a significant role in maritime trade throughout the twentieth century. Today, both these and smaller barge canals are enjoying new investment and forecasting future traffic growth.

The components that support these canals have changed little over the centuries. Almost all canals are dependent on a system of locks, or chambers, that raise or lower ships to accommodate grade changes in the surrounding land. Most are fixed, with only the water level within them moving up or down. However, some locks operate more like boat lifts and incorporate a chamber that itself rises and falls to meet requisite height differentials. A few employ marine railways, which move boats up and down a slope on rails and carry the boat "dry." Some combine technologies: the Three Gorges Dam on the Yangtze River in China, for example, contains five large ship locks with a boat lift, which if needed can raise a 3,000-ton ship vertically.

HOW A LOCK WORKS

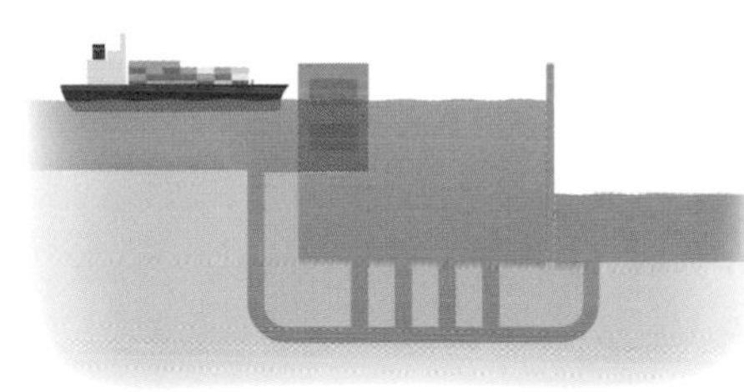

Unlike the Suez Canal, which is entirely at sea level, the Panama Canal relies on a series of three Atlantic locks that raise westbound ships 85 feet (26 m) to Gatun Lake and then three Pacific locks that lower them back down to the slightly higher Pacific.

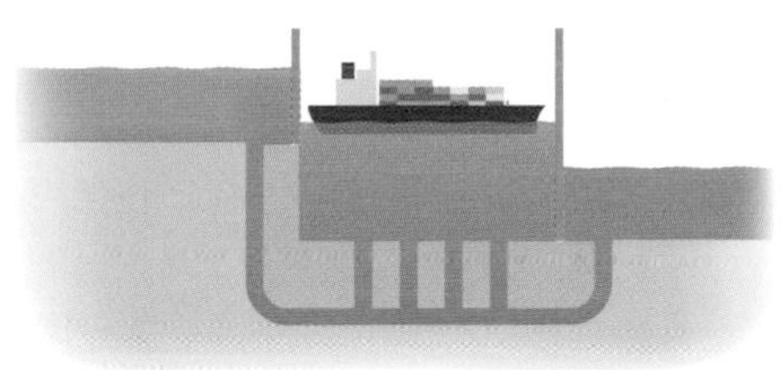

Each lock requires nearly 27 million gallons (101,000 cubic meters) of water to fill it from the lowered to the raised position. In each of its side walls there are three large water culverts used to carry water both from the lake into the chambers to raise them and from each chamber down to the next, or to the sea, to lower them.

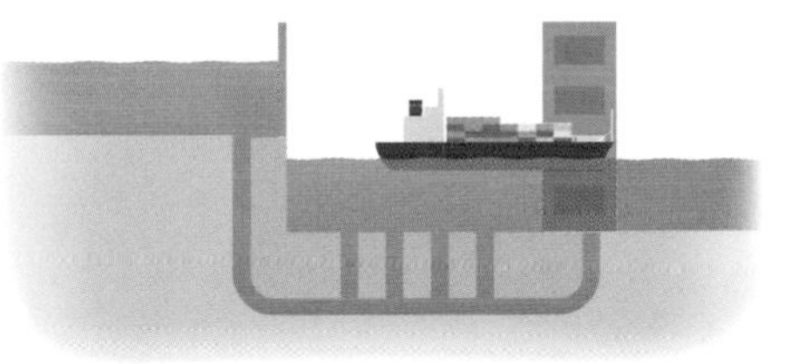

The water is moved by gravity and is controlled by huge valves in the culverts. Each 110 feet by 1,050 feet (33.5 m by 320 m) lock chamber can be filled in as little as eight minutes.

PANAMAX PLUS

PANAMAX
(965 FEET LONG, 105 FEET WIDE)

NEW PANAMAX
(1,200 FEET LONG, 161 FEET WIDE)

SUEZMAX
(UNLIMITED LENGTH, 164 FEET WIDE)

The Panama Canal opened in 1914 to much fanfare and had a significant impact on the world's shipbuilding industry. "Panamax" vessels—the largest ships able to transit the canal—became and remained for many decades the shipping industry standard.

Ultimately, the demand for containerized cargo gave rise to post-Panamax ships (those too big to transit the canal) and subsequently to pressure to widen the canal. In 2006, a plan to develop a new, third lane of the canal—along with larger locks and wider channels—was approved by national referendum.

With the completion of the project in 2015, the dimensions of the ships able to pass through the canal will increase significantly, doubling its annual throughput capacity from 300 million tons to 600 million. But the growth in capacity will not come cheaply: the estimated cost of the expansion is $5.25 billion.

STOPPING

Unlike a car or plane, which needs tremendous force to move even a short distance, an untethered ship will drift aimlessly with the current if left to its own devices. To hold it in place, a variety of anchors and mooring fixtures—often referred to as ground tackle—have evolved.

Temporary mooring fixtures, or anchors, have metal flukes that hook on to rocks or bury themselves under sand on the ocean floor. Anchors are stored within the ship's hull and are lowered on a chain through the hawsepipe, or hole in the hull, to the ocean's bottom.

A large anchor, with its accompanying chain, can weigh an enormous amount. For example, each anchor on the aicraft carrier U.S.S. *Truman* weighs 30 tons; a link alone on its anchor chains weighs 360 pounds. Anchor chains such as this can extend to 1,000 feet or more, offering the ability to stay put in fairly deep water.

Moorings, by contrast, are a more permanent fixture on the ocean floor. They can take many forms—from a dead weight like a concrete block to an upside-down-mushroom-shaped weight that buries itself in sand or silt.

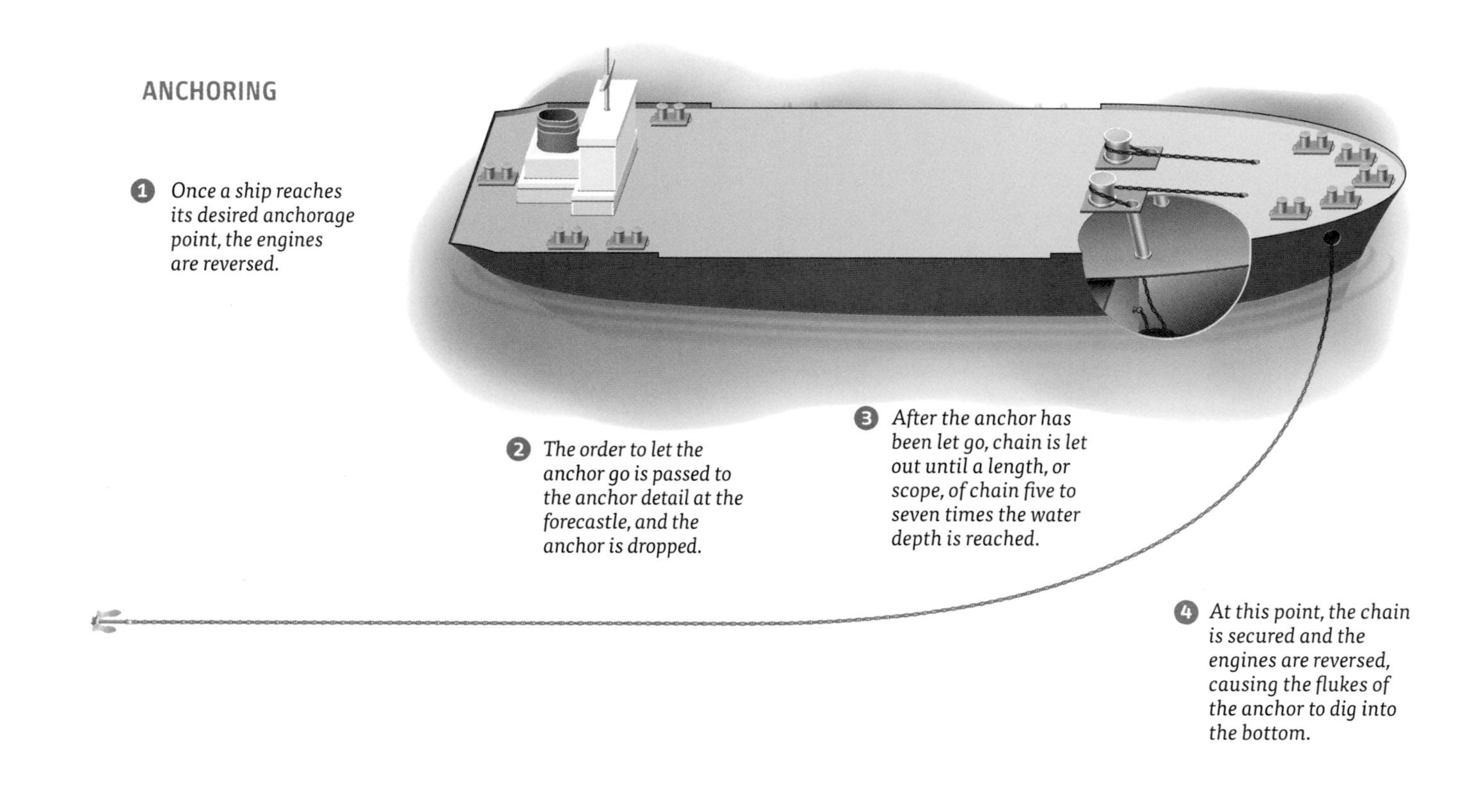

ANCHORS AWEIGH

The earliest anchors consisted of nothing more than heavy rocks. Today's anchors are somewhat more sophisticated, holding anywhere from 10 to 200 times their weight. Fluke and plow models are most popular, though the choice of anchor style is determined by a variety of factors: the type of seabed (for example, rock, grass, mud, or sand), the size of the vessel, and the state of both the seas and the wind.

MUSHROOM

SUBMARINE

STANDARD STOCKLESS

FISHERMAN

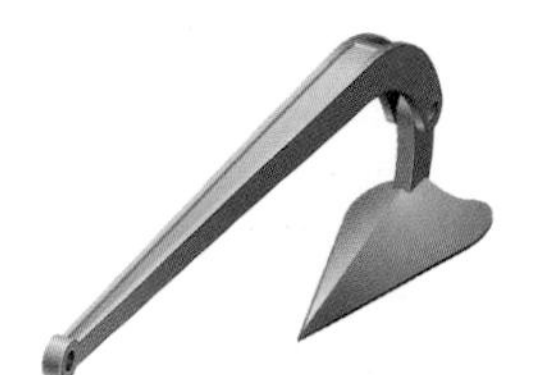
CLYDE QUICK RELEASE

DROGUE

BRUCE CLAW

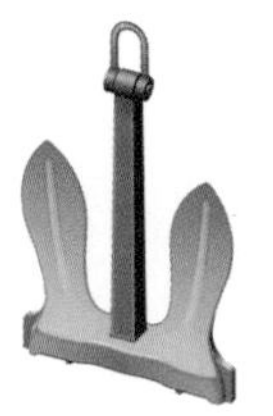
DANFORTH

Screw-in moorings also exist: in this case the blades spiral down and embed themselves in the sea floor. Typically, a floating marker identifies the mooring and can be hoisted aboard ship to secure the vessel at the mooring location.

"Mooring systems" (as opposed to "moorings") is a term used to refer to a wide range of equipment that secures a ship to a pier or wharf. Mooring systems generally include a set of ropes to secure the ship, fenders or bumpers to ensure that it does not rub directly against the dock during berthing, and a variety of other specialized equipment to prevent damage to both ship and dock while it remains at berth.

In many ports, large vessels do not park themselves for long periods of time at a dock: dock space is expensive and generally wants to be allocated to ships actively loading and unloading. Instead, they will tie up at anchorages—designated parking spaces for oceangoing vessels, located at suitable points throughout a harbor or channel. Control over these anchorages is often vested in the Coast Guard or local maritime authority.

MOORING

Breasting dolphins (as well as camels and fenders) help distribute the force of the ship evenly against the pier and ship structure.

Bow, stern, and spring lines prevent forward and backward movement while breast lines help keep the ship close to the pier.

Bitts, mostly found near the bow and stern, are used to secure lines on ships. Bollards, shaped like mushrooms, might be used in conjunction with bitts.

AT THE ZOO

A lot of mooring equipment sounds like it comes right from the zoo. "Dolphins" are a sort of floating structure used to extend a berth ("berthing dolphins") or moor a ship ("mooring dolphins"). Often they cushion a ship's impact, like a sort of extra fender.

"Camels" are also used in berthing ships. Typically, they have a rectangular shape and are set between the ship and pier to distribute a ship's weight to a variety of wooden or other piles with which it will come into contact while at berth.

Even the lines used in mooring ships carry animal monikers. To add weight to a line being thrown from ship to shore (or vice versa) during tie-up, a "monkey's paw" or "monkey's fist," shown here, might be added to the line: it's a type of knot wrapped around a weighted object that makes it easier to hurl the line a significant distance.

THE CONTAINER REVOLUTION

The biggest ships that regularly move through ports, and the ones that often require the deepest water, are containerships. Measured by their carrying capacity in what are known as twenty-foot equivalent units (TEUs), these functional-looking vessels now carry up to 8,000 or 9,000 forty-foot-long aluminum containers on their decks, each containing up to 20 or 25 tons of cargo.

The marine container is commonly eight feet wide and eight feet tall and comes in lengths of 20, 40, 45, 48, or 53 feet. Each one has a unique code painted on its side for identification and tracking and sports a novel twist-lock feature on its corners that allows it to be connected to adjacent containers for stability at sea. As a result, these containers can be stacked tall—up to seven high—both on land and afloat, with little risk of cargo loss due to weather or high seas.

In 2011, the World Shipping Council estimated that the number of ocean shipping containers in use around the globe was 18.6 million units. Most of these are standard 8 x 8-foot boxes, although "high-cube" containers—which measure 9 feet 6 inches or 10 feet 6 inches in height—are occasionally used. There are also a variety of specialized containers geared toward particular commodities: for example, reefers are refrigerated containers that carry perishable cargoes such as fruit and meat, tanktainers have a tank inside a standard container frame and carry liquids, and bin liners are designed to move rubbish from cities to recycling or dump sites.

THE PATH OF A CONTAINER

The movement of a shipping container today looks nothing like traditional longshore work. It's a highly automated, precise process, involving sophisticated computer programs and cranes and little if any manual labor. A form of intermodal transport, it involves both sea and land segments—the latter undertaken by either rail or truck.

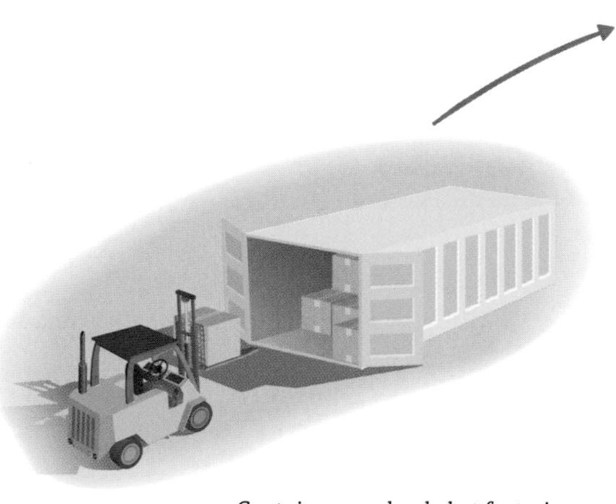

Containers are loaded at factories or warehouses. Forty-foot-long containers are most popular, although volumes are measured in terms of twenty-foot equivalent units (TEUs).

Once loaded, the container is put on a truck or a railcar for the journey to the port from which it will be shipped.

MCLEAN'S MAGIC

The invention of the container changed the world—making it affordable for people on one side of the globe to purchase items from another and leading to an explosion in maritime trade. Yet the man who invented the container was in no way related to the sea.

Malcom McLean was an American trucker, a scrappy Southerner who'd built a business moving other people's freight—on land. While delivering a shipment of cotton to a port, he noted how inefficient it was to unload the entire contents of the truck piecemeal onto a ship—only to reload them into a truck again at the other end.

Malcom devised a cargo box that could be detached from a truck chassis, placed securely on a ship, and then reattached to a second chassis for delivery to its destination. The labor saving was enormous—resulting in a drop in transportation costs of up to 80 percent. But perhaps his cleverest move was not to patent the technology: he shared it with the industry, and as a result it quickly became the international standard.

Since the invention of the container in the midtwentieth century, the system for moving containers on and off ships at ports of call has become fairly sophisticated. Because time is money in maritime trade, a range of specialized equipment has been devised to move these boxes as rapidly as possible to their destinations. The most visible of this equipment are towering container cranes, which line certain berths at major ports and can reach across more than 20 rows of containers on a ship to raise or lower a particular box.

Containers move inland from the port by truck, train, or barge. Over the last half century, the vast majority has been handled by truck: drivers arrive at the port with only a cab and connect with a chassis and container or, alternatively, arrive with a container to drop off for loading onto a ship. However, increasing numbers of containers are now moving inland from ports by rail, thanks partly to the development of new on-dock rail terminals featuring rail spurs adjacent to the docks.

The way that containers move inland is in part determined by geography. Many rail movements in the United States take the form of double-stack service: containers are placed on top of each other and thus efficiently carried two-high across long distances. In Europe, where distances are shorter and neither tracks nor rail routes are configured to take the extra weight and height required for double-stack moves, barge movements are a viable alternative to traditional truck and rail carriage.

At the port, the weight of each container is assessed and it is assigned a particular location on the ship. Careful balancing of cargo is critical to maintaining the stability of the ship at sea.

Container cranes with an extended reach lift the container and load it into the designated cell on the containership. It is then locked into place by its corners to prevent movement at sea.

After arrival at its destination port, the container is removed from the ship and placed by a crane either on a railcar or on a truck chassis for delivery to its intended recipient.

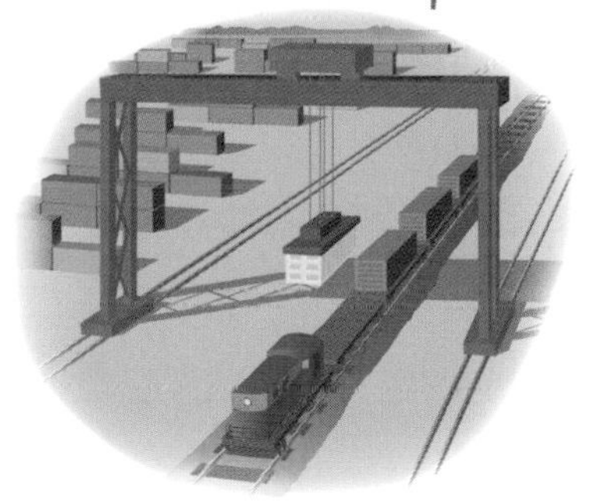

At the destination, the container is removed from the truck chassis or railcar and typically moved to a warehouse for unloading and processing.

SAFETY AT SEA

Security on the seas has been an issue for as long as commercial ships have plied the world's oceans. Largely out of sight of governmental authorities, pirates have historically found it relatively easy to hijack a merchant ship without consequence—either to take it captive or simply to steal desirable cargoes like opium, gems, or slaves.

Today, piracy at sea remains a concern to mariners from all countries. Hundreds of attacks on big and small boats are reported each year. Most of these occur between the Red Sea and Indian Ocean, off the Somali coast and in the Straits of Malacca—places where ships have to slow down to ensure clear passage. Though the losses to the industry from piracy are significant (an estimated $13 to $16 billion each year), there are few ways to pursue and even fewer to prosecute: most incidents take place outside of territorial waters.

Security in port raises a different set of issues for shippers and regulators than safety at sea. Waterfront theft and crime still exist in ports in developing countries, but are less of an issue in developed ones thanks to containerization. Instead, the focus of many port security teams is on smuggling and terrorism.

PIRATE PATCHES

BARBARY

Among the most notorious sea criminals were the Barbary pirates, who operated off the coast of North Africa between the sixteenth and nineteenth centuries. So established was the Barbary gang that rulers of places like Morocco, Tunisia, Algeria, and Libya took a cut of the profits from their regular attacks upon European ships and from the tribute payments exacted by the pirates in return for not attacking ships passing through the area.

SOMALI

In eastern Africa, pirates have proved that even the largest ships are vulnerable. A small band of pirates attacked the containership Maersk Alabama off the coast of Somalia as it headed for Mombasa, Kenya, in 2009—holding the ship's captain hostage until they were killed by U.S. Navy personnel.

PRIVATEER

Authorized by governments under letters of marque, privateers were permitted to attack foreign shipping during wartime from the sixteenth through the nineteenth centuries. Used successfully by the British against the Spanish and the American colonists against the British, these private citizens were able to keep their gains from captured goods and vessels.

MALACCA

The Straits of Malacca remain one of the most dangerous passages in the world as far as piracy goes. The 550-mile (885 km) channel separating the island of Sumatra from the Malay peninsula saw several hundred attacks between 2002 and 2007—leading Lloyds of London to classify it as a war zone.

JOLLY ROGER

The curiously named Jolly Roger has been a calling card of pirates as far back as the early eighteenth century. A white skull set above a crossed pair of bones on a black background was perhaps the most recognized (though not the only) flag used to signify the pirates' intention to fight to the death against anyone who failed to yield to their demands.

Flown by "Black Sam" Bellamy and Edward England

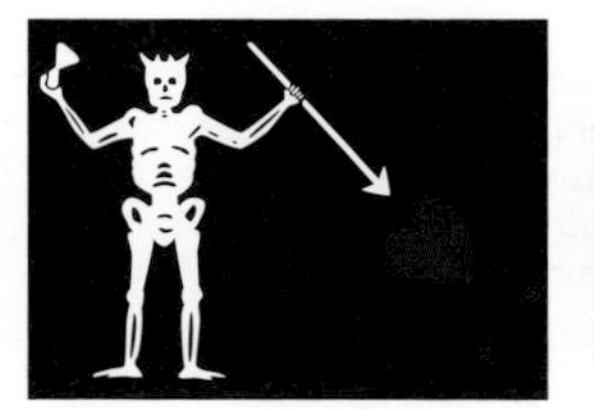

Flown by "Blackbeard"

Flown by "Calico Jack" Rackham

With much of maritime trade containerized, it is not always easy for customs or local port authorities to identify smuggled commodities, such as drugs and nuclear material. In some ports, dogs and strategically placed cameras can identify unusual cargoes and behaviors. In others, officials rely on expensive sniffers and radiation detection equipment that can more reliably detect illegal cargoes.

Regardless of technology, it remains all but impossible to properly check each of the thousands of containers moving through the world's largest ports daily: the cost of procuring sufficient manpower and equipment to do so would be more than ports or customs agencies could afford. Instead, officials typically rely on a virtual review of all cargo data and educated and random sampling of suspicious or high-risk containers.

In recent years, a new threat—terrorism—has loomed large on port security agendas around the world. Considerable effort now also goes into the vetting of port personnel, including terminal workers, truck drivers, and longshoremen. In many ports, the informal shape-up for work on the docks has given way to a system of background and fingerprint checks and identification tags.

PORT SECURITY

CONTAINER SCANNING

The first line of defense involves using automated targeting tools to identify containers that pose a potential risk for terrorism and then prescreen and evaluate those containers before they are shipped.

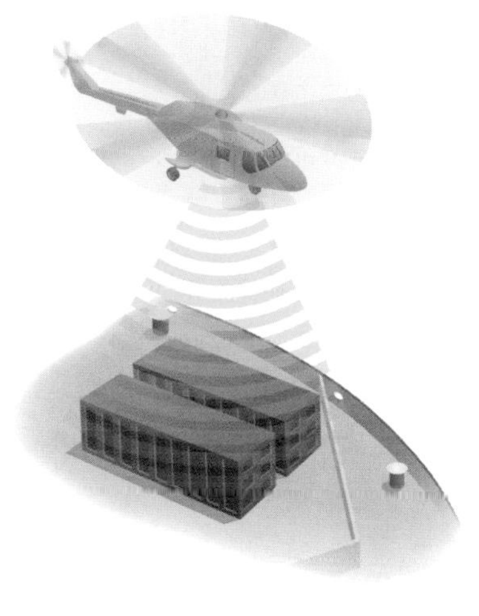

HELICOPTER SCAN

Prior to allowing suspect vessels or cargo to enter ports, law enforcement helicopters outfitted with radiation detectors can be used to inspect the vessel from the air.

BOATS

Port security vessels outfitted with a multitude of detectors can scan the hulls of ships entering the port complex, detect traces of weapons of mass destruction materials, and transmit real-time data to land-based labs.

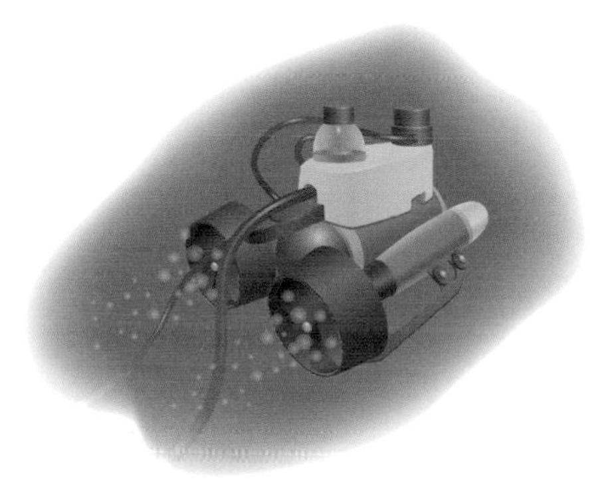

SUBMERSIBLES

Ports may employ remotely operated vehicles, or ROVs, equipped with video cameras, sonar, radiation detectors, and ship hull inspection systems to inspect ships from the water.

TRAINED DOGS

Thousands of highly trained dogs are used worldwide to detect myriad contraband ranging from chemical and biological weapons to illegal drugs to human smuggling.

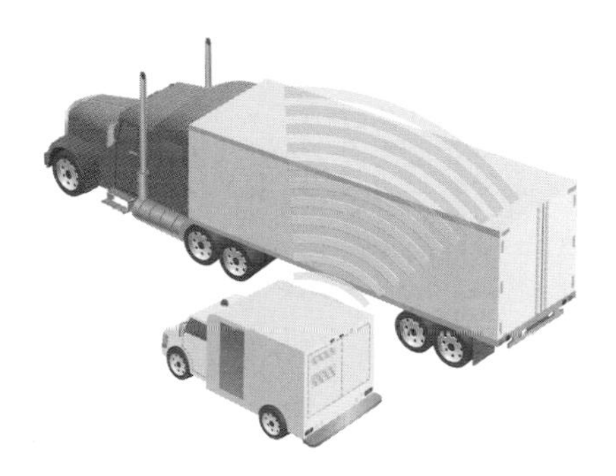

MOBILE SCREENING

Ports may use large-scale X-ray and gamma-ray machines and radiation detection devices to screen cargo at its destination. Radiation portal monitors and thousands of handheld radiation screening devices are used to discover fissile material.

HOW A DRY DOCK WORKS

Prior to entering the dock, a series of keel blocks uniquely laid out for each vessel are placed on the dock floor to support the ship.

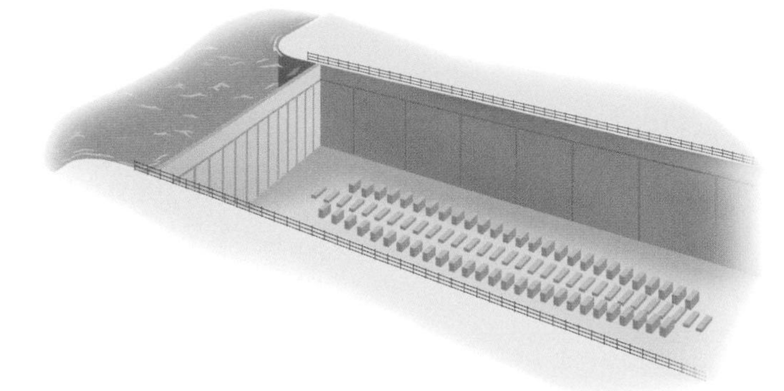

The dock doors (if installed) are opened and the dock is flooded (or submerged if floating).

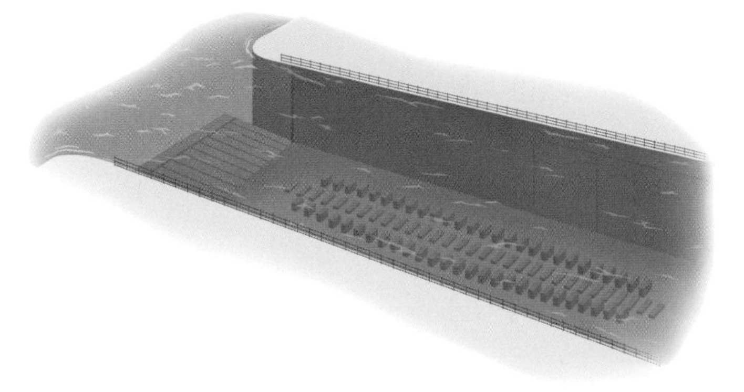

HIGH & DRY

Regular maintenance and repair of ships is important in keeping them gainfully employed at sea, but it is also critical to keeping these valuable assets insured. Classification societies like the American Bureau of Shipping or Lloyd's Register have systems of required maintenance at specific intervals based on the size of the ship and its manufacturer. Some of this maintenance can take place at sea, but much of it must take place at a shipyard located in a port.

The technology that allows ships to be repaired and maintained in the water has remained the same over thousands of years, with relatively minor improvements or innovations. Dry-dock technology has been traced back two thousand years to Ptolemaic Egypt as well as to the seafaring states of ancient Greece, Rome, and China. The oldest surviving "graving dock," where dry-docking occurs on land, is made out largely of wood and stone and is located in Portsmouth, England—where Henry VII constructed it in 1495.

Though much the same in function, twentieth-century graving docks look quite different from their more rounded predecessors due to the boxy shape of modern oceangoing vessels. Large graving docks today can be over 3,000 feet (900 m) long and 150 feet (45 m) wide, with very thick walls—a far cry from the slender, ramped structures that existed to repair sailing or rowing ships in the ancient world. The largest such dock in the world is found at Harland & Wolff's shipyard in Belfast, Northern Ireland, and boasts two enormous gantry cranes—Samson and Goliath—that reach over 300 feet (900 m) into the sky and have become civic landmarks since their erection in 1969 and 1974, respectively.

THE GLOMAR EXPLORER

Among the most unusual users of dry-dock technology at sea was the *Glomar Explorer*, a so-called deep-sea mining ship constructed by the Hughes Company for the Central Intelligence Agency in the early 1970s at a cost of $500 million. Her real mission was to recover a sunken Russian submarine lying under 16,500 feet of water off the coast of Hawaii.

Known as Project Jennifer, the 63,000-ton vessel and its claw-equipped barge arrived on the scene in 1974—six years after the sub had gone down. Although the public would not know of the incident until later, the *Glomar* was able to recover a sizable portion of the submarine—including several nuclear-tipped torpedoes and, sadly, eight crewmen, who were subsequently buried at sea.

The ship is pushed and pulled into the dock with the help of tugs and lines from the dock wall.

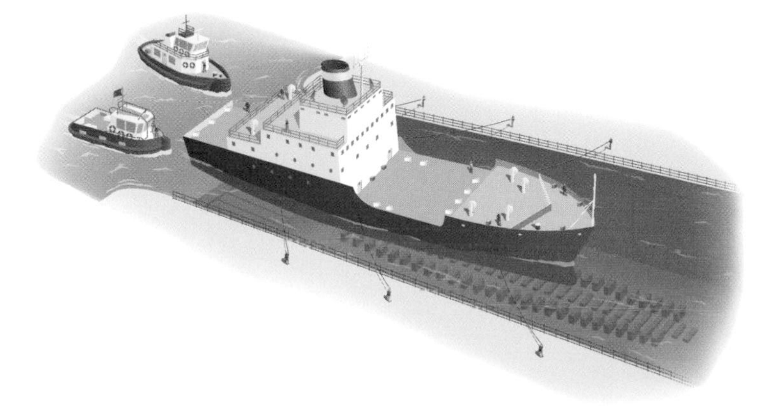

The dock door is shut (if applicable) and water is pumped out of the dock. As the water level lowers, divers ensure the ship is properly resting on the blocks before the ship reaches the critical draft where it can become unstable and capsize.

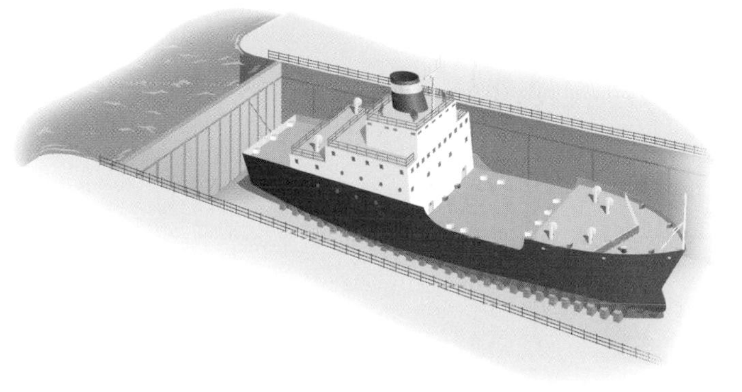

The operation of dry-docking a vessel in a graving dock is fairly straightforward. Constructed as a large basin in the water, a swinging gate opens to allow a vessel needing repair to float in from the sea. Water is drained out of the basin, leaving the ship supported on blocks resting on the floor of the now-dry structure. Once maintenance or repair work has been completed, the ship is refloated and the dry-dock gates opened to allow it to return to sea.

To ensure that no harm comes to the ship when its weight is no longer supported by water, each ship carries a docking plan that indicates where the blocks within the dry dock need to be placed and at what angle; typically, they are inserted to support the key structural members of the vessel. The docking plan also lists where the various component parts of the ship are located so the maintenance crew has no surprises.

Not all maintenance or repair work occurs in graving docks along the shore. Floating dry docks are a form of pontoon that can be maneuvered easily to reach a variety of locations. By opening valves, the main chamber of a floating dry dock will be inundated with water and sink low enough to allow a ship to float onto its deck. Once in place, the deck can be lifted to allow the water to run out and the ship's hull to be easily accessed for maintenance or repair.

Naval dry docks work on the same principle as commercial dry docks, but may be somewhat more specialized. Among other features, some boast a roof atop the structure so that spy satellites can't easily take pictures of the submarines or ships being repaired. Certain floating dry docks are also submersible and might be used for sensitive military repair work.

BREAKING A SHIP

Where do ships go when they die? For most, the answer is the beach.

Thanks to low labor costs and minimal environmental regulation, the center of the world's ship-breaking industry has, since about 1970, shifted to a handful of ship-breaking yards on beach destinations in India, Bangladesh, and Pakistan. Vessels are literally run onto the beach at high tide under their own power and then dismantled for scrap over a period ranging anywhere from two to six months.

Breaking a ship inevitably involves handling a variety of hazardous substances: asbestos, PCBs, lead, and other heavy metals. In the West, workers dismantling a ship would need specialized training and equipment and be able to claim damages in cases of personal injury. On the Indian subcontinent, in contrast, large volumes of untrained and largely unequipped workers dismantle vessels as large as supertankers or containerships at rates as low as four dollars a day.

LAND

BEHIND THE WHEEL

While travel by water has changed little in the last hundred years, almost nothing about modern road transportation resembles what it was a century ago—before the invention of the automobile. Today's cars and trucks bear little resemblance to the wagons and coaches that transported people and goods during the nineteenth century and road networks worldwide have been transformed beyond recognition.

Notwithstanding their modern appearance, highway systems built to accommodate cars have their origins in ancient history. Think back to the Romans, whose achievements in terms of roadway construction and drainage were impressive by any standard. Unfortunately, that level of sophistication would not return for two thousand years: until the early nineteenth century, urban roads in preindustrialized cities were often congested and filthy, and country roads tended to be primitive—if they existed at all.

There were some exceptions. In parts of England and in some areas of the United States, private toll road companies received franchises from the government to construct, improve, and maintain new roads to connect important destinations. In return, they were able to levy a fee from users—much as public agencies do today on many bridges and highways. The result was a patchwork of well-maintained roads, used only by the small section of the population that could afford them.

In most places, however, good roads did not exist and shippers looked to the water as their highway. Canal building received a boost in the nineteenth century due to this inability to move cargo reliably between points by land—a demand made more urgent by the Industrial Revolution and the growth in manufacturing that accompanied it. But the Industrial Revolution that inspired the construction of new canals also sowed the seeds of their demise—giving rise first to the railroad and soon after to the motorcar.

The first patent for a motor vehicle, in 1886, was awarded to a German named Karl Benz, whose "motorwagen" would revolutionize the way people across the globe lived and worked. Benz's new invention had three wheels but, unlike other early experiments based on steam or electricity, it relied on gasoline. His four-stroke single-cylinder engine would become the prototype for cars to come.

The earliest motorized vehicles were complex and expensive to produce. But the development of the Model T changed that completely: by simplifying the production process to make a car through the introduction of assembly-line techniques, Henry Ford and his company precipitated a dramatic drop in the price of the motorcar. Between 1908 and 1915, a car that had previously cost $850 dropped to $290 in price.

HISTORY OF THE WHEEL

The basic concept of the wheel dates back thousands of years. With a means of propulsion—animal or mechanical—and a way to attach the wheel to an axle, or whatever needs to be moved, a wheel reduces ground resistance and facilitates forward motion. But a wheel will be only as successful as the terrain it is riding along: until the advent of roads, wheels were rarely as efficient as sail as a transport technology.

EARLY WHEELS

Rollers comprised of wheels date back much further, but the earliest wheeled vehicle is generally believed to have appeared around 3500 B.C. in what was then Mesopotamia.

Though cars became more affordable with new manufacturing techniques, the earliest owners of cars were aficionados with money to burn: they loved both the technology and the driving experience. Car travel was primarily a weekend hobby rather than a reliable means to get from point to point. Many enthusiastic motorists joined clubs that took responsibility for promoting and maintaining roads outside of urban areas—ensuring themselves a pleasant weekend driving experience.

The nature of car travel would change quickly and radically over the next 50 years, moving from a hobby to a necessity. By the middle of the twentieth century, governments were planning and funding new networks of highways to connect major population centers. Motor vehicles reduced reliance on public transportation and shifted commuting habits dramatically from rail to road, giving rise to new suburban communities outside of city centers. The result was an explosion in road travel.

It would be another 50 years before the environmental implications of that explosion would be understood. Today, manufacturers and policy makers are devoting huge sums of energy, time, and money to finding new ways to manage both the numbers of cars on the roads and their environmental impacts—including promoting hybrid and electric cars, new ways of tolling and congestion mitigation, and fuel-efficiency standards.

WHICH SIDE IS RIGHT?

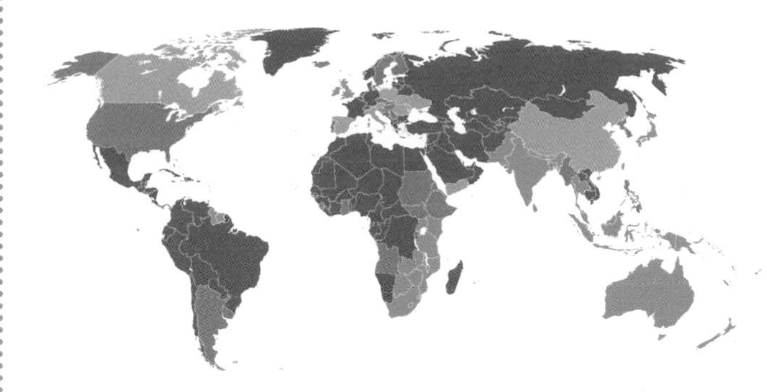

Has always driven on the right.
Originally drove on the left, but now drives on the right.
Once had different rules of the road, but now drives on the right.
Has always driven on the left.
Originally drove on the right, but now drives on the left.

Why do some countries drive on the left and others drive on the right? No one fully knows. To make it even more confusing, some countries drive on both sides: mainland China drives on the right while Hong Kong and Macau drive on the left. Similarly, the U.S. Virgin Islands drive on the left—which somehow seems not very American.

None of the seven European countries that drive on the left shares a land border with a country that drives on the right. But there are indeed places where drivers must swap sides at the borders—particularly in parts of Asia, South America, and Africa. Thailand, for example, drives on the left—but almost all of the countries that border it drive on the right.

Typically, countries that drive on the left rely on vehicles that put the driver in the right-hand seat of the car, and vice versa, though there are exceptions. Some countries permit both types of vehicles to be operated, and a number of countries in the Caribbean operate left-seated driving vehicles on left-driving roads.

Curiously, not all of the former British colonies continued to drive on the left after independence. At least half a dozen—Gambia, Ghana, Sierra Leone, Nigeria, Canada, and of course the United States—now drive on the right side of the road.

SPOKED WHEELS

Spoked wheels are attributed to the Egyptians, who found them strong enough to be attached to their chariots.

WHEELS WITH IRON RIMS

The first wheels with iron rims, Celtic chariots, date back to 1000 B.C. But the concept was a good one, the iron rim giving strength to the wheel and otherwise supporting the weight of whatever was being moved.

PNEUMATIC TIRES

The earliest tires, comprised of hard rubber, appeared on the bicycles of the Victorian era. The inflatable structures we know today, made of cords and wires set within a softer rubber, were developed by John Dunlop, a Scottish inventor, in 1888.

ENGINE

An internal combustion engine converts gasoline into motion by burning oil. A small amount of gasoline is put into a cylinder (a small, enclosed space) and ignited by a spark to release energy in the form of expanding gas. A piston moving up and down resets the cylinder for the next explosion. Hundreds of these explosions per minute in multiple cylinders produce sufficient energy to move a car forward. Cooling systems, which typically involve a liquid coolant that passes through a radiator, ensure that the heat produced by the combustion is removed to keep the engine from overheating.

STEERING

Two different forms of steering predominate today—recirculating-ball and rack-and-pinion mechanisms. In the former, found on most trucks and many larger cars, the steering wheel turns a bolt, which moves a block connected to a gear that turns the tires in a specific direction. In a rack-and-pinion system, found on most cars, the steering shaft is connected to a pinion gear on a rack that connects through a tie rod to the spindle on the wheel mechanism itself. Suspension systems, which include a variety of springs and shock absorbers, connect the car body to the wheels in a way that minimizes vibration but allows the wheels to hug the road.

ELECTRICAL

Electrical energy needed to start the engine is stored in the battery. While the car is running, the alternator provides a continuous voltage to replenish the battery. The battery powers all the electrical components, including fans, windshield wipers, headlights, and power windows and seats.

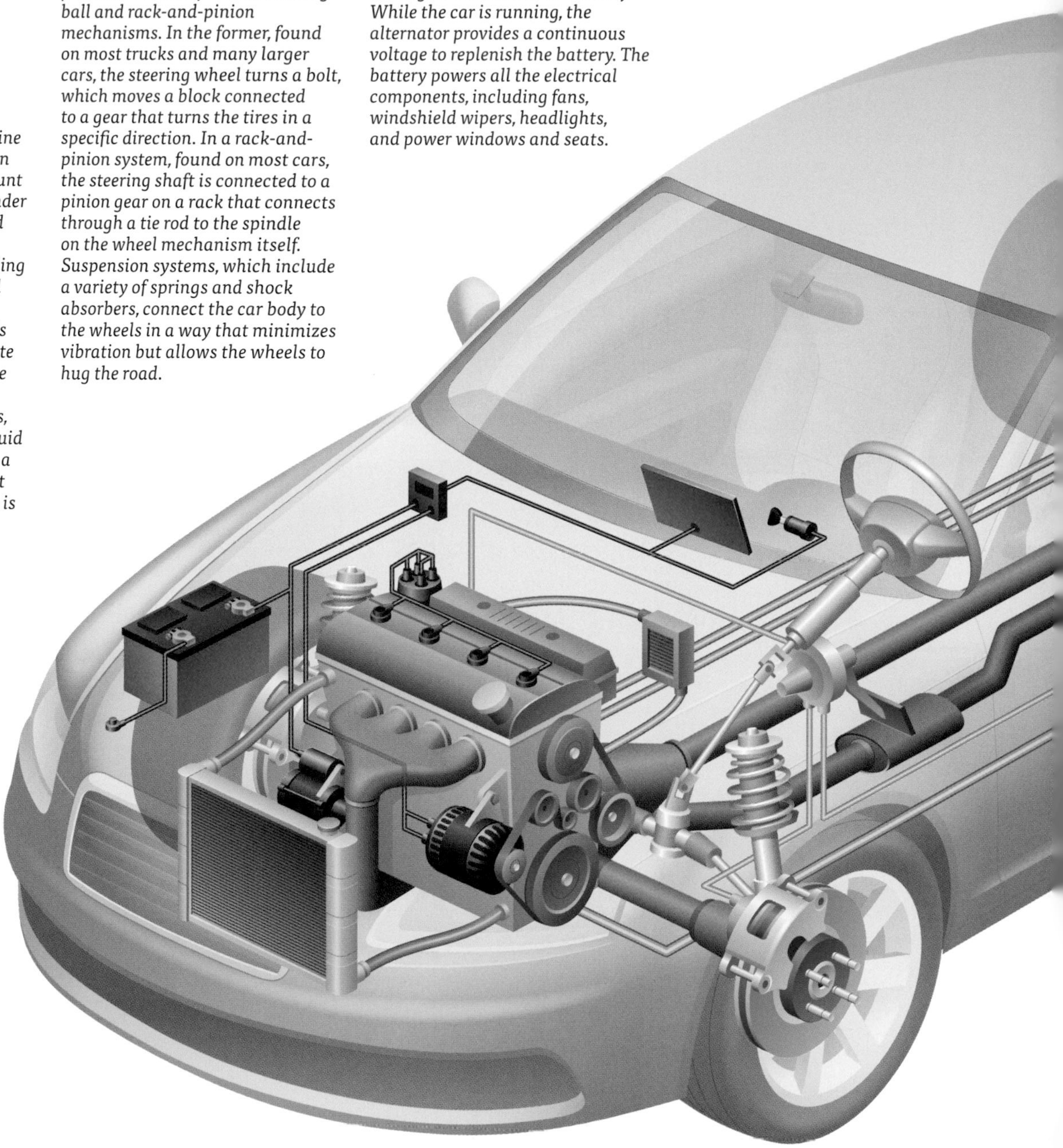

ANATOMY OF A CAR

Millions of people around the world rely on cars for their daily transportation needs. Only a small minority of them actually know how they work.

Most cars rely on the internal combustion engine to convert high-energy forms of fuel into motion. It does so by pushing a combustible mixture of fuel and air flowing into an enclosed space (also known as a cylinder), where it explodes (or combusts) and transmits energy to the piston in the form of expanding gas. At any point when a car engine is running, dozens of these explosions will be under way in individual cylinders within the engine—the more cylinders, the more power is delivered to the vehicle.

Four stages, or strokes, are required to convert gas to motion: intake, compression, combustion, and exhaust. To move the wheels, the motion that comes out of the engine must be rotational rather than linear. The pistons are connected by a rod to a device known as a crankshaft, which turns in a circular motion as a result of the pistons' moving up and down in a continuous cycle of explosions.

EXHAUST

Exhaust from the engine moves through a muffler, or silencer, located between the engine and the tailpipe. Mufflers are metal cases containing baffles, which form a series of acoustical chambers that help to reduce engine noise. In many places, catalytic converter mechanisms, which reduce emissions by breaking up nitrous oxides and oxidizing carbon monoxide and other unburned hydrocarbons, are required components of exhaust systems.

BRAKES

When a driver depresses the brake pedal, the force exerted by the driver's foot is multiplied hydraulically via a fluid, usually oil, running through hoses to pistons on each wheel. The pistons exert sufficient pressure either on the brake shoes, which in turn push against the brake drum, or on a brake pad, which pushes against the brake rotor, to stop the car.

TRANSMISSION

Car transmissions rely on sets of gears to allow the engine's power to be translated into a variety of different speeds. In manual transmissions, gears of different sizes are locked and unlocked via a clutch pedal operated by the driver. In cars with an automatic transmission, only one set of gears—called a planetary gear set—is needed to achieve all of the different combinations to convert engine torque into the desired level of forward motion.

HORSEPOWER

Why do we measure the strength of engines in horsepower?

Horsepower owes its place at the center of transportation technology to James Watt, a Scottish engineer who lived at the turn of the nineteenth century and is known for his work on steam engines. Watt estimated that the amount of work one horse could do in a minute was roughly 33,000 "foot pounds"—that is, moving 330 pounds of coal 100 feet, or 100 pounds of coal 30 feet, or some other pair of numbers that multiply to 33,000. This measurement of engine power has been adopted across modes as an industry standard.

Engine power alone is not a measure of a vehicle's performance: that depends on how much weight the engine is pulling at a given moment. But broadly speaking, the higher the horsepower, the higher the performance—and the faster the vehicle—be it a train, a car, or a plane—can get moving.

Other major components support the combustion process directly: the fuel system pumps gas from the tank and mixes it with air in just the right proportion; the sparkplugs provide the spark to initiate combustion; the battery provides energy for the spark as well as electricity for the car's radio, windshield wipers, lights, and onboard computers. Valves open and close at the precise time to inject the right amount of fuel and air into the pistons. The alternator works to recharge the battery itself while the car is running, and oil serves as a needed lubricant to ensure that key components (including the crankshaft, pistons, and bearings) move smoothly.

Not all cars follow the same formula: some, like trucks, rely on diesel fuel instead. Because diesel fuel has a higher density of energy, no sparkplug is required for combustion. Instead, in diesel engines air is injected into the cylinder, where it is pressurized and heated; fuel is then injected and ignites solely as a result of the heat and pressure in the cylinder. The result is typically better mileage and greater reliability.

REFINING OIL

Modern-day road transport would be impossible without fossil fuels and the energy-packed hydrocarbons they contain. Freeing those hydrocarbons and turning them into products that trains and planes and automobiles can use is the job of the refinery.

Crude oil contains hundreds of hydrocarbons mixed together in a way that isn't particularly useful for transportation. Refining separates the various grades or products by heating up the crude oil in a process known as distillation: parts of the crude, or fractions, vaporize at different temperatures; upon condensation, they become a variety of products ranging from gas and diesel to kerosene, heating oil, wax, and asphalt.

Traditional gas is among the lightest of the products produced at a refinery; kerosene and diesel are slightly heavier; with the heaviest being fuel and lubricating oils,

Crude oil, or petroleum, is comprised of decaying plants and animals dating back to ancient sea beds. Once a deposit has been found offshore, a drilling rig is constructed and the petroleum is pumped to the surface.

Double-hulled tankers and supertankers carry large amounts of petroleum to refineries typically located along the coast. Depending on where the refinery and oilfield are located, pipelines on land may perform the same function.

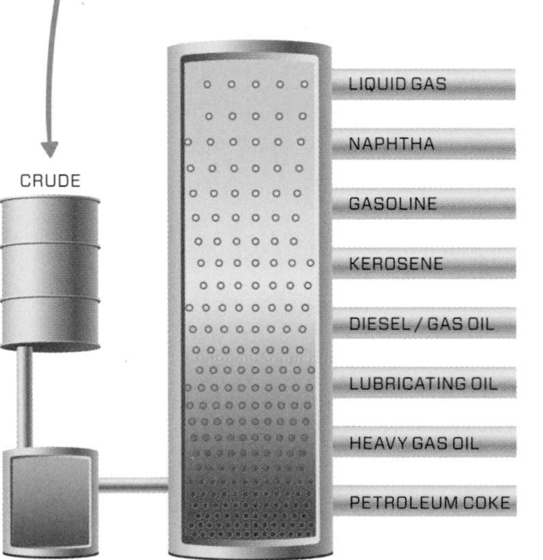

At the refinery, the crude oil is heated to a very high temperature—giving off a vapor that is channeled through a column. As the vapor rises, it cools—with certain parts (or fractions) condensing to liquid before others. These liquid fractions are pulled out of the column and either go directly into storage or are processed further to constitute a specific product—ranging from kerosene, gasoline, and diesel to gas, naphtha, and heating oil.

From the refinery, refined product is moved to end users. In the case of motor fuel, the end user is often a gas station. Tankers carrying multiple grades in segregated compartments are used to make deliveries to stations.

wax, and asphalt. Although gasoline represents only about 40 percent of the output of the crude oil distillation process, demand for it typically outstrips demand for other products. As a result, refineries rely on various forms of processing to convert heavier distillates, such as diesel fuel, into gasoline when demand requires. Different combinations or mixtures of distillates will create gas with different octane ratings with different retail prices.

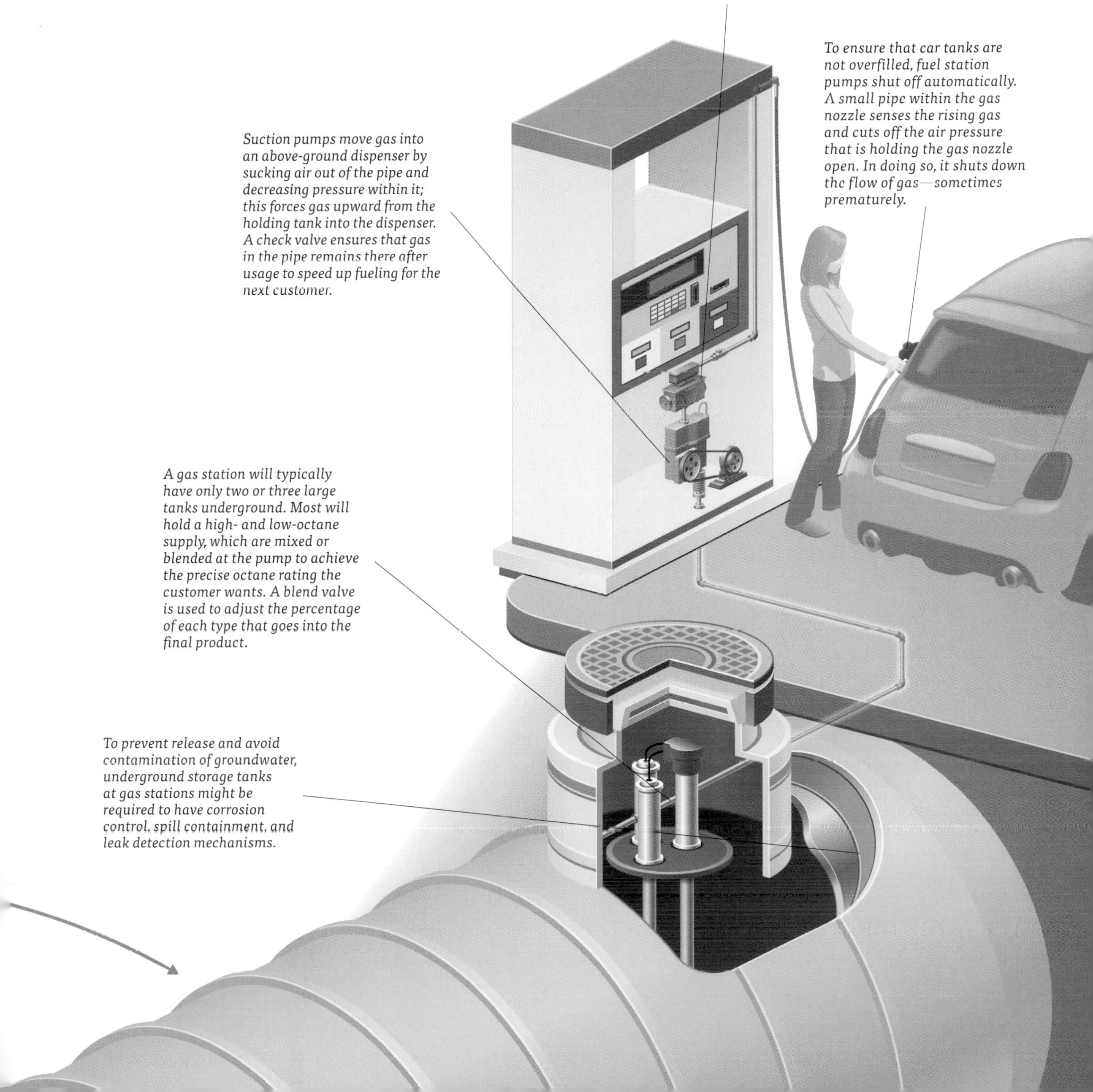

As gas travels through the pump, a flow control valve regulates its speed and a meter measures gas flow volume. This information is transmitted to a computer, which measures it in tenths of a gallon and calculates the price. Because the density of the gas changes with outside temperature, the computer evaluates temperature and modifies the flow meter (and hence the price) accordingly.

To ensure that car tanks are not overfilled, fuel station pumps shut off automatically. A small pipe within the gas nozzle senses the rising gas and cuts off the air pressure that is holding the gas nozzle open. In doing so, it shuts down the flow of gas—sometimes prematurely.

Suction pumps move gas into an above-ground dispenser by sucking air out of the pipe and decreasing pressure within it; this forces gas upward from the holding tank into the dispenser. A check valve ensures that gas in the pipe remains there after usage to speed up fueling for the next customer.

A gas station will typically have only two or three large tanks underground. Most will hold a high- and low-octane supply, which are mixed or blended at the pump to achieve the precise octane rating the customer wants. A blend valve is used to adjust the percentage of each type that goes into the final product.

To prevent release and avoid contamination of groundwater, underground storage tanks at gas stations might be required to have corrosion control, spill containment, and leak detection mechanisms.

TRUCK BASICS

Trucks come in two basic configurations: those that have a cab and trailer that are two distinct units and those for which these two parts are integrated. The former are generally known as semitrailers (also tractor trailers, semis, or articulated trucks). Their trailers have wheels only at the rear end: the forward axle is provided by the towing cab and carries roughly half of the weight of the load.

Connections between the trailer and the cab are critical to the functioning of tractor trailers. Trailers are attached by a kingpin to a horseshoe-shaped coupling device, known as the fifth wheel. Additional connections, called glad-hand connectors or palm couplings, contain air hoses and provide air pressure from one part of the truck to another. Electric cabling, often referred to as the pigtail, carries power to the trailer for lighting in a coiled shape—so that it doesn't snap when the truck turns corners.

Aerodynamics is critical to trucks. The large quantity of air that passes over the silhouette of a truck at high velocity creates vacuum pockets and differences in pressure that spawn eddies of air. These eddies create back-and-forth motion, or swaying, and can destabilize a truck. This resistance can also be costly: a truck moving at 55 miles per hour will displace roughly 18 tons of air for each mile traveled. Overcoming this resistance can require somewhere between 50 and 75 percent of a vehicle's horsepower if the truck is not designed in a streamlined fashion.

Weight is also a critical factor in the design of trucks. Roads often carry weight specifications based on the number of axles a truck contains, with each axle permitted to carry a designated weight. To provide maximum flexibility, some trailers have axles that can move along a track on the underside of the truck body and are secured by pins; this allows the truck's loaded weight to be distributed evenly so as not to exceed axle weight limits on roads.

AERODYNAMICS

Roughly half of a cruising truck's horsepower goes to overcoming the aerodynamic drag generated by the contrast between the high-pressure zone at the front of the truck and the partial vacuum created behind it. The large alternating eddies and turbulent air created by this condition can cause truck trailers to sway. Anything that can smooth the air flow over the trailer without diminishing its carrying capacity, such as rounding corners, closing the gap between cab and trailer, or adding a "tail" to ease the flow of air past the end of the trailer, reduces drag and helps to improve mileage and performance.

5TH WHEEL

The "fifth wheel" is not a tire-type wheel but a horsehoe-shaped coupling device that connects the cab and trailer. Devised in the early nineteenth century, it replicates a wheellike device found on horses and carriages that allowed for easier turning of corners. In the modern version, a kingpin inserted in the center of the device allows a robust connection but permits the two parts of the truck to be released from each other quickly.

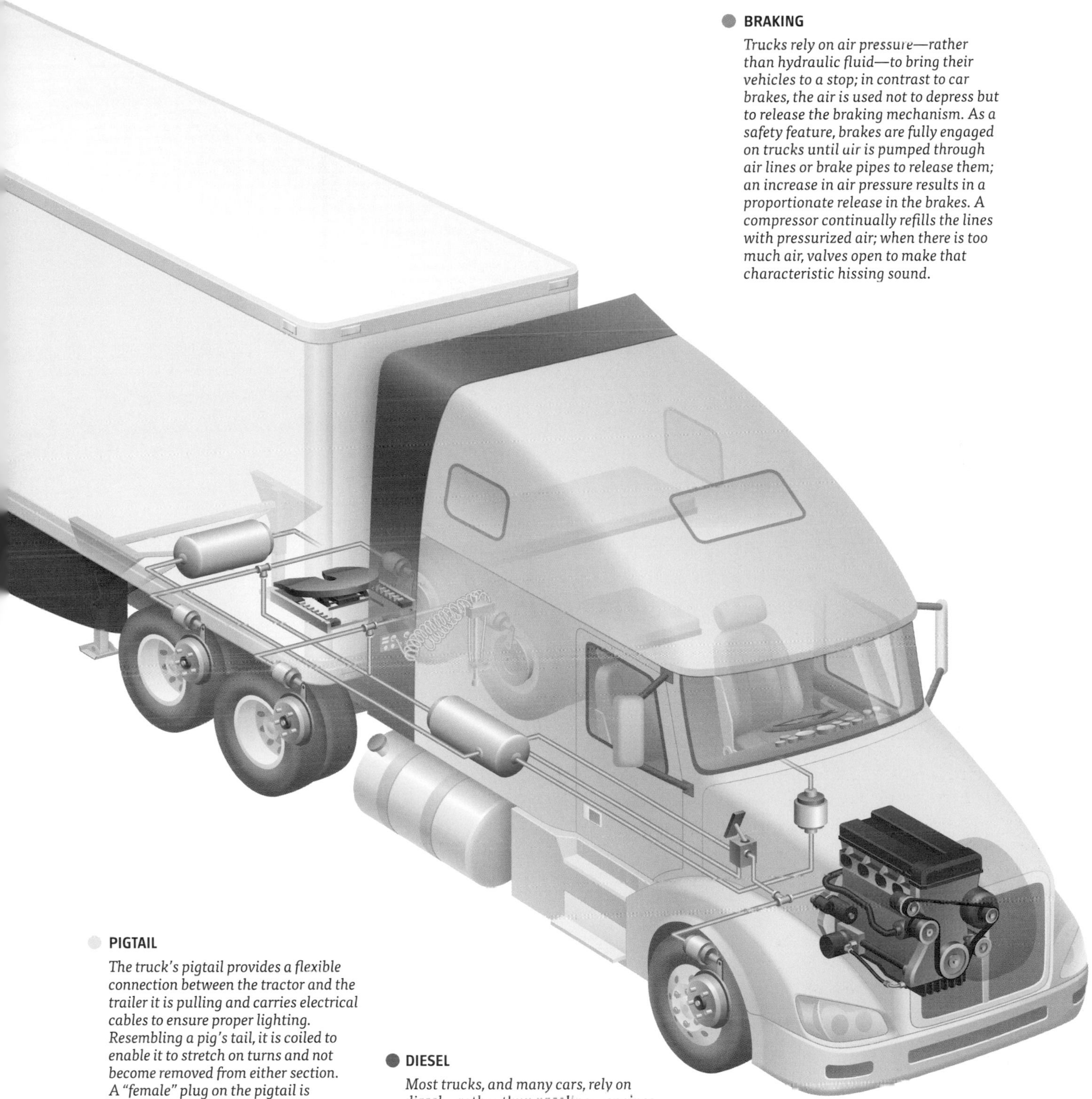

BRAKING

Trucks rely on air pressure—rather than hydraulic fluid—to bring their vehicles to a stop; in contrast to car brakes, the air is used not to depress but to release the braking mechanism. As a safety feature, brakes are fully engaged on trucks until air is pumped through air lines or brake pipes to release them; an increase in air pressure results in a proportionate release in the brakes. A compressor continually refills the lines with pressurized air; when there is too much air, valves open to make that characteristic hissing sound.

PIGTAIL

The truck's pigtail provides a flexible connection between the tractor and the trailer it is pulling and carries electrical cables to ensure proper lighting. Resembling a pig's tail, it is coiled to enable it to stretch on turns and not become removed from either section. A "female" plug on the pigtail is connected to a "male" outlet on the back of the tractor.

DIESEL

Most trucks, and many cars, rely on diesel—rather than gasoline—engines. Similar to what occurs in gasoline engines, diesel fuel is injected into an engine cylinder—but no spark is required for ignition. The pressure of compression is sufficient to cause the diesel fuel to ignite. Among other advantages, diesel has a higher energy density than traditional gasoline, so it gets better road mileage.

MOVING CARGO

There are almost as many types of trucks on the road as there are commodities that they carry. Many of them have been purpose built, either to accommodate certain types of commodities or to meet the height, weight, or length regulations that governments impose on road users. These weight and length restrictions vary greatly from place to place. The Netherlands, Sweden, Finland, and Australia, for example, allow very long trucks of up to 80 feet in length. Places like the United States and UK regulate both length and weight more tightly, with weigh stations dotted along major highways for enforcement.

The body combinations and number of axles and wheels that trucks utilize vary enormously from place to place and country to country. In the United States, semitrailers usually have two axles and eight wheels located at the rear of the trailer and rely upon the tractor for its forward support. Some of these, particularly those involved in long-distance haulage, are enormous. Weighing in at up to 80,000 pounds, American 18-wheelers are roughly 16 times the weight of the average car and have twice as many gears. As a result, they are often regulated more heavily than smaller commercial vehicles and passenger cars.

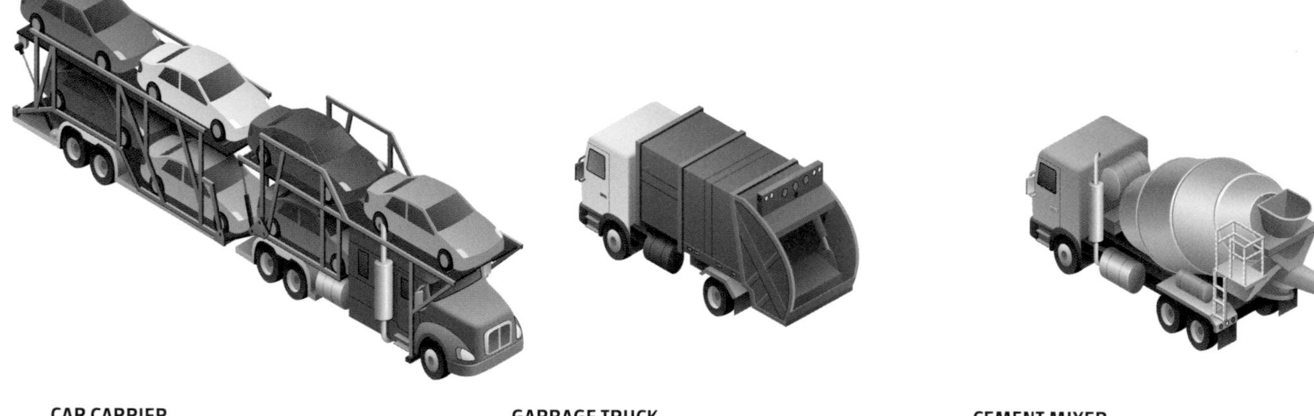

CAR CARRIER

Car carriers (sometimes known as vehicle transporters) carry millions of new cars each year from manufacturing plants to dealers to ensure they arrive undamaged and unused for the customer. Used cars might also be shipped on car carriers, either to or from auctions held by the dealers who sell them.

GARBAGE TRUCK

Garbage trucks not only transport waste; they also crush and compress it while on the move. These purpose-designed vehicles can nonetheless prove useful in other situations: many cities attach plows to them in wintry weather and send garbage fleets out to plow streets.

CEMENT MIXER

Cement mixers consist of an engine, truck frame, and rotating mixer that keeps premixed concrete from hardening en route from plant to job site. Some contain separate holding tanks for the aggregate, cement, and water and mix the materials on site to the customer's specification.

LOW BOY

Low boys, also known as low loaders, are semitrailers with a platform section dropped in height to allow carriage of heavy machinery that would otherwise extend beyond height restrictions on roads. On some versions, vehicles with wheels can be driven onto the truck deck for road transport.

REEFER

A reefer, or refrigerated trailer, relies on frozen carbon dioxide (dry ice) or liquid nitrogen to maintain refrigeration for highly perishable items. As the gas evaporates, it provides coolant for the cargo. Full-size reefer loads can last anywhere from two to four weeks without an outside power supply.

TANKER

Tanker trucks move liquids such as gas, oil, and milk, but can also carry gases and some solids. Carrying anywhere up to 9,000 or so gallons at once, sometimes refrigerated or pressurized, they often have multiple axles to distribute weight and absorb shock.

MOVING PEOPLE

Not all commercial vehicles move goods: many move people. Manufacturers of trucks are often also manufacturers of buses—some that travel long distances and some that confine themselves to the boundaries of an urban area. Similarly, manufacturers of automobiles are also manufacturers of small vans and taxis—which are usually just retrofitted or purpose-built cars.

In parts of Western Europe and the United States, traditional people-moving vehicles such as buses and taxis are increasingly being joined by a variety of smaller vehicles. Pedicabs, for example, long found on the urban streets of South America and Asia, can now be seen regularly on the streets in London and New York—cities historically foreign to human-powered travel.

Among the most recognizable people carriers in the world is the North American school bus. Unlike elsewhere, school buses there must be painted "school bus yellow" to identify themselves and ensure they are visible. Not surprising, they are heavily regulated for safety and warning devices must form part of the vehicle's standard operation.

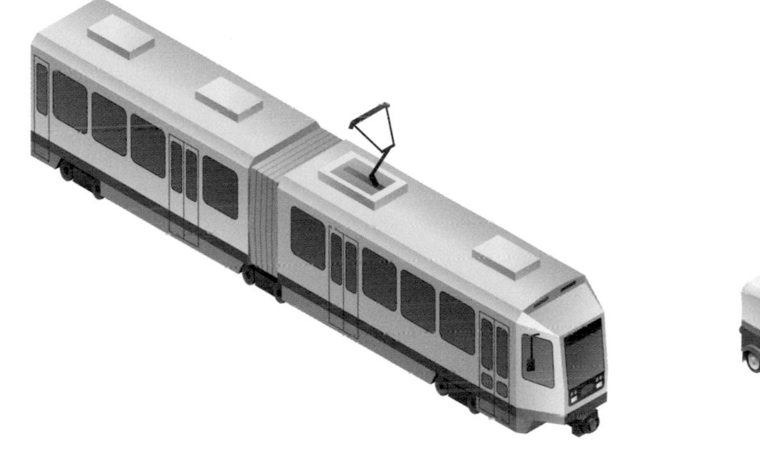

TRAM

Trams, also referred to as streetcars or trolley cars, come in many flavors and sizes. Most run on electricity provided from a pantograph above, but some move under their own diesel power and others draw power from a third rail. Most run on streets and are not segregated from other traffic.

TUK-TUK

Also known as auto rickshaws, tuk-tuks are three-wheeled motorized vehicles that provide local transport in developing countries (and in some Asian cities). Like traditional rickshaws, the carriages are not fully enclosed.

DOUBLE-DECKER

Double-decker buses, long a national symbol in the UK, are increasingly found in other countries as efficient forms of urban transit. Their shorter length and single, rigid chassis makes them easier to maneuver and their height provides a nice vantage point for users—a fact that hasn't been lost on the tourist and sightseeing industry, which now employs them heavily in global capitals.

ARTICULATED BUS

Comprised of two separate rigid sections, articulated buses swing around a pivot in the middle—allowing them to maneuver more easily in cities. Also known as bendy buses in the UK, their capacity (120 people to a double-decker's 85) makes them popular with bus rapid transit (BRT) operators.

TAXI!

The earliest taxis weren't motorized: they were pulled by horses. The Hackney Carriage Act, passed in England in 1635, offered the first control over the horse-drawn carriage-for-hire industry. The word "taxi" emerged over two centuries later—in 1891—with the invention of the taximeter—a device that could measure the time and distance a person traveled and ascribe a fair price to it. Motorized taxis soon appeared on the streets in urban centers: the first yellow cabs appeared in New York City in 1905 and the famous black cabs in London soon after.

PAVING A ROAD

The making of roads has progressed a lot since the days of the Appian Way. Roads are no longer constructed by soldiers, prisoners of war, or slaves but instead by government transportation agencies or by contractors hired by those agencies. While more skilled labor is involved in roadway design and construction than ever before, modern roads are probably no better than they were in the past: indeed, ancient roads that consisted of a drained stone surface required considerably less maintenance than the tarmac roads we drive on today.

The oldest paved roads date back 4,600 years to Egyptian civilization. Most were built with slabs of sandstone or limestone, although some included petrified wood. Stone sleds moved on rollers above them. Some

1 Prior to paving, earthmoving and grading equipment will create a level surface suitable for paving. Embankments are constructed, major bumps are leveled, and depressions or holes are filled. Drains and sewers will be inserted where necessary.

2 Road scrapers and graders are used to scrape the dirt below vegetation levels to ensure an even base. The center of the road must be left higher than the edges to facilitate drainage. The screened dirt or gravel fill is sprayed and compacted repeatedly until it is even.

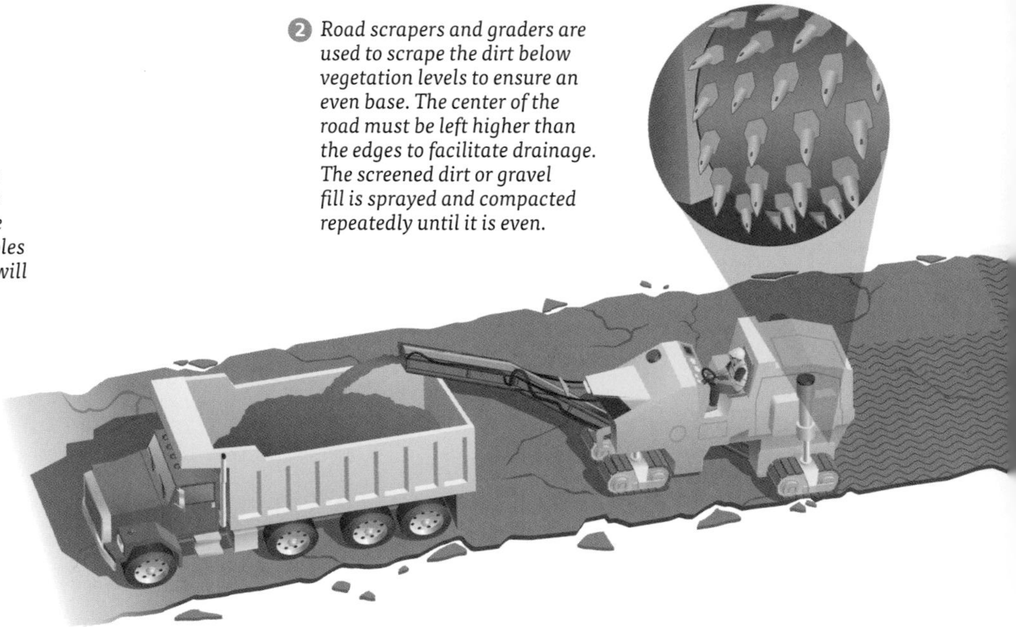

ROAD ANATOMY

Major roadways are normally constructed of asphalt or concrete paving. Asphalt is a petroleum product that can be used to bind sand and crushed rock and can be poured in place to create a hard surface. Though cheaper in most places, it is not as strong or long wearing as concrete—which is made of cement and water poured into steel molds and formed into large roadway slabs.

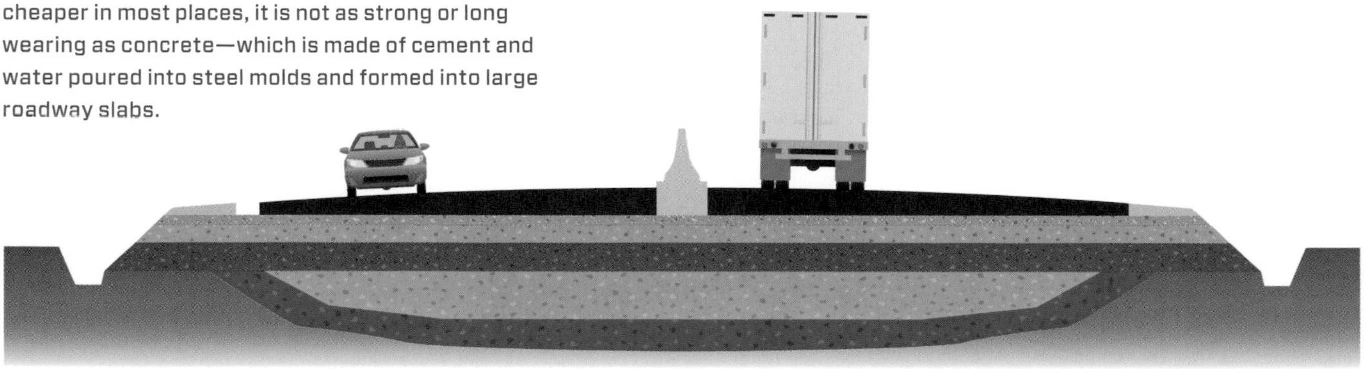

Embankments are constructed in the earth to provide the bed for the roadway. Ditches are created on either side of it for drainage and other purposes.

Subgrade layers consist of a variety of fill material used to bring the roadway to an even level.

The subbase layer extends across the roadway and the shoulders on either side.

A base course is spread across the roadway. It usually consists of gravel or a combination of Portland cement and lime.

Asphalt is poured on top of the base course and may or may not be extended to the road's shoulder.

six and a half feet wide, these roads stretched from quarries to ancient lakes that would meet the Nile at flood; from there the blocks of basalt could be floated to Giza.

Though technologies for moving materials have gotten much more sophisticated, the process of developing a road is not much faster today than it was in the ancient world thanks to the bureaucratic approval processes that characterize modern democracies. Planning and designing a major road can take up to two years; acquiring the right-of-way and constructing the road can take anywhere from three to five additional years—assuming no legal challenges to its construction delay the process. Complex or controversial roads can easily take a decade.

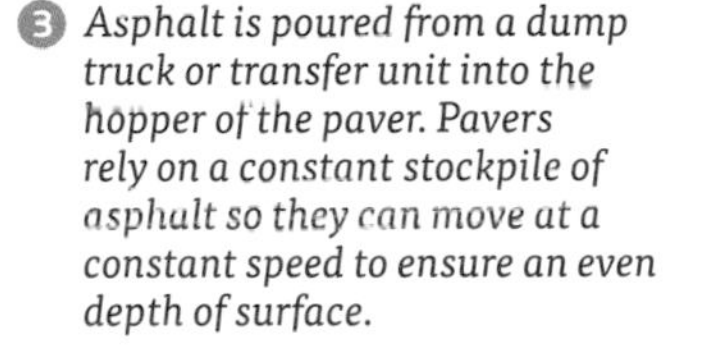

❸ *Asphalt is poured from a dump truck or transfer unit into the hopper of the paver. Pavers rely on a constant stockpile of asphalt so they can move at a constant speed to ensure an even depth of surface.*

❹ *Material is conveyed into a screed, which contains an augur that moves through the material for even distribution on the roadway. Flames heat the bottom of the chamber to prevent the asphalt from sticking to the screed.*

❺ *As the new asphalt layer is dumped on the prepared roadway, workers smooth it out in preparation for compaction by a roller.*

❻ *Paint stripers can carry up to hundreds of gallons of paint, though many are smaller. The paint is run through hoses and applied under pressure through one or more gun carriages to the newly paved roadway. In some cases, striping involves the placement of long strips of adhesive rather than actual paint.*

RECYCLED MATERIALS

Recycling has literally hit the streets. Rubberized asphalt is now commonly used to supplement traditional petroleum-based materials as a road surface. It's a form of concrete that's made by blending rubber from ground-up tires with asphalt and traditional aggregate material. The result is stronger than conventional roadway material: on average, it lasts 50 percent longer. A two-inch-thick road surface uses up to 2,000 old tires per mile of car lane constructed.

Other recycled materials also appear increasingly on roads. "Glasphalt" relies on crushed glass as an additive and has been used widely in the United States and Canada for city roads and driveways. Fly ash, shingles, and coal-mine refuse have also been used in roadway concrete mix.

SYNCHRONIZATION

Traffic-light synchronization is an increasingly popular way to reduce urban congestion and emissions. By synchronizing lights at intersections along a specific corridor to maximum green-light timing, individual cars can move at a constant speed through multiple intersections before having to stop. The speed at which a vehicle can travel without being caught by a light varies from place to place depending on traffic conditions (at one time they were reported to be 28 miles per hour on major north-south avenues in Manhattan).

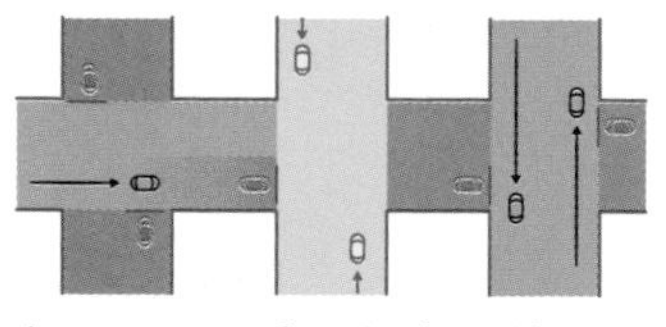

A car on a synchronized corridor approaches an intersection with a red light, which is permitting traffic on the crossing street to clear.

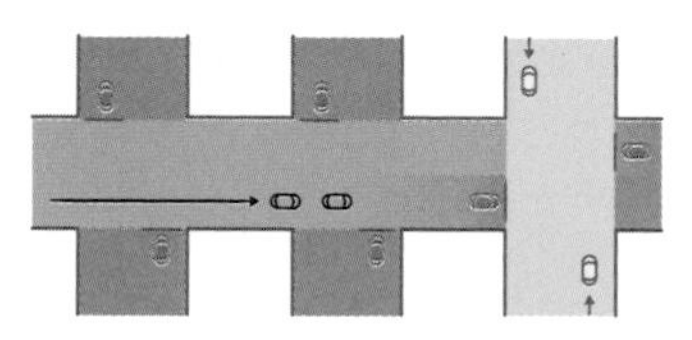

As the car reaches the intersection the light turns green.

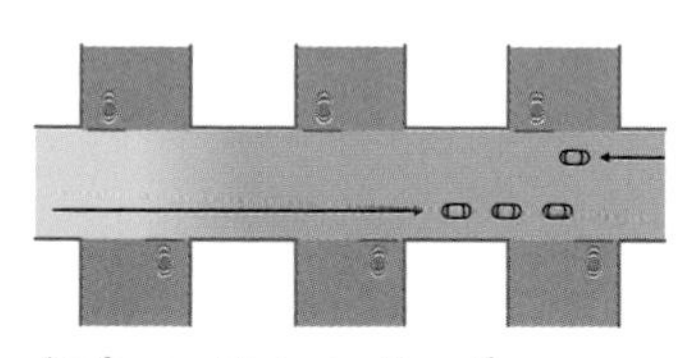

At the next intersection, the same thing happens, and the car (and cars directly behind it) proceeds through without stopping.

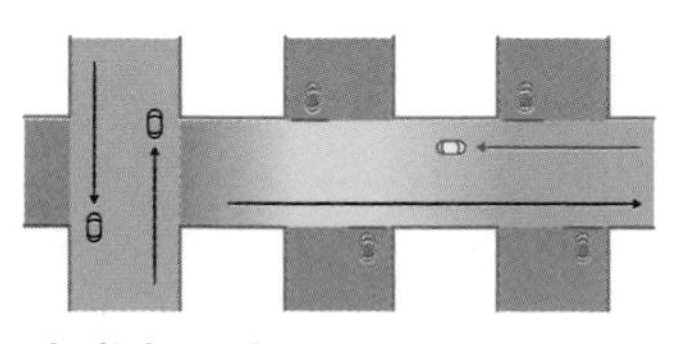

The light at the original intersection returns to red while cars continue to move forward through multiple intersections uninterrupted.

CROSSWALK

Crosswalk markings can take many forms. In some places, white lines demark both the mandatory car-stopping points and the areas where pedestrians should cross; in others, black and white stripes constitute what are known as zebra crossings.

SIGNALS

Flashing signals increasingly rely on symbols, often with a countdown timer, rather than words to tell pedestrians when to cross.

SENSORS

Sensors underneath the pavement might be used to trigger the light's change. These consist of multiple turns of insulated loop wire nestled in a shallow slot in the roadway. Relying on a form of metal detector to detect the presence of one or more automobiles at the intersection, they can be used for measuring speed and counting vehicles in addition to recording congestion and activating a change in signal.

CONTROLLING TRAFFIC

Traffic lights on roads evolved a century ago in response to the increasing popularity of the automobile. Today they are more or less universal around the world, with most relying on three colors: red, amber, and green. Unbeknownst to most drivers, these colors are not pure: the red light actually contains some orange, and the green light contains a touch of blue—primarily to make sure that color-blind people can differentiate between red and green.

The great majority of these bulbs show up as a vertical set of round lights. In Quebec, however, where traffic signals are mounted horizontally, the red light is a square, the amber a diamond, and the green a circle. In many places, the three traditional signals are being supplemented by arrows indicating turning options.

The operation of these lights is similar around the world. In general, a light must be amber at least one second for every 10 mph

RED-LIGHT CAMERAS

Today, police less frequently catch motorists running a red light than cameras do. Red-light cameras snap pictures of violators and timed photos are sent with a ticket to violators. While hundreds of cities across the United States have used this technology to bolster income to their traffic enforcement departments, the cameras have not led to a decrease in accidents. One study found that the cameras resulted in a smaller number of right-angle crashes, but a larger number of rear-end collisions.

EMERGENCY SIGNALS

In cases of preemptive signaling, equipment inside an emergency vehicle—such as an ambulance—allows it to change the traffic light ahead so that it has a steady stream of green lights.

AUTOMATED PHASING

In a traditional traffic light, electromechanical controllers wired to the signals rely on dial timers; the length of the cycle is determined by the size of the gear within it. Ranging from half a minute to two minutes, these cycle periods are fixed. In modern lights, dynamic control—in which signals are programmed to adjust their timing to actual traffic flows and condition—is common. Many can be programmed remotely to take account of commuting patterns or special events.

of speed limit to give time to drivers to brake. "All red time," when lights are red on both roadways of an intersection, can range from three to five seconds to allow cars to clear the intersection. These times, as well as the time of both red and green signals, can be varied to accommodate local traffic conditions and avoid gridlock.

In many places, detectors in the pavement—either wires or magnetometers that measure changes in the earth's magnetic field—provide information about traffic conditions to remote traffic managers. Other nonpavement invasive approaches, such as microwave radar, infrared technologies, or video image processors, can also be used to detect automobile volumes and flow. Once information from a local intersection is received, a remote traffic controller will evaluate it, compare the data with preset parameters, and decide whether to adjust the intersection's light phasing and timing.

THE FIRST TRAFFIC LIGHT

The origins of traffic lights can be traced back to railroad signaling systems, which relied on a palette of colored lights to indicate when trains should or should not move forward. Railway signals were configured with green on top and red on bottom, precisely the opposite of how traffic signals evolved—no doubt to avoid confusion between the two.

The earliest traffic signals were not really traffic lights at all. They resembled railway semaphore signals, with arms that reached out horizontally and featured colored lights on top for night traffic. They were operated by police to manage and direct horse-drawn traffic; one of the earliest was installed outside the Houses of Parliament in London in 1868.

The earliest traffic lights, like the earliest signals, relied heavily on human intervention: police judged traffic and decided when it was time to change the light (they blew a whistle beforehand). The modern tricolored traffic light dates back almost a century to its reputed invention by a police officer in Detroit in 1920. Like other early experiments in Cleveland, Salt Lake City, and Houston, it was a manually controlled signal—but one that relied on the red, amber, and green system that we know today.

INTERSECTIONS & INTERCHANGES

Roadway intersections around the world take a variety of forms, from simple intersections governed by stop signs to complex highway overpasses involving on and off-ramps. All are testament to the fact that basic roadway design can go a long way to keeping traffic moving where multiple roads intersect.

Although all traffic engineers aim to avoid accidents and gridlock at crossings, there is no single way to design a meeting of two or more roads. Instead, there are a series of options that traffic engineers will consider in planning an intersection. Design options range from the simply named (e.g., continuous-flow intersection) to the descriptive (e.g., jughandle or hook turn) to the uniquely place-based (e.g., Michigan left).

The physical design chosen for a given intersection will be a function of many things. The first consideration will usually

JUGHANDLE

Jughandle turns are largely the province of one American state; New Jersey boasts about 600 of them at present. They require cars traveling on a divided highway that want to turn left at an intersection to exit the highway on the right side, and then be fed onto the crossing road—where they can continue their journey across the intersection once they clear the light.

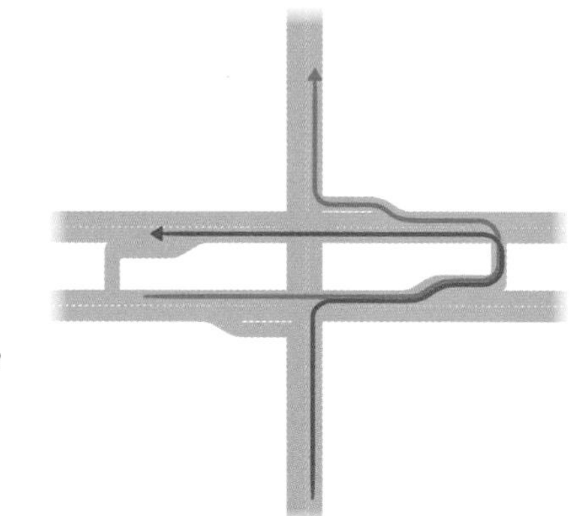

MICHIGAN LEFT

Michigan lefts first appeared in Michigan in the 1960s. Cars turning left onto a divided highway are instead required to turn right and do a U-turn to join the major roadway. Michigan lefts have been shown to reduce collisions in intersections, speed up traffic (there is no left-turn phase at the traffic light), and decrease pedestrian accidents.

ROTARY

Rotaries, also known as roundabouts or traffic circles, can be found around the world. Traffic entering the circle, either in free flow or directed by signals, must give way to traffic in it as well as to exiting traffic. Because vehicles in a roundabout move more slowly and in one direction, they have been proved safer than traditional intersections—resulting in fewer collisions and less severe passenger and pedestrian injuries.

CONTINUOUS FLOW

Continuous flow intersections, also known as displaced left-turn intersections, involve vehicles crossing over the counterflow lanes and using a special ramp that allows them to make a turn while traffic is flowing in both directions at the major intersection. Signal timing at the intersection ensures that the cross-over and the subsequent turn can be made while there is no oncoming traffic.

THE STOP SIGN

The red and white octagonal stop sign is recognizable to drivers worldwide. Originating in Detroit in 1915, the octagon shape was selected in 1922 by roadway engineers who reportedly felt that the more sides a shape has the more danger it connotes. In the 1930s, reflective lettering was added for better visibility. But the background was yellow until 1954—when the familiar red field with white lettering was established.

Since then, the stop sign's height has been increased from two to five feet above the roadway and its size from 24 to 30 inches across. And its popularity has spread: it is now an official road sign around the globe, including the European Union. Not everyone relies on the octagon, however: Japan's stop sign is an upside-down triangle with a red background and white lettering.

be the volume of traffic moving through an intersection on the roads that comprise it and the directions that these vehicles most frequently take (straight, left, or right turn). The presence of pedestrians, speed of traffic, and the nature of visibility at the intersection are also important factors.

In many countries, traffic circles (also known as roundabouts or rotaries) have taken the place of traditional intersections. The idea of a traffic circle dates back to the early twentieth century, when places like the Place d'Étoile in Paris and Columbus Circle in New York were built to accommodate multiple streets meeting in one place. Interest in using them as a form of traffic control evolved in the midtwentieth century in Britain, where they continue to proliferate. France too is a fan of the roundabout: half of the roundabouts in the world (roughly 30,000 of them) are there.

INTERCHANGES

Because intersections are prohibited on interstate highways, traffic engineers rely on a variety of configurations to allow drivers to move from one highway to another. They feature a series of exit and entrance ramps, occasionally with overpasses, designed to maintain speed and minimize weaving between those getting on and off a particular road.

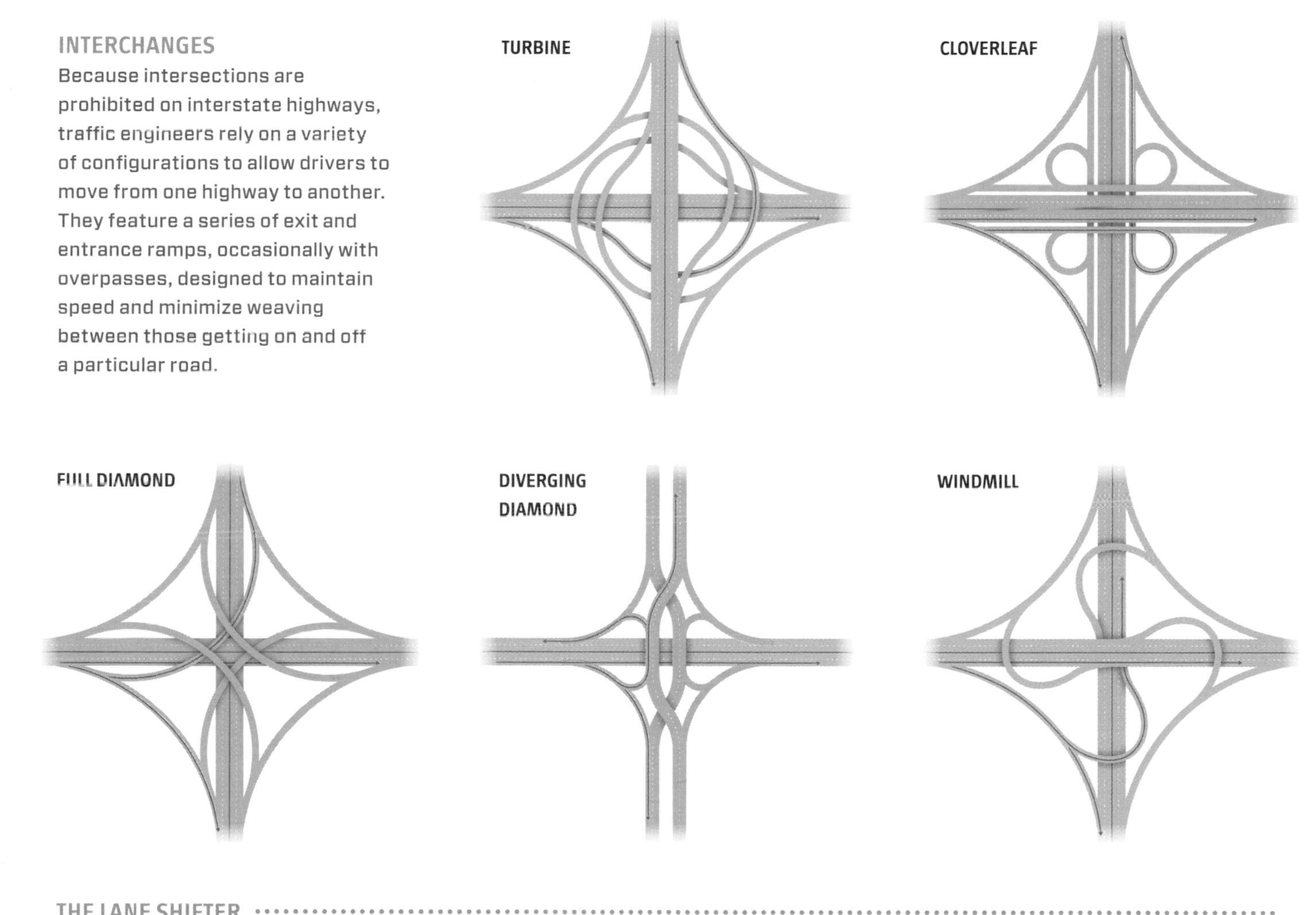

THE LANE SHIFTER

One way to manage the flow of traffic is to expand the number of lanes handling peak travel on the road during the morning or at night. Movable barriers, made up of three-ton sections of concrete hinged together with a steel pin to form a chain, are now regularly used to expand or contract the number of lanes moving in a particular direction on a multiple-lane roadway.

The barrier is moved by a self-propelled barrier transfer machine (BTM) anywhere from four to 24 feet in a lateral direction. Conveyor wheels grip the T-shaped top of the barrier, lift it several inches off the ground, and then move it through an S-shaped line within the vehicle to reposition it—giving rise to the more common name for the device: the zipper machine.

HIGHWAY NUMBERING

Highway numbering is something of a science. Each country has its own approach to the numbering of both highways and the more minor roads that intersect or parallel their path. In the United Kingdom, the numbering system is also a lettering system: major roads typically begin with an A or B and are followed by a series of anywhere from one to four numbers. Roads with an M in front of them are major motorways.

In the United States, a similar system of national, state, and local roads exist. Interstate highways have one system of numbering while state highways and routes often carry the name of the state first followed by a route number.

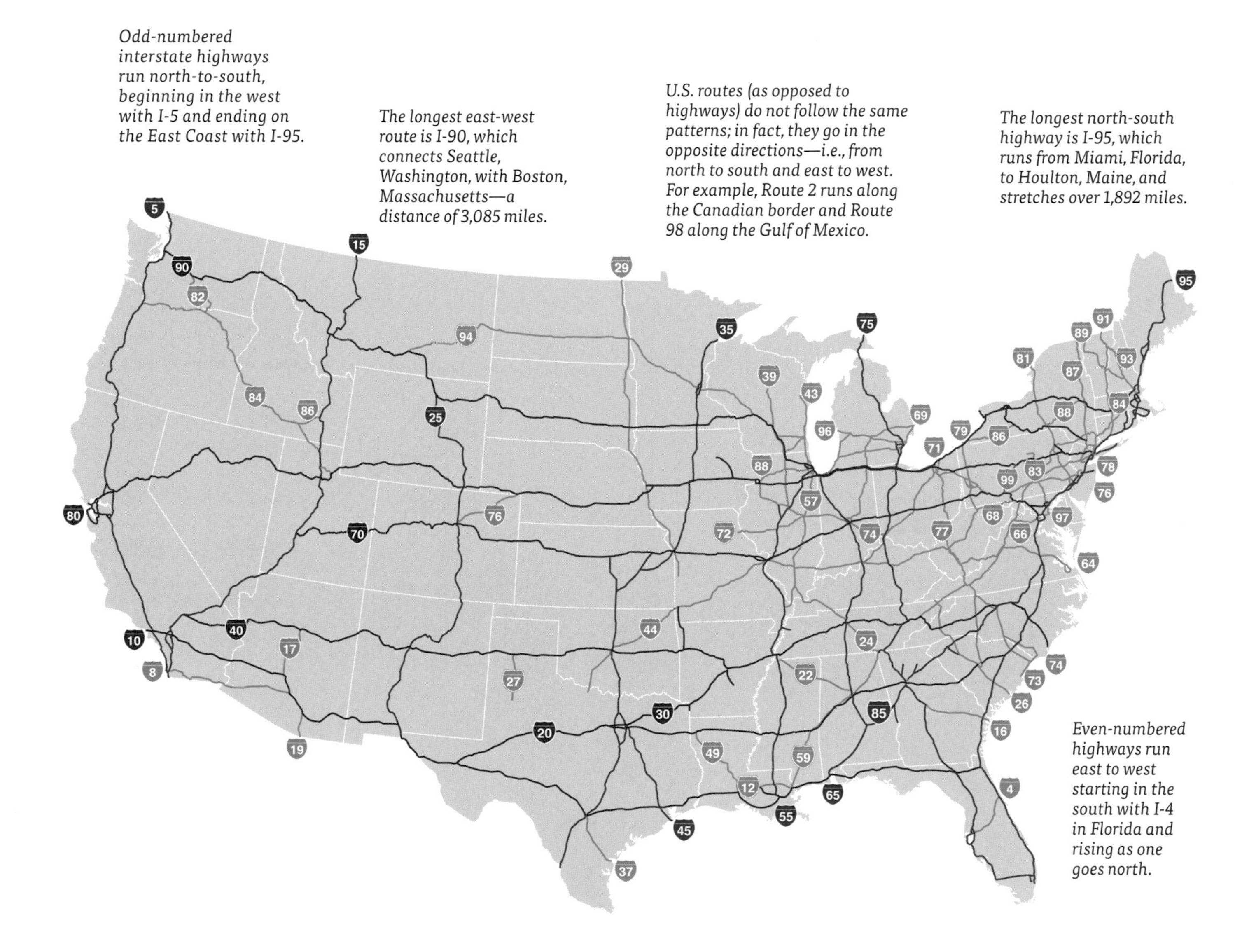

IKE'S ROAD TRIP

The U.S. interstate highway system, originally known as the National System of Defense and Interstate Highways, is among the biggest capital projects ever underwritten by the federal government. Between 1957 and 1969, 41,000 miles of limited access highway were built to tight standards: lanes were 12 feet wide with 10-foot shoulders; bridges had 14-foot clearances, grades were less than 30 percent, and roads would accommodate speeds up to 70 miles per hour.

The program was the brainchild of many, including President Dwight Eisenhower. As part of a transcontinental military convoy in 1919, the abysmal roadway conditions in the American west convinced him of the need for an integrated network of high-speed roads—and he acted decisively once in office.

TOLLS

Although road tolls existed as far back as ancient Babylon, modern tolls date back to eighteenth-century turnpike trusts established to construct and maintain roads in parts of Great Britain. Although Britain abolished private tolling when roadways became public, countries such as the United States later set up dedicated highway corporations to build new toll roads and reinvest the proceeds in future maintenance.

Today, there are many different schemes for implementing tolls to finance and manage roadways. HOT, or high-occupancy tolling, schemes encourage car pooling by offering cheaper or free travel to high-occupancy vehicles (HOVs) but also permit single-occupancy vehicles (SOVs) to access these lanes at a higher price. Likewise dynamic tolling, in which roadway charges are based on car volumes, can be used to keep traffic flowing at a designated speed and discourage travel at congested times. Based on supply and demand algorithms, dynamic rates can rise and fall frequently—as often as every five to fifteen minutes.

As the setting of toll prices has gotten more sophisticated, so too has the method of toll collection. Fewer and fewer highway tolls are collected by human beings in small booths. Electronic tolling systems, which rely on signals from transponders located in cars to debit a preexisting account, are now widespread. These transponders continuously emit radio signals, which are picked up by antennas located above or adjacent to each toll lane. The normal detection zone is between six and 10 feet wide and about 10 feet long.

Increasingly, cars are moving at speed through boothless toll areas that feature antennas located over the roadway. At some, drivers can maintain highway speeds; at others, they must slow significantly.

Manual toll collection is still done in places, but labor costs can be high—up to one third of toll revenue—and the amount of pollution created by congestion at booths can be significant.

As a car approaches the toll, the radio frequency emitted by the lane antenna or electronic reader activates the car's transponder, which returns a signal with the driver's unique information.

At some tolls, light curtains—beams of light shined across the road—detect the presence of a car. In others, treadles—sensors embedded in the pavement—calculate the number of axles.

A transponder is a battery-operated two-way radio, also known as a radio frequency identification unit, that transmits signals from its location in the car and contains identification data as well as account information.

In high-speed and automated tolling booths, video cameras record vehicles without transponders—taking a photo of the license plate so that a payment notice can be sent to the owner.

KEEPING CARS OUT

Since 2003 in London, some 1,500 fixed cameras have monitored vehicles moving in and out of the city center. With some exceptions, those penetrating a specified ring cordon demarking the central zone during weekday hours must pay a 10-pound ($15) charge in advance.

The system relies on camera and infrared technology to record vehicle registration numbers (license plates). A computer compares these numbers to lists of those paying in advance. Violators are issued a penalty notice of up to 120 pounds ($180).

London is not alone in its attempt to manage weekday congestion. Singapore, Milan, and Stockholm have also adopted congestion pricing to deter car traffic in their centers, though their programs have different rules and penalties.

HYBRID CARS

The term "hybrid car" describes a range of vehicles that rely on at least two different sources of power. Most on the road today are hybrid electric vehicles (HEVs), which have standard internal combustion engines as well as one or more electric motors that complement them. Some rely on fuel cells, natural gas, or other forms of energy for one of their sources of power.

The balance between power sources varies from one hybrid to another. Some feature a small electric motor that assists the primary combustion engine in acceleration and charges itself during deceleration. Others, like the Prius, have two motors that share the task of providing power to the wheels via a power splitter: electric power serves as the sole source of power when the vehicle is stationary or moving slowly but is shut off at higher speeds. The combustion engine is used to run a generator only when the primary battery power drops too low.

There is some debate as to whether hybrid cars are really any more efficient than a lightweight conventional car with a small engine—particularly on highways. Regenerative braking technology used in hybrids adds little value in high-speed driving situations, where the brakes are not being actively used. Most pundits agree that the economic and environmental advantages of hybrid electric vehicles are primarily limited to lower-speed city driving, where the electric motor takes over completely.

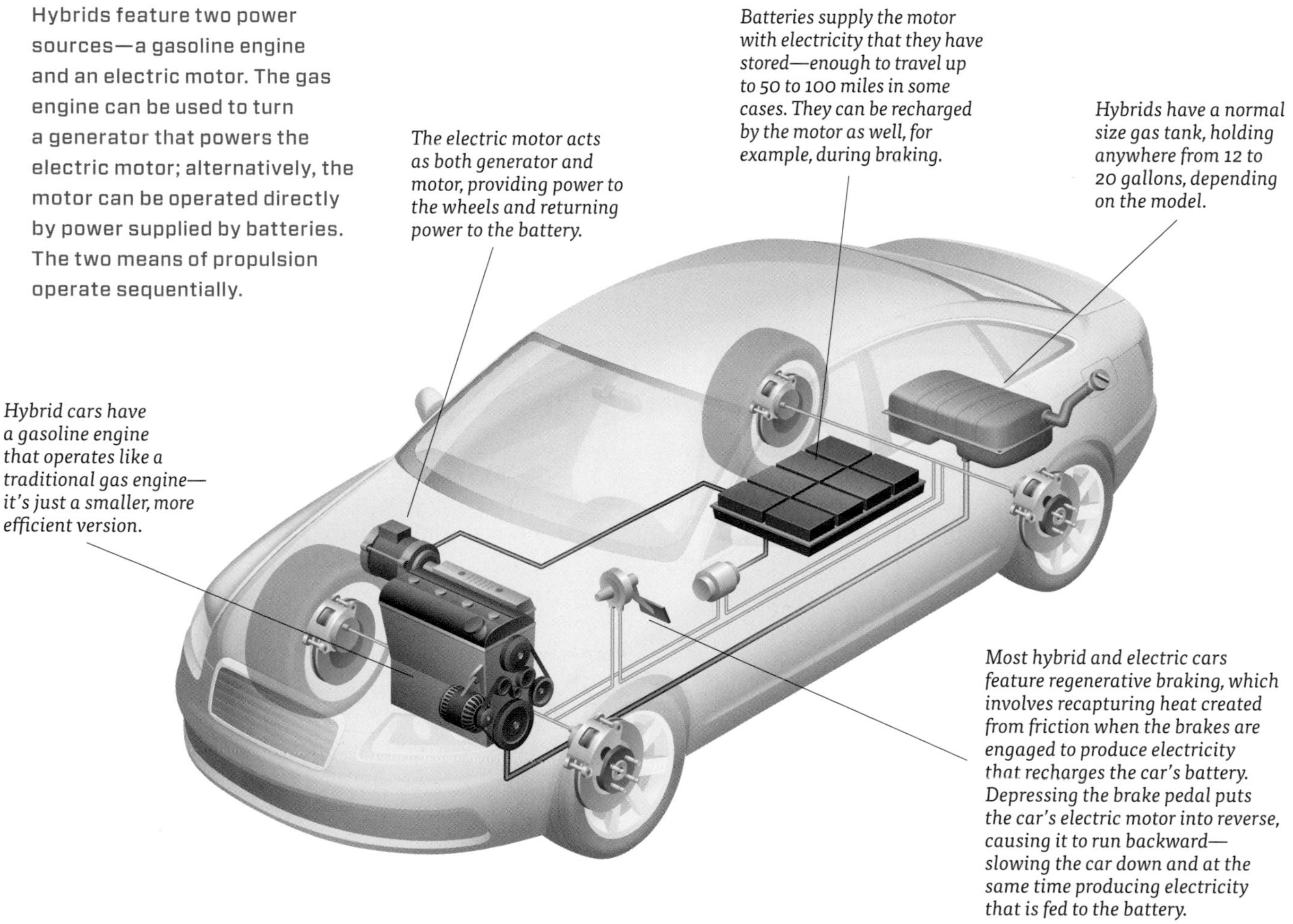

Hybrids feature two power sources—a gasoline engine and an electric motor. The gas engine can be used to turn a generator that powers the electric motor; alternatively, the motor can be operated directly by power supplied by batteries. The two means of propulsion operate sequentially.

The electric motor acts as both generator and motor, providing power to the wheels and returning power to the battery.

Batteries supply the motor with electricity that they have stored—enough to travel up to 50 to 100 miles in some cases. They can be recharged by the motor as well, for example, during braking.

Hybrids have a normal size gas tank, holding anywhere from 12 to 20 gallons, depending on the model.

Hybrid cars have a gasoline engine that operates like a traditional gas engine—it's just a smaller, more efficient version.

Most hybrid and electric cars feature regenerative braking, which involves recapturing heat created from friction when the brakes are engaged to produce electricity that recharges the car's battery. Depressing the brake pedal puts the car's electric motor into reverse, causing it to run backward—slowing the car down and at the same time producing electricity that is fed to the battery.

ARTIFICIAL NOISE

Hybrid cars are too quiet. Without the combustion operation of an engine, which kicks in at higher cruising speeds, hybrids are deadly silent as they move along local roads. Because of the danger they pose to blind, old, and young people, systems that produce warning sounds are now being introduced into their design.

The move to make electric cars noisier is most prominent in the United States, where automobile manufacturers have been given until 2017 to come up with solutions to the problem of noiseless cars. Several sounds are being considered, among them a whir, a melody, and a chirp. The last was incorporated into the 2011 Chevy Volt as an optional feature that drivers turn on and off—though how they know when to use it is unclear.

In countries that have embraced electric car technology, charging points often feature an indicator light so drivers can see whether it is available for use. Many have a sensing mechanism that disconnects power from the unit when a car is not charging—to prevent exposed wires should a car accidentally drive away while connected.

Two speeds of charging exist: normal (240vAC) or high-voltage (500vDC). Normal charging can take up to eight hours while high-voltage recharging can take significantly less. However, the latter remains risky as fast charging can put additional stress on the car's battery.

Electric cars have a battery charger built into the car; it is connected to the electrical network via a charging cable.

Charging units along the side of the road are typically connected to the electrical grid.

ELECTRIC CARS

To many, hybrid cars are only a stop along the path to an all-electric car future. Electric cars are generally accepted to be more environmentally friendly than hybrid cars because there are no tailpipe emissions whatsoever. Even accounting for the emissions attributed to the power plant producing the electricity used to charge them, the carbon dioxide resulting from use of an electric car is estimated to be between one third and one half that of a conventional car.

All-electric cars rely on special, and quite expensive, batteries for their power needs. These batteries must be charged regularly and replaced periodically. Otherwise, they have very low maintenance as compared with a traditional internal combustion engine—whose many parts alone lead to an increased risk of failure.

They're also more efficient. In engine lingo, efficiency is determined by how well an engine does at converting the energy from fuel combustion into propulsion—in other words, how little or much waste heat is lost in the process. In conventional gas-powered cars, roughly 20 percent of fuel energy is used to provide forward movement or power the car's accessories. In an electric vehicle, roughly 60 percent of that fuel energy is used.

WHO KILLED IT?

In 1990, the State of California ordered automakers to make electric vehicles available in the state for the model year 1998—and GM, Ford, Chrysler, Honda, Nissan, and Toyota did so. Yet within three years, nearly all had been taken off the road—bought back by manufacturers and literally crushed in an apparent attempt to stave off enthusiasm for the novel technology.

But several hundred Toyota RAV-4 vehicles somehow avoided the chop shop and were sold to their users. Many are still on the road today. Toyota's vehicle has a range of 100 miles, a recharge time of five hours, and a top design speed of 80 miles per hour. Scarcity has added to their value, and they are now worth more than their owners paid for them in the early 1990s.

SAFETY

As car ownership has grown and roads have gotten more crowded, driving has nonetheless gotten safer. This is in part a function of technological development, which has provided a multitude of new systems for cars and roadways. But it is also due to government regulation—of car and car part manufacturers, of roadway speeds, and of human behavior inside the vehicles.

Cars are undoubtedly safer than they have ever been. Some safety features, like the seat belt, are termed "active" rather than "passive." The first patent for a safety belt dates back to 1855, but the retractable three-point seat belt as we know it did not even appear until the late 1950s. Installation of seat belts in new passenger vehicles became mandatory in the United States in 1968 but Australia was the first to make its use compulsory in the front seat in 1970. Airbags, which grew out of World War II aviation technology, were added as an alternative to active seat belt systems in 1989 but did not become mandatory for new cars in the United States until 1998.

Roadways too are safer. A variety of safety features are now built into roads as they are paved or constructed, including rumble strips, which create noise and vibration as a car tire brushes across them, and safety edges, which replace the sheer roadway edge drop-off with a more gradual slope. "Cat's eyes," reflected glass spheres affixed to the roadway, and concrete Jersey barriers also help demark lane lines or the edge of a roadway.

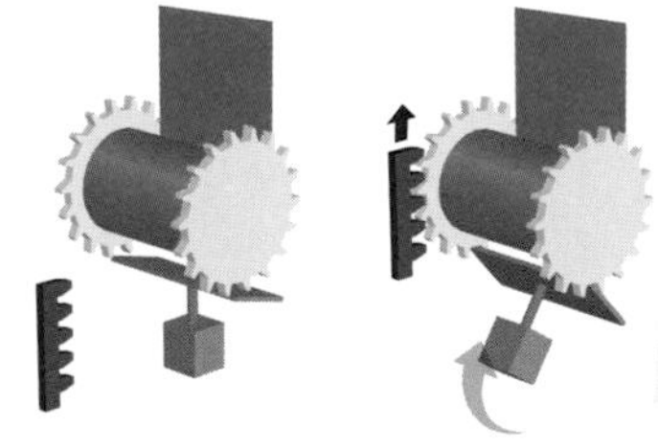

Pretensioner
Locking mechanism

HOW A SEAT BELT WORKS

Pulling a seat belt forward unrolls a spool located on a spring mechanism. The spool resists the unrolling motion and tightens the belt when the pressure is released. During a sudden deceleration in speed, the belt stretches slightly before locking to lessen and spread the impact on the passenger's body. Newer seat belts also have a pretensioner that pulls the belt in to protect the passenger in the event of a crash. Like airbags, pretensioners are explosive charges electrically triggered by impact and located under the seat; slack is taken up before the body has a chance to move forward, eliminating the trauma of hitting the belt.

HOW AN AIRBAG WORKS

An airbag consists of a thin nylon fabric folded into the steering wheel, dashboard, or side door. It inflates like a solid rocket booster: when a sensor is triggered, a solid propellant is ignited and produces blasts of nitrogen gas that inflate the bag at over 200 miles per hour. The whole event takes no more than one twenty-fifth of a second.

CRUMPLE ZONES

Crumple zones are areas at the front and rear of the car that collapse relatively easily on impact with another object. They are purposely designed to absorb some of the impact so that only part of it is transmitted to the passengers. The main cabin portion of the car is built more solidly.

U.S. SPEED LIMITS

The earliest roadway speed limits date back to around 1900. A century later, these limits vary widely from country to country. Even within the United States there is great variation: although all states adopted the 55 miles per hour speed limit during the oil shortages of the 1970s, by 1987 they were again setting their own limits. Rural interstate limits now reach 80 miles per hour in parts of Texas and Utah, but rarely top 65 miles per hour in the Northeast.

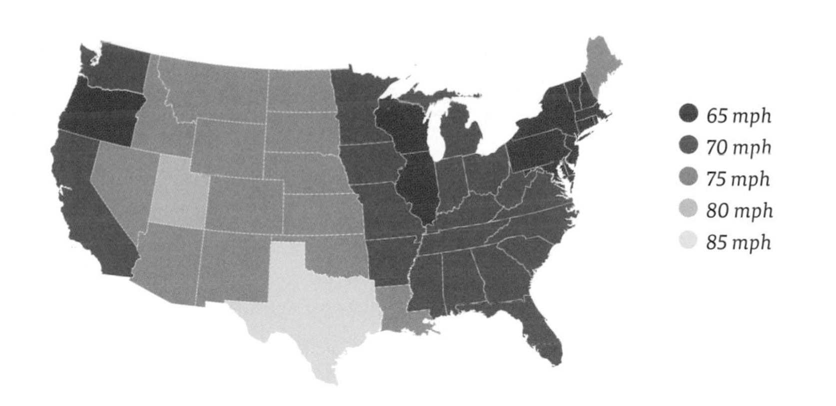

RUMBLE STRIPS

Rumble strips cause the car to vibrate and make a loud noise, letting the driver know that he is leaving the appropriate lane at the road's edge or center. They can be rolled in at the time the roadway is poured or added later by machines that can meld hot plastic to a preexisting surface.

JERSEY BARRIERS

Jersey barriers are steel-reinforced concrete modular barriers used to separate lanes of traffic. Devised at Stevens Institute of Technology in New Jersey, their sloped shape acts to lift the colliding vehicle off the ground and pivot it away from incoming traffic, absorbing energy that would otherwise go into a vertical crash.

"CAT'S EYES"

"Cat's eyes" consist of two pairs of reflective glass spheres cased in a rubber dome and attached to a road or highway. Invented and used extensively during the World War II blackouts in Britain, they are used today to mark traffic lanes and islands.

CRASH TESTING

Before a car can be sold commercially, it must go through extensive crash testing—complete with substitute humans. Constructed to resemble human sizes and weights, crash dummies are made up of metal, rubber, and synthetic materials.

Buried within their frames are numerous devices to measure the impact of a crash on humans—including load sensors (to measure the force with which a particular part of the body is hit), movement sensors (to measure how much the chest moves), and an accelerometer (to measure how fast body parts move in a variety of directions).

In addition, paint of varying colors is applied to body parts likely to be in contact with the car—normally the head, leg, and arms. After the test (usually at around 35 miles per hour) is complete, paint stains will be left on the locations that the dummy's body has hit—offering car manufacturers important data they can use to design safer cars and new protective systems.

ON THE RAILS

Compared with sea travel, trains are a relatively new form of transportation. Only 200 years have passed since the earliest experiments with locomotive engines, and only 50 since the debut of high-speed trains.

The earliest trains were not what we recognize as trains at all—they were wheeled wagons pulled along tracks by men or animals. Among the earliest was the Diolkos wagonway—a passage across the Isthmus of Corinth used to transport ships and cargo from the Aegean to the Ionian Sea 2,000 years ago. The wagonway catered to wheeled wooden sleds, pulled both by men and animals, and ran in grooves dug into limestone.

Over time, grooves gave way to stone and then wooden tracks. Wood remained in use well into the nineteenth century; the earliest locomotives ran on timber tracks. Iron appeared as a material first as a flange, or rib edge, on wheels, and then as a top layer or plate to protect the wooden track below. By the midnineteenth century, iron had become the material of choice for the entirety of the rail track.

As tracks evolved, so too did the cars that were rolled along them. The earliest horse-drawn railways were often sanctioned by government franchise and were only somewhat more efficient than traditional horse-drawn road coaches. But with the evolution of the steam locomotive in the early nineteenth century, the competitive advantage of railways would take an enormous leap forward.

Early steam trains were built for moving freight—coal in particular. George Stephenson, considered the father of the first steam locomotive, made history in Britain in 1825 when his locomotive pulled six loaded coal cars 9 miles (though it took him two hours). The ability to carry human loads was soon recognized, and scheduled passenger rail service was established in the UK by the 1830s.

The efficiency of railroads as a means of transport would grow significantly in the second half of the nineteenth century, largely due to the introduction of steel rail made cost effective by the new Bessemer process. Steel's performance as a track material was far superior to that of wood, which wore out, or iron, which had a tendency to fissure or crack. Early steel tracks lasted upward of 15 years, far longer than competing materials.

Even before steel came to the rails, the growth of the railroads was dramatic—particularly in places that waterways couldn't reach. In the United States, in 1840, some 3,000 miles of track existed across the country; by 1860 the amount had grown tenfold to 30,000. It was also important to nation building: the completion of the Transcontinental Railroad in 1869 marked a milestone in the westward expansion of the country.

Steam would not endure as a mode of transport for long-distance rail travel, due to the invention of the diesel-electric–powered locomotive in the early twentieth century. Its diesel engine produced energy to drive generators that powered a traction motor connected to the train wheels. The new locomotive didn't need to stop for water along the route and demanded less maintenance than an average steam engine.

Nor would steam survive for more localized commuter travel once electricity was invented. In New York City, for example, a particularly bad accident at the turn of the century caused by a steam-obscured signal resulted in commuter steam trains being banned from downtown areas. Electrification was not only cleaner but also offered the ability to travel underground—freeing up large swaths of city land for development.

Above ground, electrification in city centers gave rise to new forms of trolleys and streetcars—less polluting, more regular and more efficient than either horses or steam. Underground, new railways could silently carry large numbers of people between central city points on subways, or metros, consisting of electrified rail and trains. London's and Paris's successful experiments with subway trains soon transferred to other cities.

By the first decade of the twentieth century, today's rail technologies had largely been established. Trains relied on diesel fuel or electricity and the principle networks of steel rail track in developed nations had been laid. Improvements over the next 50 years or so would be around the margins: in the signaling that controls trains, in the appearance of rail cars, and in the speed and quality of the ride. None of these innovations provided a real foundation for rail to compete with the two big inventions of the twentieth century, the automobile and the airplane, and by the mid-1960s rail travel had lost much of its audience to them.

In the last several decades, however, the world has witnessed a return to the rails. Greater speeds than ever are being reached by a combination of high-speed trains, continuously welded steel rails, and new technologies like magnetic levitation. Improved signaling and communications have led to unprecedented levels of automation, including driverless trains. Combined with concerns about road congestion and air pollution, these improvements have underpinned a growing demand for rail travel across the globe.

FROM SEA TO SHINING SEA

TRANSCONTINENTAL

Among the earliest railroads to unite a continent was the Pacific Railroad or Transcontinental Railroad. Meeting up with eastern networks in 1869, the railroad opened huge swaths of the American continent to settlement.

TRANS-CANADA

The first Canadian transcontinental railway, the Canadian Pacific, connected Montreal with Vancouver when it opened in 1885.

TRANS-SIBERIAN

The Trans-Siberian Railway stretches almost 6,000 miles from Moscow to Vladivostock in Russia. Completed in 1916, it is not the longest continuous railroad in the world, but is perhaps the most famous—covering seven time zones.

TRANS-AUSTRALIAN

The first east-west transcontinental railway in Australia opened in 1917 and ran from Port Augusta to Kalgoorlie.

TRANS-ISTHMUS

The 48-mile Panama Railroad opened in 1855, connecting the Atlantic Ocean with the Pacific across the Isthmus of Panama, then part of Columbia.

MAIL TRAIN MEMORIES

Mail trains in the United States and Britain have a rich and storied history. Bags of mail initially rode with other rail freight, but by the midnineteenth century highly trained mail clerks rode aboard special cars to sort mail en route—up to 600 pieces per hour.

The mail cars were designed to maximize space inside for the sorting process. A mail-catching unit allowed the trains to collect mailbags at stations without stopping. These mail-on-the-fly contraptions were configured as steel catcher arms or nets that would snare a mailbag hanging alongside the station tracks.

By 1950, mail trains faced stiff competition from highway and air networks. The United States established the Highway Post Office Service in the 1950s to serve 150-mile routes by specialized mail-sorting buses, but the idea was abandoned in the 1970s. Britain's Royal Mail trains endured longer: the last one ran in 2004.

BELOW THE TRACK

Railway tracks represent the top layer of a complex and precise sandwich of materials. They typically sit on a track bed made of crushed stone ballast, which in turn sits on earth that is carefully shaped for the purpose. Often a mat of plastic or other waterproof material is placed between these levels to ensure that loose or wet clay or silt does not penetrate the tracks above.

LASER DOZER

Track beds are often graded with a special bulldozer known as a laser dozer, which incorporates special 3-D controls and may feature a GPS guidance system to direct the design of the ballast bed.

COMPACTOR

A compactor then moves over the ballast, laid to a depth of roughly 12 inches (30 centimeters), to ensure the foundation is even.

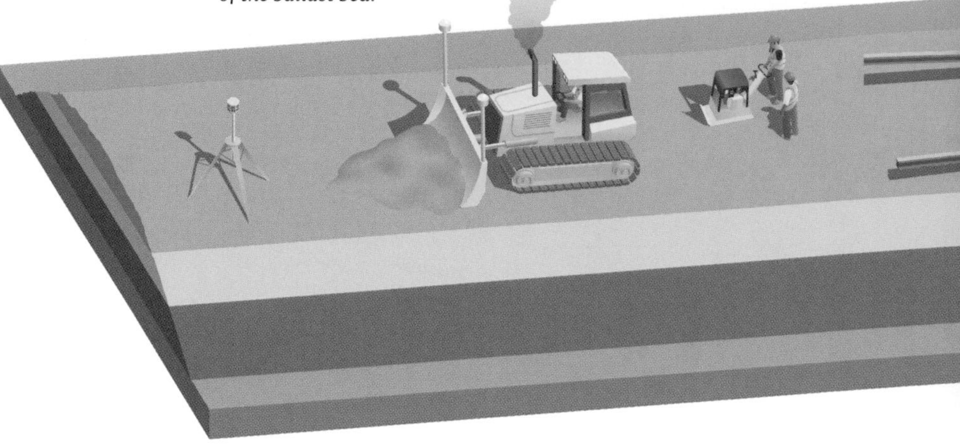

LAYING TRACK

Trains can't go very far without tracks—parallel rails separated at a consistent gauge, or width. Originally, sections of rail were bolted together via a variety of mechanical devices. The concept of welded track—or as it was originally known velvet track—first emerged in the United States in 1837. Since then, welded track has become the norm in the industry—particularly for trains moving at any speed.

The welding of rail was originally a labor-intensive business. Today it is done by machine in a process known as flash-butt welding. The ends of the rails are heated by an electric current and then pressed together enough to take account of buckling in summer and pulling apart in winter. Rails of roughly 60 feet long can be welded to form a continuous surface up to a mile in length.

Rails stay in place thanks to "sleepers," as known in the UK, or ties, as referred to in the United States. These ties, usually made of wood or concrete, transfer the load of the train from the rail to the gravel below it. They rest perpendicular to the rail on the bed below, held in place by tie plates and spikes or clips.

Underneath the rails and the ties is a layer of ballast, or stone. The name "ballast" came from the broken stone and gravel used as a ship stabilizer on open ocean journeys and subsequently repurposed inland for rail use. Ballast allows small amounts of movement while facilitating drainage. Because it provides a natural home to moisture, herbicides are used to regularly clear the ballast of weeds.

Not all rail systems rely on ballast. Subway systems, prone to much less expansion and contraction as they are less exposed to the elements, often have a rigid floor made out of concrete. Pumps are used to remove water from the system when it finds its way in during storm events.

Maintaining the integrity of the track is critical to the performance of rail networks. Historically, special inspection trains were used to monitor the condition of track along a route. Today, sensor-based systems supplement these visual inspections to ensure the integrity of the track.

GAUGES

Over half of the world's railways rely on a distance between tracks of 4 feet 8 ½ inches—the gauge George Stephenson used for his early locomotive. But gauges vary from over five feet in India and Spain to under four in Brazil and Japan. Occasionally, rail cars must switch gauges as they cross borders; they are lifted off the track and their bogey (underside) is replaced. Certain tracks are built with dual gauges to accommodate multiple bogey widths.

Standard rail gauge (4 feet 8½ inches) is found across North and Central America, Europe, China, and the Middle East.

Wide-gauge railroads (over 5 feet in width) remain in parts of India, Spain, Portugal, and Ireland.

The narrowest rail gauges (under 4 feet) can be found in southern and central Africa and many parts of Asia, including Indonesia, Taiwan, and the Philippines.

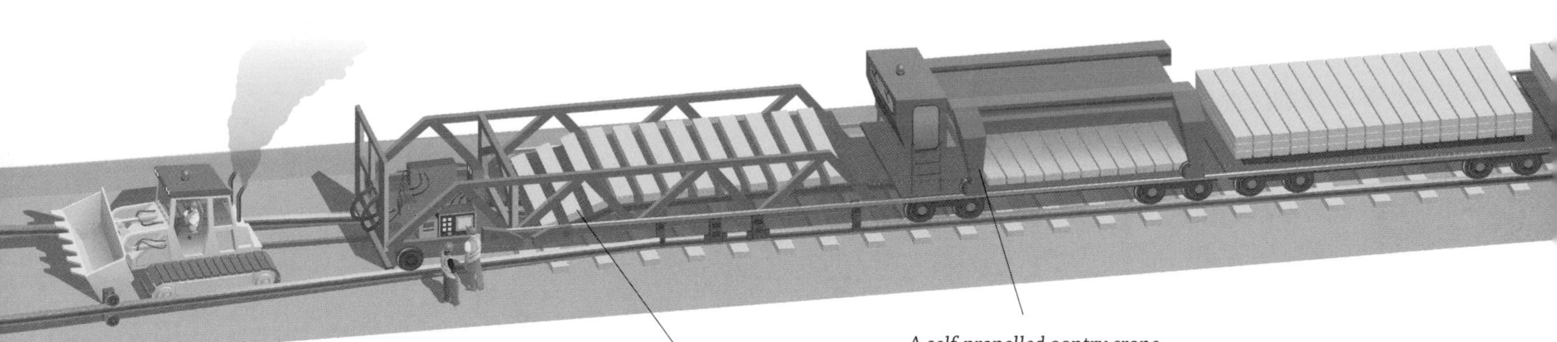

First the ties move along a conveyor to a tie-drop area. Rail lying along the line is threaded through guides at the rear of the unit, guided into place on the ties, and connected with a special device known as a nipper clipper.

A self-propelled gantry crane keeps ties supplied to the conveyor system.

NEW TRACK CONSTRUCTION MACHINE

A track construction machine installs the rails and ties in one operation.

BALLAST FILL

Machinery moves along the track to place additional ballast between the sleepers, or ties, once they are in place.

TAMPER

A ballast tamper, or tamping machine, aligns the track (ensuring that the two lines are parallel) and vibrates it to settle and stabilize the ballast beneath it. This ensures that the weight of the train will fall evenly across the rails, avoiding strain on any one portion.

TRACK STABILIZER

The track stabilizer applies force onto the newly laid track to ensure that the ballast beneath it is totally stable and will not move up and down or side to side.

SWITCHES

Trains depend on a sophisticated system of switches to ensure they pick the right set of rails when faced with a choice of directions at a junction. The switches do the physical work of guaranteeing that the train's wheels are directed along the correct set of tracks as part of what's known as an interlocking—a section of the network that controls the switches and signals in its area.

Rail switches consist of a pair of tapered rails, known as switch points, or switch rails, that are connected to a motorized mechanical assembly and shift in parallel to direct a train along a specified path: the train will follow the pair of rails that are fully connected at the moment that it passes through the junction. The place at which the two switch points cross is known as the frog and is designed to make sure that the train will always be supported by at least one rail and will not fall into the gap between sections of steel.

Early switches were operated from signal boxes adjacent to the track and were manually controlled by a system of rods that physically moved the points into place. Today's switches are typically moved by electric or hydraulic action and are tied directly to the signaling system so there is no confusion about which position a given switch is occupying.

Not all switches are alike. There are right-handed and left-handed ones, each with different lengths and curvatures. The shape of the switch is critical to determining the maximum speed at which a train can cross the junction: longer switches and shallower crossing angles permit higher speeds at the junction.

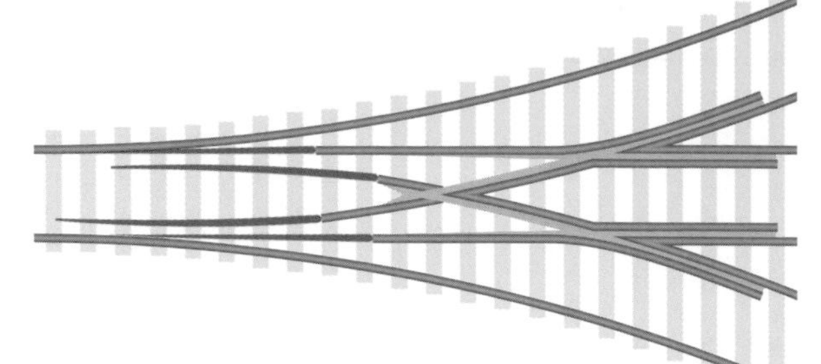

THREE-WAY

Three-way switches may be found at junctions where rail track separates into three directions. More complicated than the typical two-way switch, the additional switch points at a three-way junction require trains to move slowly as they navigate the switch.

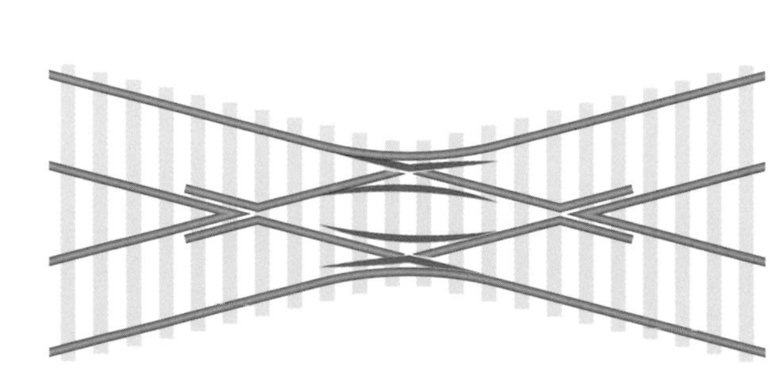

SLIP

At a single-slip switch crossing, trains coming from one direction have the option of going on one of two paths, but trains coming from the other must pass the switch and then reverse to use the other branch. Slip switches are often found at rail sidings, where trains serve users on a less frequent basis.

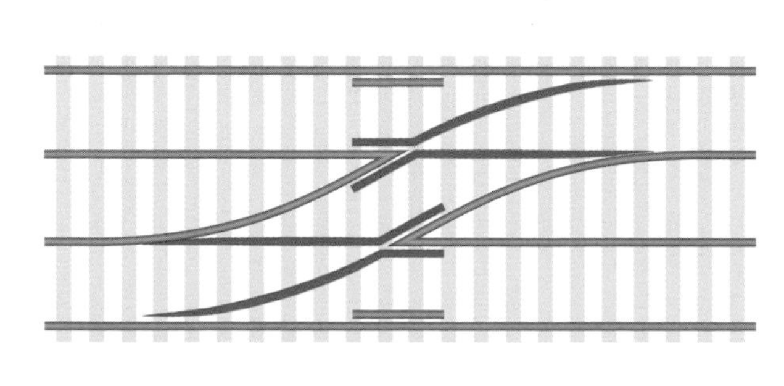

CROSSOVER

Crossovers are found where trains need to switch from one track to another. They can be found on parallel tracks that typically carry trains in the same direction, such as local and express trains on a subway, or on tracks moving in different directions (also known as a double, scissors, or diamond crossover).

MANY LANGUAGES

Signals function as rail traffic lights, ensuring that a stretch of track is occupied by only one train at a time. At one time, railroads relied on semaphore arms—mechanical blades that were lowered to indicate occupied track. While semaphores are still found today, modern railways typically rely on a system of colored lights similar to those at road intersections—green to signal proceed, amber to signify caution and slower speed, and red to convey the need to stop.

But there is no universal signal language. Individual railways developed their own systems over time, and systems vary widely between countries. Even within countries, signal systems can differ: Germany still relies on the two systems that were in place when it was unified in 1989.

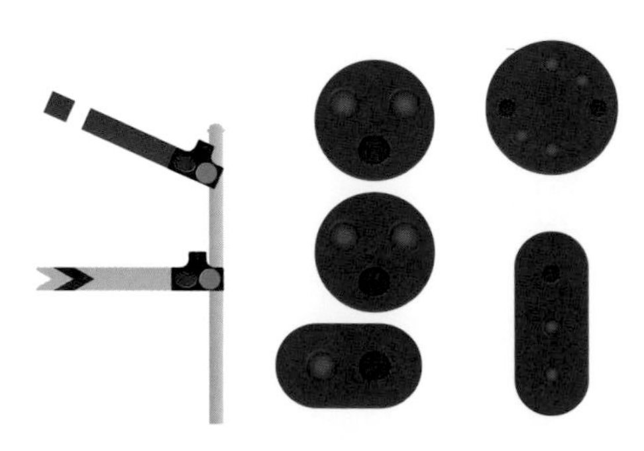

SIGNALS

Signals convey a variety of information to the train driver. In Britain, a system of route signaling, or directional signaling, is used. A route indicator on a junction signal tells the driver which track to follow; his training provides him with the route knowledge to travel at the correct speed for that stretch of track. Like commercial airplane pilots who are eligible to captain specific models of aircraft, train drivers under this system must be qualified to drive a particular route.

In most places, however, a system of speed signaling is used, which tells the driver at what speed to proceed but does not convey directional information. Speed signaling relies on a series of signals, often conveyed by lights but occasionally by semaphore blades, which tell the driver whether and how fast to proceed. The specific language of signal systems varies from place to place.

Information to program speed signals often depends on an automatic block signal (ABS) system. "Block" refers to the section of track between two points—often between two signals. Its length depends on the volume of traffic, speed, and length of trains; stopping distance; and sight characteristics. When one track block is occupied by a train, a signal will be tripped automatically and trains in front of or behind it will not be allowed to move onto the block. Although historically these blocks were fixed, real-time information about train locations now allows moving block signal technology that keeps track around a moving train clear.

FIXED BLOCK

In a traditional fixed-block system, occupancy of a section of track is detected via a low-voltage current that comprises a circuit between two signals. The train's metal wheels, which convey current from rail to rail, will short the circuit upon entering the block; this indicates the presence of a train there and is a signal to display an amber or red signal on either side of it.

Under this system, higher speeds (with longer stopping distances) require longer blocks—reducing the number of trains that can be run on a line at a given point in time.

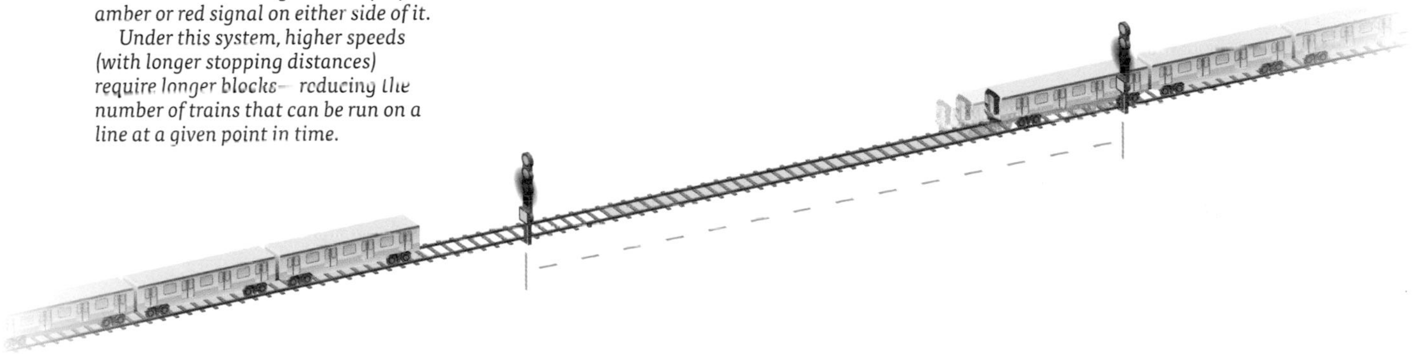

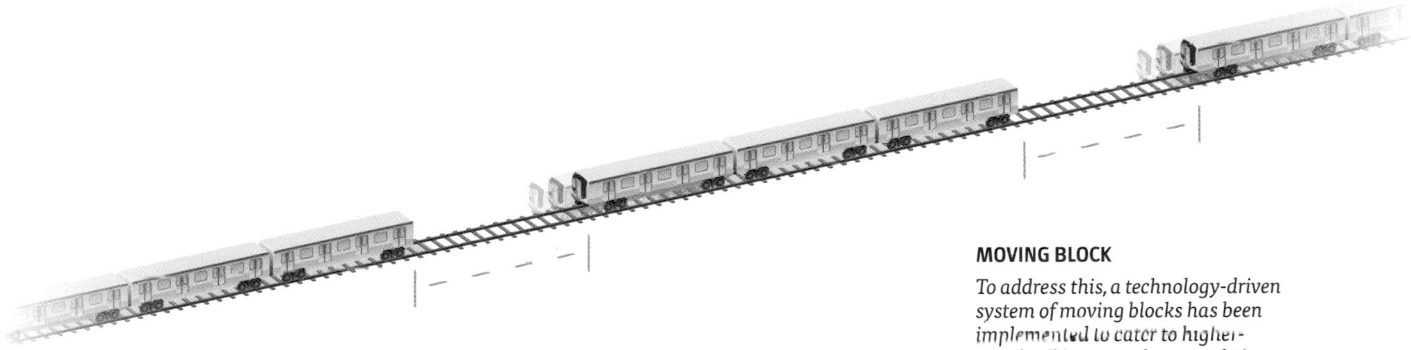

MOVING BLOCK

To address this, a technology-driven system of moving blocks has been implemented to cater to higher-speed rail in some places: real-time information about the location of the train provided by a cab-based signal defines a no-go zone around a moving train. The system calculates the stopping distance of each train and adjusts separation between trains accordingly.

MOVING FREIGHT

Though it seems something of a throwback in a world of space travel, rail freight continues to provide a vital lifeline to most of the world's population. Many commodities moved by rail are simply too heavy to travel on conventional highways or too low in value to make higher-cost transportation by truck worthwhile. To keep the cost of transportation low, freight trains tend to be long and carry things like grains, aggregates, coal, and other bulk commodities. Often these are configured as "unit trains"—made up of car after car of exactly the same product.

Like trucks, rail freight cars are often designed to carry specific commodities. Open wagons are used for bulk products and minerals. Hopper cars might be covered, for things like sugar or grains, or uncovered, for things like coal; both allow the car to be emptied automatically. Refrigerated cars are used for perishable commodities that otherwise would not be able to survive journeys of up to a week. And a variety of configurations are used to transport containers and truck trailers, including double-stack trains, which carry containers two-high along their length.

Rail freight journeys of any distance are not the fastest means to move cargo from point to point: moving even perishable items like vegetables can take days. Loading and unloading is time-consuming and the speeds at which freight trains travel will generally be lower than that of passenger trains.

The time involved in rail freight delivery is not solely a function of train speed. Because a long-distance shipment will be handed off between routes and railroads, during parts of its journey the rail car will not be moving at all. It will instead be in a hump yard or other facility getting disconnected from one group of cars and attached to another. Rarely do freight trains have priority on rail networks either; that is normally reserved for passenger trains.

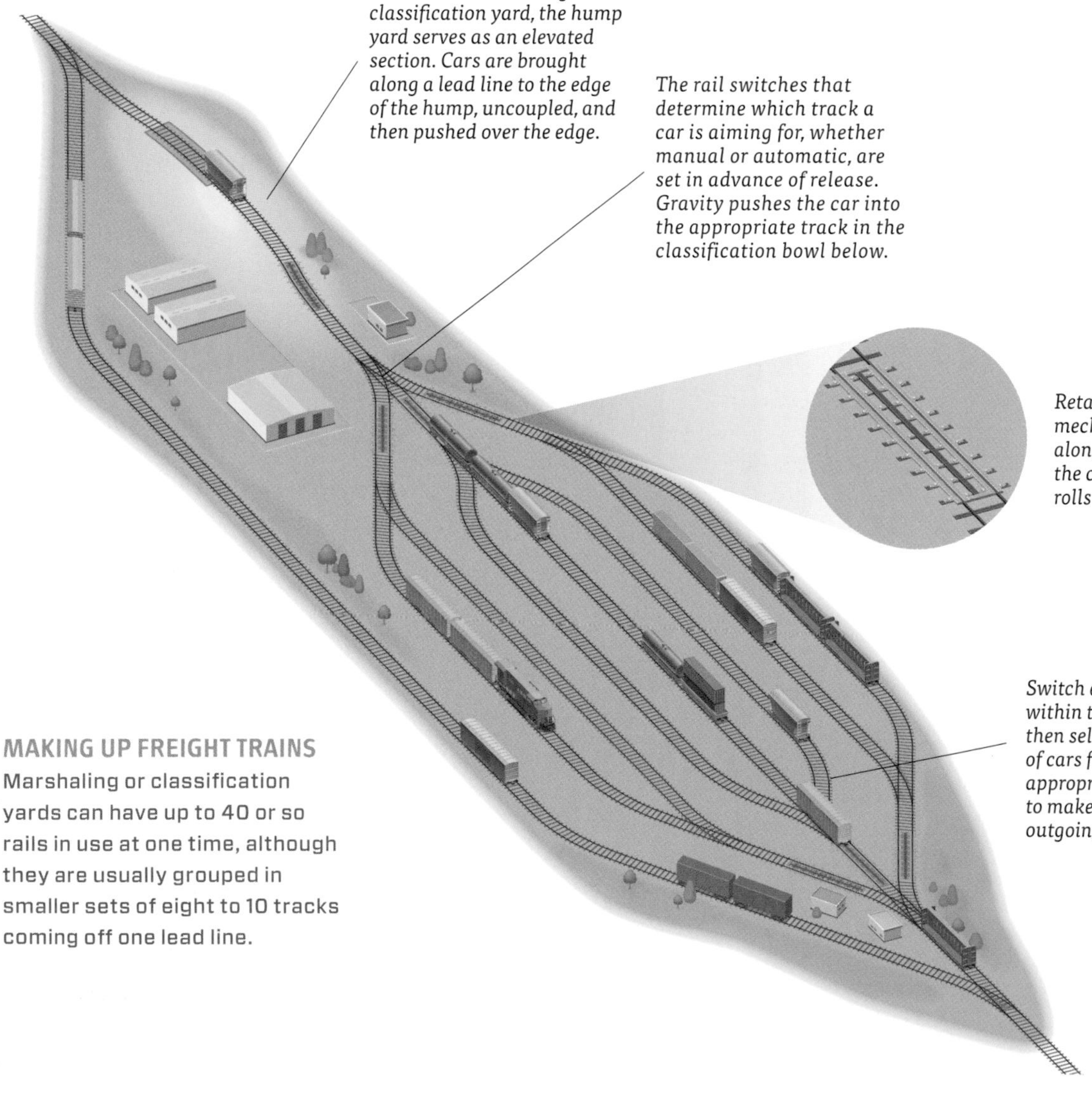

Within the marshaling or classification yard, the hump yard serves as an elevated section. Cars are brought along a lead line to the edge of the hump, uncoupled, and then pushed over the edge.

The rail switches that determine which track a car is aiming for, whether manual or automatic, are set in advance of release. Gravity pushes the car into the appropriate track in the classification bowl below.

Retarders, or mechanized brakes, along the rails slow the car down as it rolls forward.

Switch engines within the bowl then select sets of cars from each appropriate track to make up the outgoing train.

MAKING UP FREIGHT TRAINS

Marshaling or classification yards can have up to 40 or so rails in use at one time, although they are usually grouped in smaller sets of eight to 10 tracks coming off one lead line.

BOXCAR

Boxcars are the most prevalent freight cars on the rails today. Most boxcars carry freight loaded onto pallets or packed into crates. A wide variety of boxcars can be found across the world.

DOUBLE-STACK

Double-stack cars allow containers traveling in intermodal shipping to be stacked two-high. They are most often found in long-distance transport, when a large number of containers is moving to the same destination.

ROAD RAILER

Road railers are a form of regular truck trailer adapted to ride both road and rail. Built with separate sets of wheels, they travel as unit trains when they move on rails and then can be detached for onward travel by road to reach their destination.

CENTER BEAM BULKHEAD

Center beams are rail cars primarily used to carry building supplies. They feature a center partition to which the supplies can be tied so they don't shift during their journey.

COVERED HOPPER

Covered hoppers are used for dry-bulk commodities, such as grains or sand. The product is loaded from the top of the hopper at its origin and removed from the bottom at its destination.

COFC

COFC stands for "container on flat car" and looks exactly like it sounds. The car acts as a bed upon which a container used in intermodal shipping can be secured for long-distance transport.

TANK CAR

Tank cars are used to carry compressed or liquid commodities, such as oil, milk, or gas. They come in a wide variety of shapes and sizes, often unique to the commodity they carry.

REEFER

Refrigerated boxcars have their own power supply, enabling them to keep perishable foodstuffs, such as vegetables, meat, fruit, or frozen food, sufficiently cold during transport.

LOCOMOTIVE

Locomotives are among the most important of rail cars, and usually the only ones that are self-propelled. Locomotive engines provide the power to pull or push long lines of rail cars known as consists. Depending on the nature of terrain and the weight of the cargo being transported, trains may rely on one, two, or more locomotives to provide the traction needed to get the job done.

Once powered by steam, today's locomotives are powered by diesel fuel or by electricity provided either by overhead connections or by an electrified third rail. Even those powered by the same source are not really interchangeable: locomotives connected to passenger trains must be able to operate at a wide variety of speeds while freight locomotives, because of the great loads they are hauling, must have high tractive power to help get their loads moving from a stopped position.

Diesel locomotives aren't quite the same as diesel cars—in fact, they're really electric locomotives. A diesel engine on board the locomotive makes electricity, which in turn is fed to an electric motor attached to the wheels, or truck, of the locomotive to provide forward motion.

HIGH-SPEED RAIL

High-speed travel is the frontier of modern railroading. The expression "high-speed rail" generally refers to speeds above 125 miles per hour on existing track, or 155 miles per hour on new track. At these speeds, rail travel is competitive with air or road travel for journeys of medium distances—typically between 200 and 500 miles—and hence makes the most commercial sense.

Conventional high-speed rail represents an improvement, rather than a revolution, in rail technology. Continuous welded rail helps to avoid the vibrations that would otherwise result from connections between rails. Curves in the track are more gentle to accommodate higher speeds and reduce centripetal force, and level crossings are nonexistent. And most high-speed rail networks are powered electronically by overhead lines rather than by diesel.

The earliest line to implement this technology, the Shinkansen (also known as the bullet train), in Japan, is now celebrating its fiftieth birthday. Opened in 1964, it runs between Osaka and Tokyo—a distance of 323 miles (520 kilometers). Ten trains per hour, each consisting of 16 cars, make the trip between the cities. The journey time, at roughly three and a half hours, represents about half the time it would take to make the same trip by car.

Even faster than the Shinkasen is the Train à Grande Vitesse (TGV), which debuted in France in 1981. While it remains the speediest train using conventional steel-on-steel technology, the fastest trains in the world today do not use this technology at all. Magnetic levitation (maglev) trains rely on a magnetic force to repel the underside of the train from the rail line so that trains are literally floating in the air above the guideway. This decreases resistance and allows much higher travel speeds—averaging close to 180 miles per hour (300km/hr).

The first commercial high-speed maglev train appeared in Shanghai in 2002, and opened to the public the following year. Since that time, China has pushed ahead with maglev technology as part of a plan to link the country's larger cities by rail by 2020. Its fastest train, which links Wuhan in central China to Guangzhou on the southeastern coast, reaches speeds of 245 mph (394 km/hr) and has cut travel times from 10 to three hours.

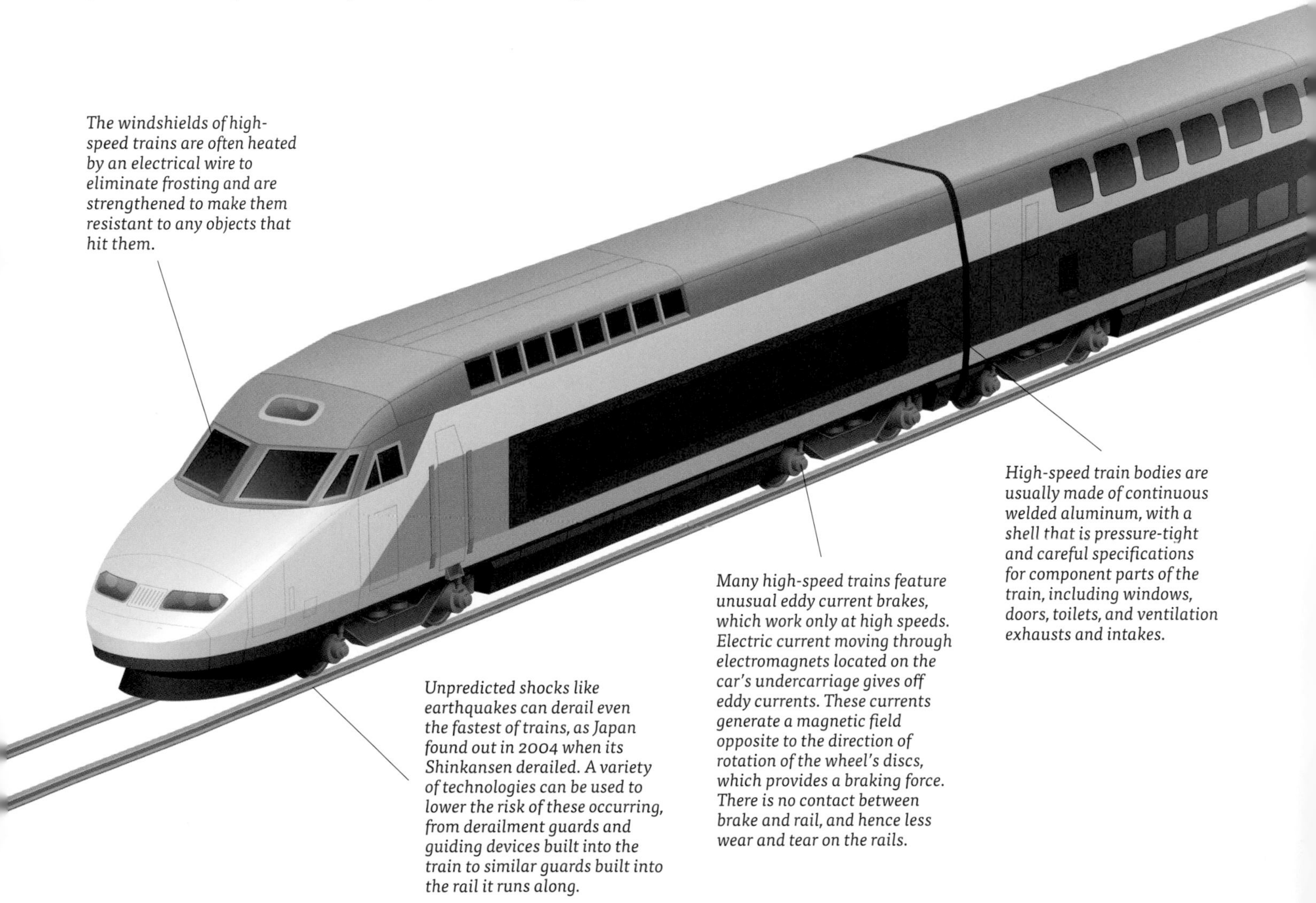

TILTING TRAINS

Tilting trains lean into their route to reduce the effects of centrifugal force on their passengers. They use a variety of mechanisms to do that: some are air powered (pneumatic) or oil powered (hydraulic) and others are powered by electricity. The mechanism is connected to a computer that indicates when the train is approaching a curve in the rail and activates the tilting mechanism well ahead of time—so that passengers feel almost no lateral forces as they round a curve.

Early experiments with tilting trains did not always work well. In the early 1980s British Rail introduced one of the first tilting trains—the advanced passenger train (APT)—on its London-to-Glasgow route. But the tilt didn't always work, and the APT was soon dubbed the "accident-prone train," prompting the withdrawal of the program and the return to conventional high-speed (intercity) service.

On a curve to the right, the train will tilt to the right to offset the g-force push on the rider to the left. Some trains rely on inertia for a passive tilt, but many have computers powering an active tilt.

MAGLEVS

Perhaps the first thing to know about maglev trains is that they don't have engines and don't burn fuel. The trains float over a guideway and rely on magnetic force for movement: the magnetic field created by the electrified coils in the guideway walls and track serve to propel the train forward.

No steel is found in maglev tracks or train wheels; instead, large magnets on the undercarriage of the train repel the magnetized coil of the guideway to lift or levitate the train up to three inches above the track. Once lifted, electric power is provided to the coils in the guideway walls and the train is pulled along the guideway toward its destination. The current alternates constantly, changing the polarity of the magnetized coils and allowing the field in front to "pull" the train while the field behind is "pushing" it.

In a prototype maglev being tested in Germany, the sides of the bottom of the train curl around the steel guideway. Electromagnets on the undercarriage raise the train roughly one centimeter (just under half an inch) above the steel rail and keep the train in a constant state of levitation.

THE WORLD OF HIGH-SPEED RAIL

High-speed rail can now be found in many places around the world, but notably not in the United States. Europe and Asia—in particular France, Italy, and Japan—served as pioneers for nearly all high-speed rail innovation in the twentieth century.

JAPAN

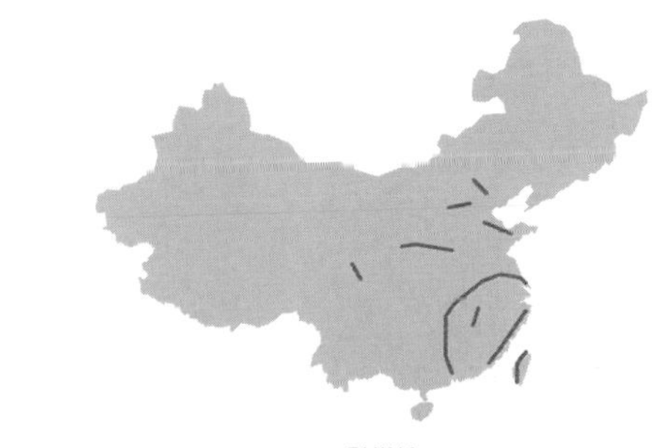

CHINA

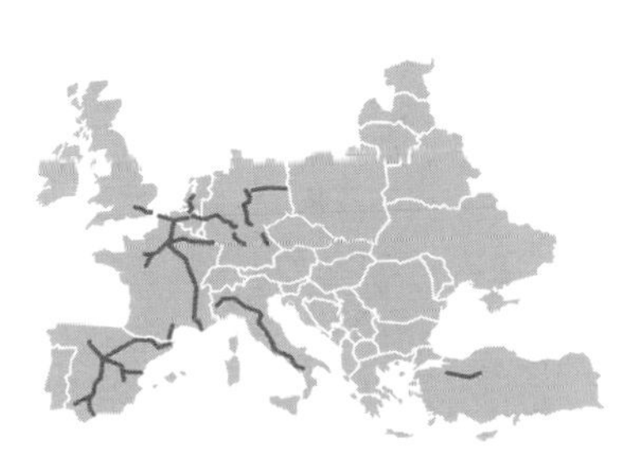

EUROPE

URBAN RAIL

Trains in the city generally operate in very different fashion from those that cross the countryside. For a start, they run more frequently and are often more crowded—standing on a city subway car can be the norm, rather than the exception. They also rely on very different technologies.

Subways or metros are the most heavily used—indeed, without them, street movement in large cities around the world would grind to a halt. Although today's underground trains are nearly all electric, the idea of the underground train predates electricity: the first London Underground train made its debut in 1863 under steam power using the name Metropolitan Railway. But it was the ability to travel deep underground without having to worry about fumes or exhaust that facilitated their growth in cities like Paris and New York, whose systems are—like London's—now more than a century old.

Many subway systems run both aboveground and below, diving under the earth as they approach the city center and emerging again across town. They often rely on conventional railway technology and track configuration, although some—including Paris, Mexico City, and Montreal—feature rubber-tired trains that run on concrete or steel rails. Both types of system use electricity to power the trains, either from a third rail or from an overhead connection.

Light-rail systems are more like trains than subways in that they spend most of their time aboveground. Having said that, very few run on diesel fuel: most are powered by electricity. They are lighter in the sense of the loads they carry, the speeds they go, and the investment required in both the track and cars that comprise them. Most travel relatively short distances within city centers and are physically separated from other forms of traffic on a dedicated right-of-way.

People movers and monorails are another form of urban travel, occasionally found in city centers though more frequently located at the airports that serve them. The term "people mover" was coined by the Disney Company in the late 1960s while designing one of its theme parks. What constitutes a people mover is not precisely defined: both automated guideway systems, like the Vancouver SkyTrain, and smaller ones, like Detroit's People Mover, fall into the category.

Cable car systems rely on two bull wheels at either end of the cable; usually one is powered by an engine and one rotates freely. Cars might be fixed or detachable; in the latter case the car is transferred to a slower-revolving mechanism at the stations.

Bogies on a monorail car connect it to the monorail beam. In straddle-beam monorails, the train straddles the beam and rubber tires on the bogie keep it stable and centered. In suspended monorails, the car is suspended beneath the monorail and the wheels are actually located within the beam itself.

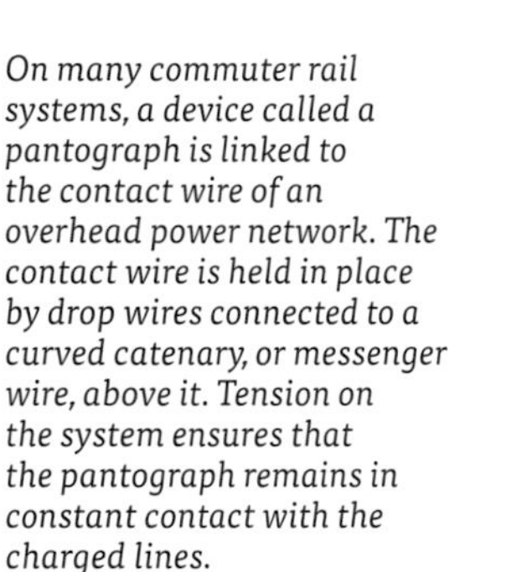

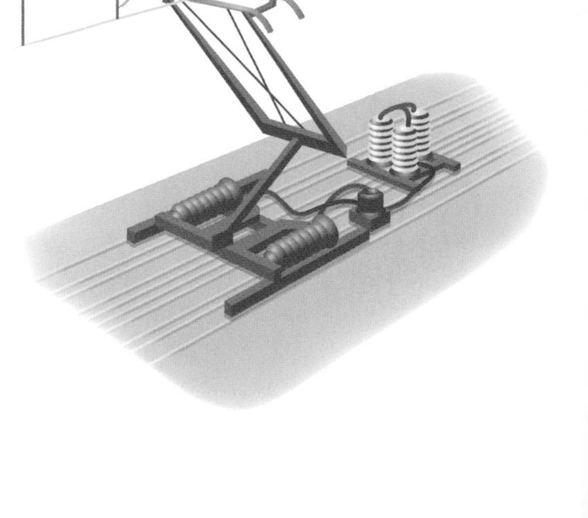

On many commuter rail systems, a device called a pantograph is linked to the contact wire of an overhead power network. The contact wire is held in place by drop wires connected to a curved catenary, or messenger wire, above it. Tension on the system ensures that the pantograph remains in constant contact with the charged lines.

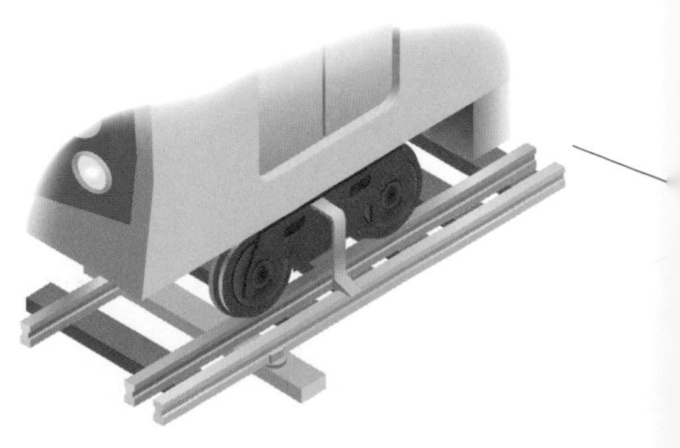

The electrified third, or conductor, rail is made of steel. The car's undercarriage is fitted with a "shoe" that connects to the top, side, or bottom of the rail. Because of the risk of electrification to people, higher voltages cannot be used and top speeds of travel are significantly lower than overhead-power systems.

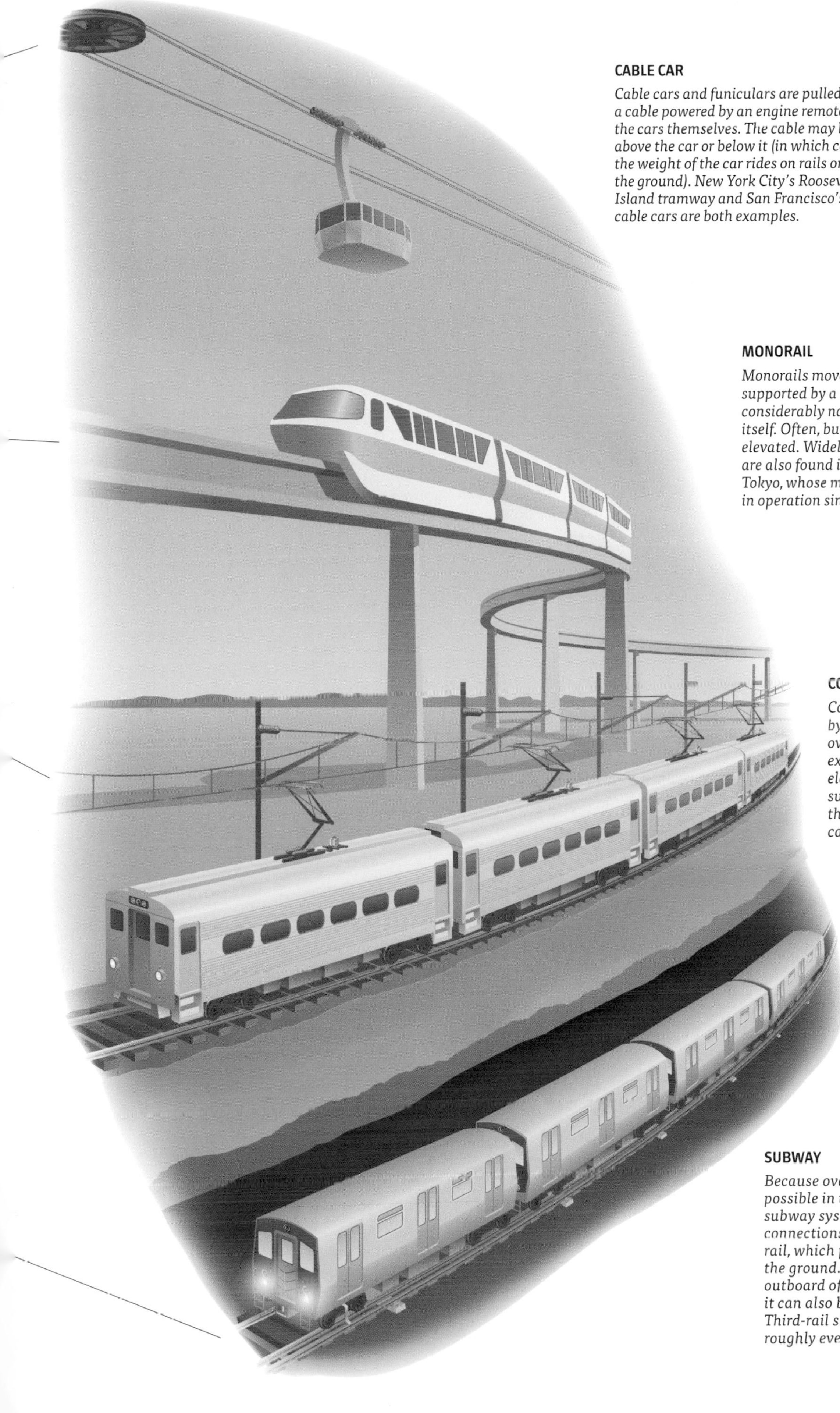

CABLE CAR

Cable cars and funiculars are pulled by a cable powered by an engine remote from the cars themselves. The cable may be above the car or below it (in which case the weight of the car rides on rails on the ground). New York City's Roosevelt Island tramway and San Francisco's cable cars are both examples.

MONORAIL

Monorails move along and are supported by a single beam, which is considerably narrower than the train itself. Often, but not always, they are elevated. Widely used in airports, they are also found in cities as large as Tokyo, whose monorail system has been in operation since the mid-1960s.

COMMUTER RAIL

Commuter trains are often powered by electricity running through overhead lines connected to an arm extending from the train's roof. The electricity flows from feeder stations supplied by a high-voltage grid through the overhead lines to each carriage's wheels.

SUBWAY

Because overhead lines are not possible in underground tunnels, subway systems often rely on connections to a third, or conductor, rail, which provides electricity on the ground. Often this rail is located outboard of the running rails, but it can also be located between them. Third-rail shoe connections occur roughly every 10 feet.

SAFETY ON THE RAILS

Among the most spectacular of transportation mishaps are train derailments—perhaps because trains look so out of place off the rails. Derailments can be caused by any number of things: broken or poorly aligned rails, faulty wheels, blockages on the track, collisions, or excessive speed.

One cause of failure stems from what's known as wheel burn. When a locomotive "slips,"or rotates on the track without actually moving the train it is hauling, the area under the wheel (no larger than the size of a dime) is superheated. Once the train moves, the steel cools down again—but the process of rapid heating and cooling can lead to molecular changes that undermine the ductility and strength of the steel.

There are several ways of detecting damage or breaks in rails early. One is simply by inspecting it regularly, or "walking the line" with what are known as detector cars. Ultrasonic testing can also be used to identify and map any irregularities in the steel. Occasionally, a track occupancy light goes off when there are no rail cars in the vicinity—another potential sign of a fault in the track.

Fractures in the rail are not the only hazard. The rail gauge (the distance between rails) needs to stay constant for top functioning of the network, yet the upward pressures put on the rails by decelerating trains can put them out of alignment. Weather can likewise upset rail equilibrium, with the freezing and thawing cycle putting additional tension or compression on the rails.

DERAILMENT

Some derailments are relatively minor, and trains can be returned to the rails by expert rerailers using wooden blocks, metal bars, and a variety of planks. But major derailments require specialized cranes, often rail mounted themselves, with the capacity to lift the great weight (up to 140 tons loaded) of rail cars.

AT-GRADE CROSSING

At-grade crossings, also known as level crossings in the UK, are generally found where roads cross rail lines. A train will trip a circuit along the line, triggering a series of actions: a gate lowers, lights begin flashing, and bells begin ringing. The train itself sends a warning as well: it gives two long toots, one short one and then another long one, to warn of its approach.

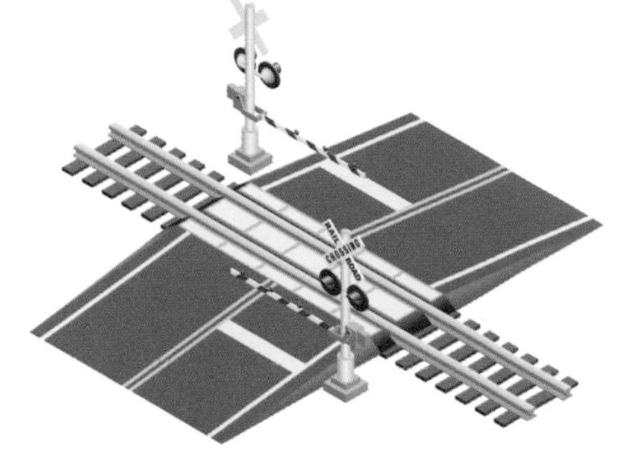

DEAD MAN'S HANDLE

The dead man's switch, or handle, must be pushed forward or engaged to keep the train in motion. In some places, for instance the New York City subway, the switch is incorporated into the train speed-control mechanism. If a train driver should suddenly become incapacitated, the switch is released and the emergency brakes are automatically applied to stop the train.

TRAIN ACCIDENTS

When train crashes and derailments occur, they are very dangerous due to the heavy machinery involved. Rail cars have been known to climb on top of one another, to fall into ditches, and to be spliced in half. Nearly always, these accidents result in injury or death.

The first recorded train mishap in the United States involving passenger death dates back to 1833, only two months after steam replaced horses on the Camden and Amboy Railroad in New Jersey. On board the train were both former President John Quincy Adams and a young Cornelius Vanderbilt.

A fire on board led to a derailment and crash, resulting in several deaths and in injury to nearly all passengers. Vanderbilt was among the severely wounded, suffering broken limbs and a punctured lung. Though he reportedly vowed never to ride the rails again, he would become the most prominent railway magnate, and his New York Central was one of the most powerful American corporations of the nineteenth century.

RAIL LIFE CYCLES

What happens to old rail cars? Most of them are pulled off the tracks, chopped up for parts, and sold to recyclers of a variety of materials. A few stay more or less intact, getting a new lease on life as part of homes, diners, or restaurants.

Some end up in a watery grave—as reefs for marine life in places distant from the tracks they once traveled. Old trains are one of the favorite forms of reef for bluefish, tuna, and flounder in places like North Carolina and Delaware, where the ocean bottom is sand or mud and cannot support the populations of baitfish necessary to attract sizable game fish. Artificial reefs are used to attract baitfish.

For the transport agency that owns the old cars, in-water disposal is an attractive option. At a cost of roughly $17,000 per car, water disposal is cheaper than removing the asbestos before processing the component parts for recycling. The price is right for more than just rail cars: a wide variety of decommissioned military vehicles are already settled at the bottom of these mid-Atlantic waters.

And what happens to old rail lines? In many countries, they are now being turned into recreational trails for joggers, walkers, or bikers. Most of them are at grade, but several are elevated. The Promenade Plantée in Paris, connecting the Bastille area with the eastern suburbs along a rail line abandoned 30 years earlier, was the first elevated park in the world when in opened in 2000. New York's High Line, opened a decade later, is among the most recent example.

PROMENADE PLANTÉE

The Promenade Plantée in Paris is generally recognized to be the world's first railway park. Opened 25 years after the Vincennes Railway Line between the Bastille and Bois de Vincennes was abandoned, it brought life both to the rail line and to the 71 arcades below it—now trendy shops and studios.

HIGH LINE

Modeled on the Promenade Plantée, New York City's High Line replaced a defunct freight rail line that had been elevated in the 1930s to relieve congestion on the avenue below it. The rail line was transferred by CSX Railroad to the city in 2005 to complete construction of a linear park.

BLOOMINGDALE TRAIL

The success of the Promenade Plantée and the High Line has given rise to imitators around the world. Like the High Line, the proposed Bloomingdale Trail in Chicago would replace a defunct freight line and cover a distance of 2.7 miles (4.3 km).

THE FIRST DINER

The rail car diner is an American icon, but few diner customers know where it came from—or how closely related it is to the cart that food vendors push along the streets of the world's larger cities.

In the midnineteenth century, a teenage entrepreneur began selling sandwiches to night workers. As his business grew, he procured a horse-drawn covered "express wagon" and parked it outside the *Providence Journal* newspaper's office. Its success inspired those who emulated him to expand their covered wagons to add places inside the wagon where customers could sit or be sheltered. To increase the attractiveness of these wagons to women, booth service, bathrooms, and longer tables were added in the 1920s.

It wasn't until the 1930s that the familiar streamlined silver rail car diner appeared. Its modern appearance suggested speed and mobility—which fit neatly with the growing popularity of the motorcar.

BRIDGES & TUNNELS

The tunnels and bridges that allow vehicular and rail traffic to pass underground or over water do more than just serve as roadway: they link cities, countries, and even continents. Spanning urban streets or highway intersections, they may at times be barely noticeable to the driver. In other places, they seem almost endless: the longest rail and road tunnels respectively, Japan's Seikan Tunnel and Norway's Laerdal Tunnel, extend for 33 and 15 miles.

While road and rail tunnels are a modern convention, bridges are not. Bridge building dates back thousands of years, with the earliest bridges made of materials found in nature: wood, vines or reeds. The Romans put bridge building on the map, both for transportation and for the delivery of water. Most of their bridge structures were arch based, involving large stone blocks wedged together with a keystone placed carefully at the top to cap the arch.

The Romans also developed a primitive form of concrete that they called pozzolana, a mix of volcanic ash and lime. This and other technological innovations made them the most prolific bridge builders in history, facilitating the construction of close to 1,000 bridges across lands that today make up part of 26 different countries. Most of these bridges were made of stone and concrete and were used for traffic, but at least 50 of them served as conduits for aqueduct systems.

The modern era of bridge and tunnel building was ushered in with the first iron bridge, built over the River Severn in Coalbrookedale, England, in 1779. Iron bridges gave way 100 years later to steel, with two of the most spectacular—the Forth Bridge in Scotland and the Brooklyn Bridge in New York—debuting within seven years of each other toward the end of the nineteenth century.

The Forth Bridge, just north of Edinburgh, was built entirely in steel—a material believed to be tougher than cast iron when it opened in 1890. Overbuilt in response to a previous railway bridge collapse, its massive towers, powerful cantilever structure, and 55,000 tons of steel sat on four separate foundations—making it strong enough to easily handle the hundreds of passenger and freight trains that have crossed it every day for over a century.

Equally monumental was the Brooklyn Bridge in New York, the first fixed structure to connect the then cities of Brooklyn and New York. At its opening in 1883, it was celebrated both for its height and its length: its towers were the tallest structures in New York while its length made it the longest suspension bridge in the world—a full 50 percent longer than its competitors. Like the Forth Bridge, it remains fully functional today.

With the Industrial Revolution came not only the ability to construct longer and stronger bridges but also the ability to tunnel under land and

ANCIENT BRIDGES

Although bridges existed before the Romans, the use of the stone arch greatly expanded their utility. In an arch configuration, downward pressure from the top of the bridge meets upward pressure from the bridge's foundations, squeezing the stone and creating a robust structure able to support considerable weight.

The Alcántara Bridge, spanning the Tagus River near Alcántara, Spain, was built by the Romans in A.D. 106. Made entirely out of stone, its Latin inscription reads "this bridge was built to last forever"—and so far it has.

water. Early railway tunnels were built under land: the first decades of the nineteenth century saw short tunnels built under parts of Liverpool and then longer stretches built through the Alps. Railway tunnels under the sea would follow: the Mersey Railway and Severn tunnels opened in quick succession in 1886, each serving briefly as the world's longest underwater transportation tunnel.

The world's first vehicular tunnel and the first to require ventilation was the privately financed Holland Tunnel, which opened in 1927 as an answer to marine congestion in the harbor separating New Jersey and New York. Its exhaust fans were an engineering marvel, replacing the 1½-mile-long tunnel's air every 90 minutes. Today, tunnels around the world rely on much the same ventilation technology to make underground or underwater driving possible over much longer distances.

With the rise of the automobile in the twentieth century, bridges and tunnels were constructed in every corner of the world. In some places, the two technologies were combined to span unusually long expanses of water or land: the 23-mile-long Chesapeake Bay Bridge-Tunnel in Virginia, for example, opened in 1964 with three bridge and two underwater tunnel segments. In 2000, the Oresund Bridge—a 7½ mile (11 km) tunnel-bridge structure connecting Copenhagen, Denmark, with Malmö, Sweden—was completed.

The Pont Saint-Bénezet, also known as the Pont d'Avignon, was built in the twelfth and thirteenth centuries as the sole crossing of the Rhone between Lyon, France, and the Mediterranean Sea. Only four of its 22 arches are still standing today.

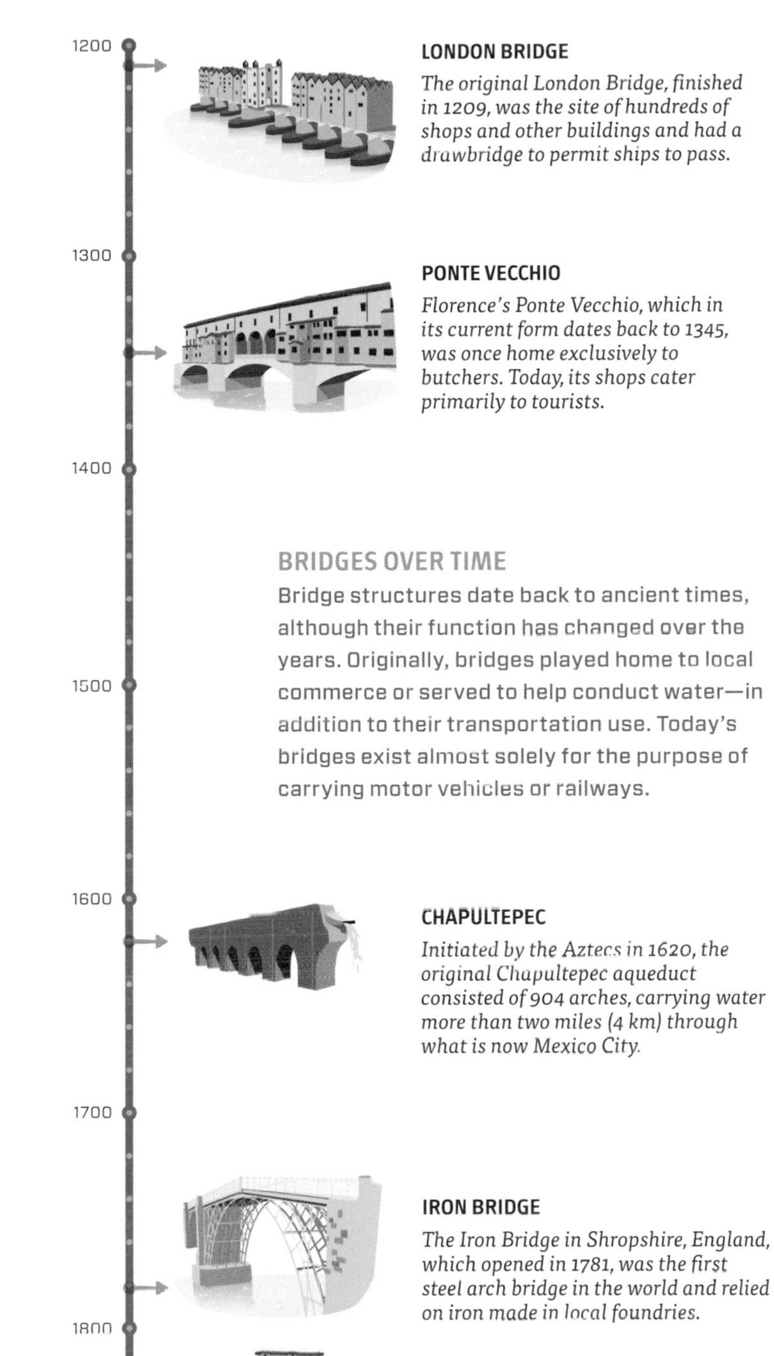

LONDON BRIDGE

The original London Bridge, finished in 1209, was the site of hundreds of shops and other buildings and had a drawbridge to permit ships to pass.

PONTE VECCHIO

Florence's Ponte Vecchio, which in its current form dates back to 1345, was once home exclusively to butchers. Today, its shops cater primarily to tourists.

BRIDGES OVER TIME

Bridge structures date back to ancient times, although their function has changed over the years. Originally, bridges played home to local commerce or served to help conduct water—in addition to their transportation use. Today's bridges exist almost solely for the purpose of carrying motor vehicles or railways.

CHAPULTEPEC

Initiated by the Aztecs in 1620, the original Chapultepec aqueduct consisted of 904 arches, carrying water more than two miles (4 km) through what is now Mexico City.

IRON BRIDGE

The Iron Bridge in Shropshire, England, which opened in 1781, was the first steel arch bridge in the world and relied on iron made in local foundries.

BROOKLYN

When the Brooklyn Bridge opened to unite the cities of Brooklyn and New York in 1883, its towers were the tallest structures in the metropolitan region.

GOLDEN GATE

The Golden Gate Bridge, renowned for its unique red color, serves as the maritime gateway to San Francisco and celebrated its seventy-fifth birthday in 2013.

PUENTE DEL ALAMILLO

The Puente del Alamillo in Seville, Spain, serves as one example of modern bridge design, in this case the work of Santiago Calatrava.

HOW BRIDGES WORK

Bridges that elevate roads or railways over water or land vary widely in both appearance and the way they work. The most significant difference is between those that are fixed and those that must open for passing ship traffic.

Fixed bridges tend to fall into four major types. Beam bridges consist of horizontal beams supported at either end. Truss structures improve upon the beam bridge by adding girders in a latticelike form to give additional rigidity and permit greater length between vertical-supporting spans.

Arch bridges operate by transmitting the weight above each arch to its sides, usually into supports or abutments placed at each end of the arch. As modern engineers strived to increase the length between abutments, the simple elegance of Roman stone arch bridges has given way to steel and prestressed concrete arch designs.

In suspension bridges, the roadway is supported by vertical suspender cables hanging from horizontal supporting cables that span the two towers. Although some suspension bridges are self-supporting, typically the supporting cables are attached to rock deep in the earth through anchorages located at either end of the bridge.

The balance of two forces, compression and tension, allows bridges to span great distances. Tension refers to the forces pulling bridge elements apart; extreme tension can lead to snapping. Compression refers to the forces pushing bridge components together; too much compression can lead to buckling. Transferring and dissipating these forces is key to bridge design.

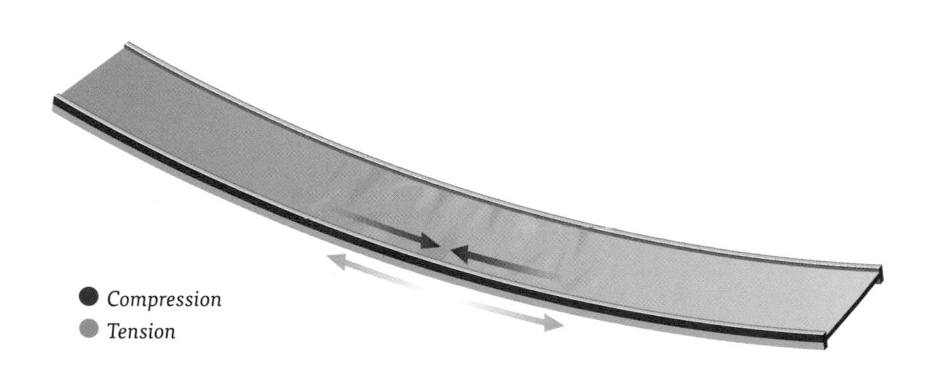

FIXED BRIDGES

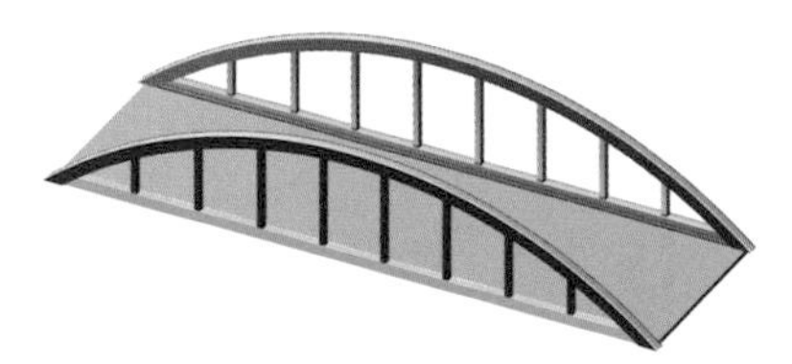

SOLID-RIBBED ARCH
Solid-ribbed arch bridges feature vertical cables that transmit the load from the roadway to the arch that supports it.

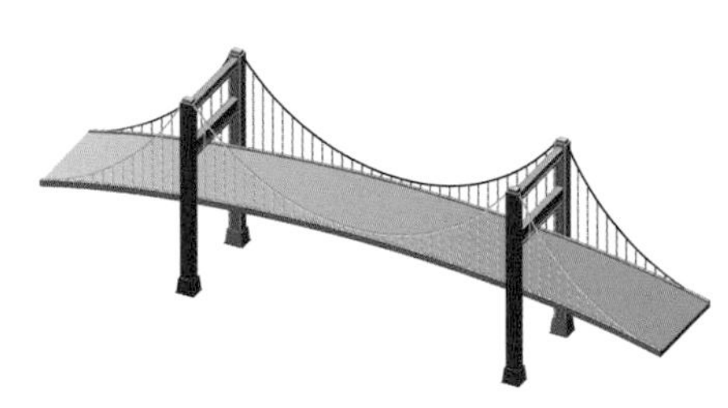

SUSPENSION BRIDGE
In a suspension bridge, the weight is carried along vertical cables from the roadway to larger cables hung across each side of the top of the towers.

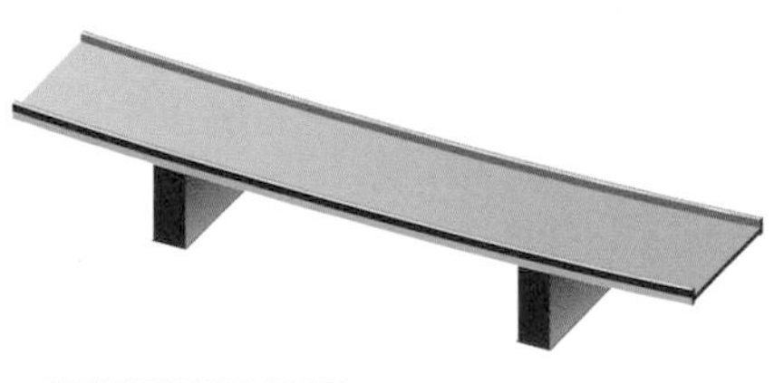

CONTINUOUS SPAN
Continuous span bridges require support, usually from below, so that the beams that constitute the bridge do not bend or bulge.

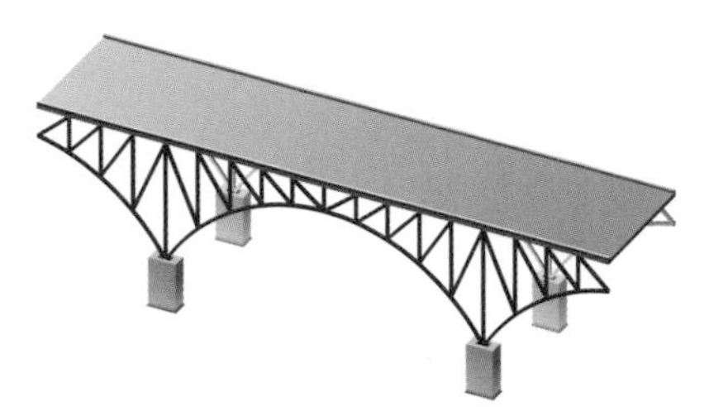

SPANDREL-BRACED
In a spandrel-braced arch configuration, the weight of the roadway and deck are carried along diagonal bracing to the arch below.

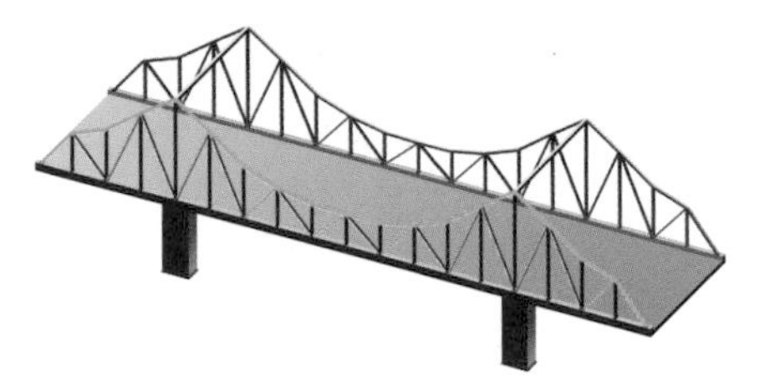

CANTILEVER THROUGH TRUSS
In a cantilever through truss bridge, the diagonal members form part of the truss structure supporting the roadway.

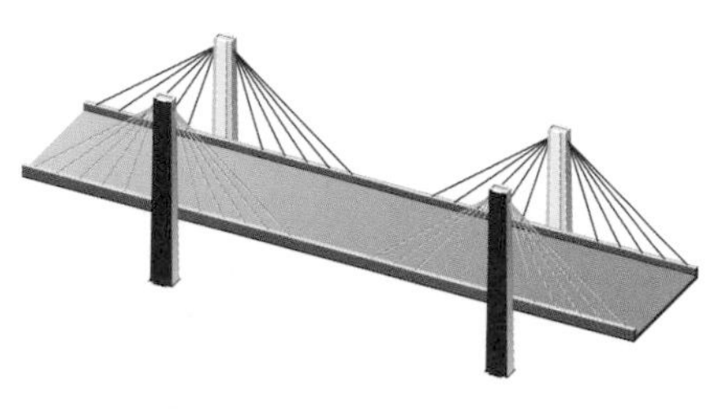

CABLE-STAYED
Cable-stayed bridges feature one or more towers that support the bridge's deck in either a harp or fan shape.

Suspension bridge decks are stiffened to mitigate the effects of sway on the hanging roadway.

Cantilever bridges rely on arms extending from support structures that project out horizontally, supported on only one end and balanced by some form of counterweight. Typically these arms are constructed from steel trusses or concrete girders.

Movable bridges are often found on waterways where maritime traffic requires access. Bascule bridges, from the French word for "seesaw" or "balance," are normally low to the water. The bridge roadway splits in two parts that are lifted to a vertical position by motors and counterweights several times heavier than the bridge itself.

Lift bridges have a center span that rises horizontally to allow marine traffic to pass. However, the counterweight need match only the bridge's weight—making lift designs suitable for heavy bridges, such as railroad crossings.

There are many other kinds of movable bridges, from swing bridges, which turn on a vertical axis, to retractile bridges, which double back on themselves. Not all are raised: pontoon or floating swing bridges sit right in the water.

MOVABLE BRIDGES

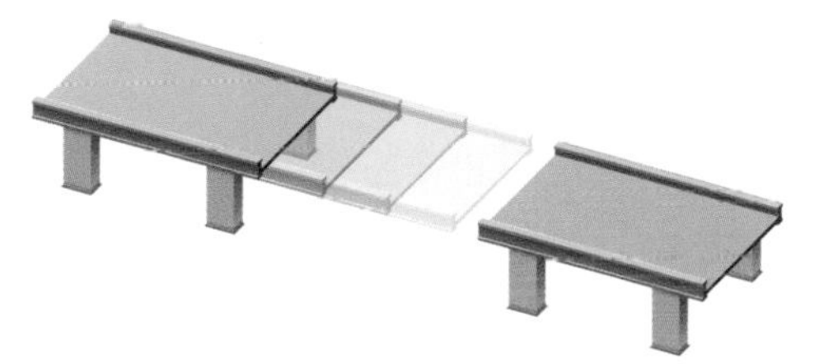

RETRACTABLE

Retractable bridges feature a roadway or deck that can slide to one side to allow ships to pass.

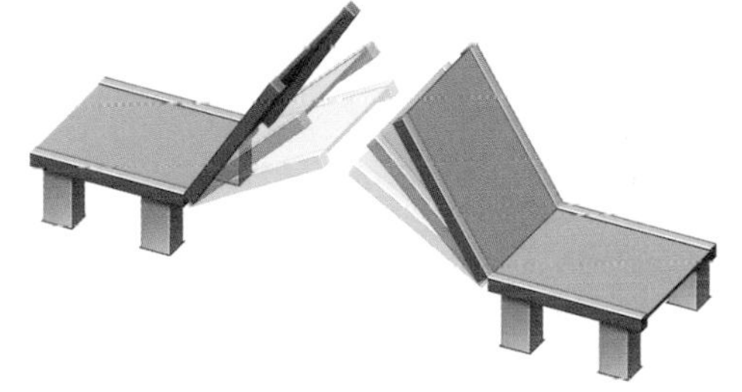

BASCULE

Bascule bridges, also known as drawbridges, rely on a counterweight heavier than the deck to maintain balance and assist in lifting either one or two leaves.

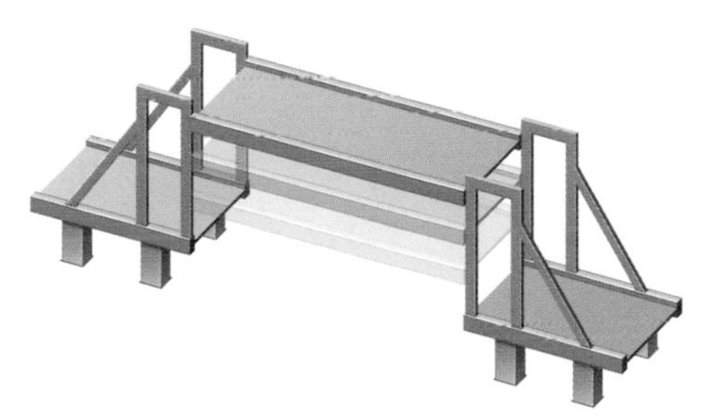

VERTICAL LIFT

The tower counterweights of a vertical lift bridge do less work than those needed to lift bascule bridges, so heavier materials can be used in the deck.

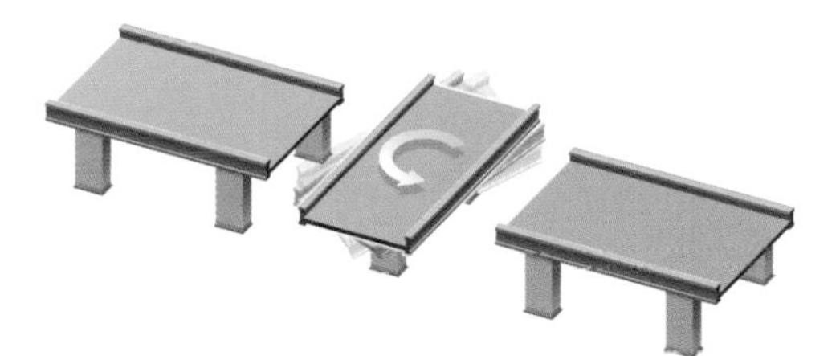

SWING

Swing bridges rotate around a central point or a pin at one end. Generally, the moving section turns 90 degrees from its original position.

ROLLING BRIDGE

The Paddington area of London is home to what is generally acclaimed to be the world's only curling bridge. Built in 2004 as part of a local regeneration and office development project in the Paddington Basin of the Grand Union Canal, the steel and timber bridge is 39 feet (12 meters) long. With the activation of hydraulic pistons, the bridge curls up slowly so that its ends join—forming an octagonal shape roughly half the width of the canal water it spans.

The bridge spends most of its time in the curled, or fetal, position. Opened at least once a week, it is more a curiosity than a precedent for bridges elsewhere—but remains a uniquely modern take on the form of retractable bridges that were once used to protect castles, forts, and other battlements across the British Isles.

SUSPENSION BRIDGE CONSTRUCTION

CAISSONS

In seabeds where bedrock is reachable, soft material sitting on the sea floor will be excavated to prepare for the lowering of the caisson—a watertight structure that allows underwater construction (in this case, the foundation for the towers) to take place. In cases where bedrock is not reachable, pilings will be driven or a large concrete pad will be placed on top of compacted gravel to support the construction of the tower.

TOWER CONSTRUCTION

Single or multiple towers will then be constructed of concrete, steel, or stone atop the caisson. "Saddles" will be placed on top of the towers to hold the main suspension cables in position once they are spun. The saddles incorporate rollers to allow the cables to move slightly once in position.

BUILDING A BRIDGE

Bridge building dates back thousands of years; indeed, the origins of many of today's most beloved bridges—including suspension ones—can be found in early bridges built by various societies out of logs, vines, and reeds. But the modern era of bridge building really began in the nineteenth century, with the need for bridges to support the great weights associated with first rail and then road travel. Though many of the bridges of that period were successful, the science of building with iron was still in its infancy and a number of bridge experiments ended in disaster.

Among the most notable failures were those in Canada and Scotland. In 1907, a huge cantilever structure being constructed over the Quebec River 9 miles north of Quebec collapsed, taking with it the lives of 75 people. Even more spectacular was the 1879 collapse of a newly completed rail bridge over the River Tay in Scotland: it collapsed in high winds with a train on it, killing everyone on board.

Perhaps the most famous bridge disaster of modern times, which was caught on film, happened in Tacoma, Washington, in 1940. The Tacoma-Narrows Bridge, constructed in only 9 months, was one of the longest and most elegant roadway suspension bridges of its time: the ratio of its depth in relation to the length of its span was 1:350. The narrowness of the bridge proved problematic only days after its opening, when the bridge began to twist and turn—and then

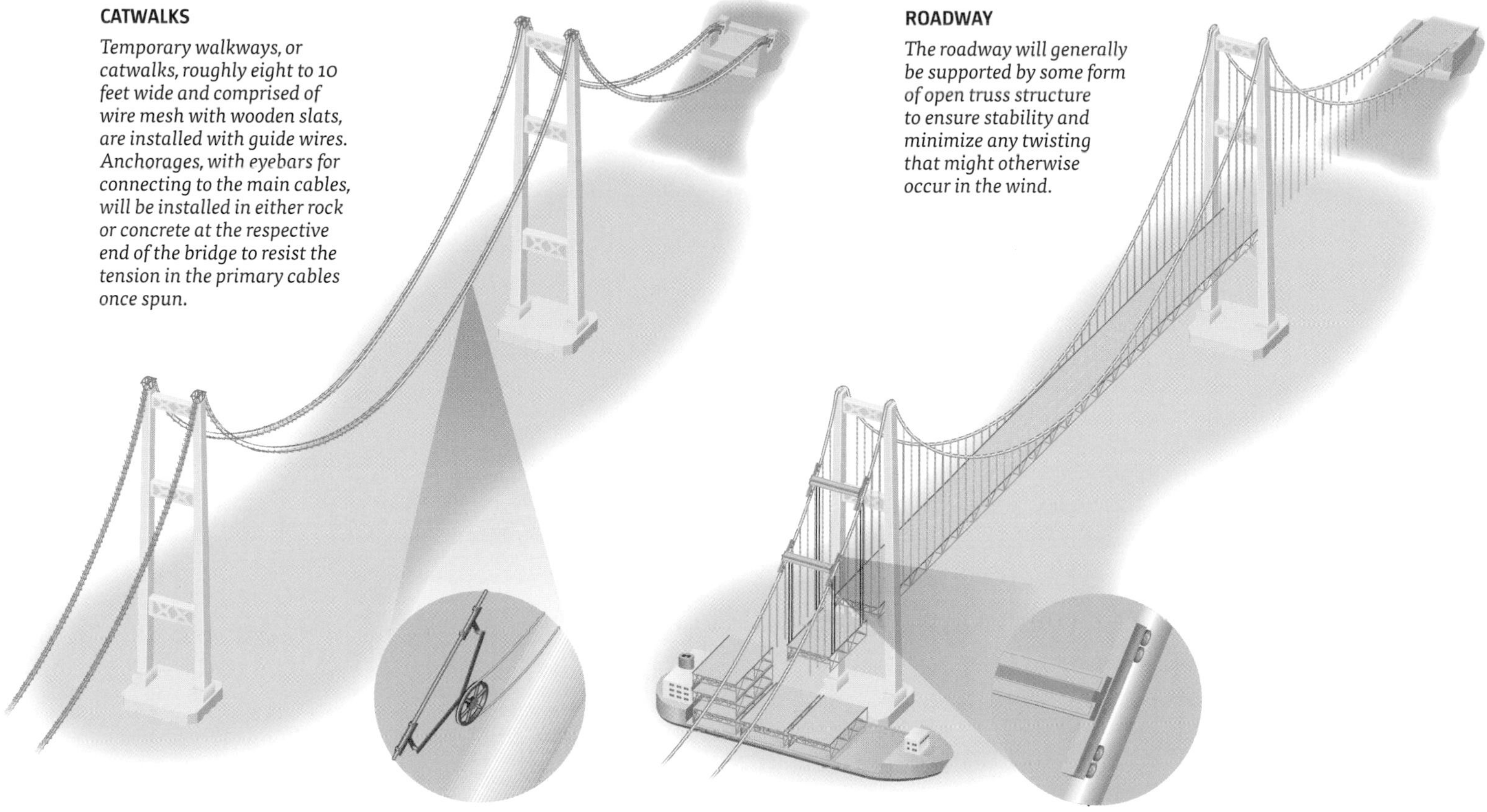

CATWALKS

Temporary walkways, or catwalks, roughly eight to 10 feet wide and comprised of wire mesh with wooden slats, are installed with guide wires. Anchorages, with eyebars for connecting to the main cables, will be installed in either rock or concrete at the respective end of the bridge to resist the tension in the primary cables once spun.

ROADWAY

The roadway will generally be supported by some form of open truss structure to ensure stability and minimize any twisting that might otherwise occur in the wind.

CABLES

Pulleys on a traveler then repeatedly loop the steel wire from one anchorage to the other, where it passes through an eyebar and travels back again. Workers pull each strand to the necessary tension and bundle the individual cables into a cable strand when the desired thickness is reached. Cable bands, to connect to the vertical suspender cables, are attached at regular intervals.

DECK CONSTRUCTION

Suspender cables are then hung to support the deck and roadway. Hoists attached to the suspenders or to the main cable lift the bridge deck into place—section by section. When possible, these prefabricated sections will be lifted from a barge below the bridge; alternatively, a cantilever crane may lift them into place from the deck itself as it extends outward from the towers.

collapsed—giving rise to its posthumous nickname: “Galloping Gertie.”

Thanks to advances in computers and load testing techniques, today’s bridges are designed with more sophisticated science than ever before. Elaborate simulations and engineering calculations are done to assure that the bridge can withstand the weights it will carry; simultaneously, wind and earthquake testing are done to simulate its performance under a variety of meteorological and seismic conditions.

Bridges must withstand terrific exposure to the elements—most prominently wind, moisture, and temperature variation. Exposure to temperature variation is particularly tricky. When heated, steel expands—which means that a steel bridge’s cables and structural elements, including parts of the roadway, will expand. In cold weather or climates, just the opposite will occur. Expansion joints must be built in to accommodate these temperature-related movements.

Because of their extended, unsupported spans, suspension bridges are notable in terms of the amount of movement that can occur as a result of such temperature fluctuation. The Golden Gate Bridge in San Francisco, for example, moves up and down within a range of 16 feet (5 m). Similar movement is found on the Verrazano-Narrows Bridge in lower New York harbor, which stands 228 feet (70 m) above the water in temperatures above 100°F (38°C) but 12 feet (3.6 m) higher when the temperature falls to 10°F (-12°C).

BRIDGE SAFETY

Bridges are subject to constant inspection so that cracks and other forms of corrosion and fatigue can be identified and corrected before they affect the bridge's structural integrity. Once done solely by bridge workers, inspection work is now increasingly being done by sensors placed on the bridge or by techniques like infrared thermography, ground-penetrating radar, or magnetometers capable of detecting voids, cracks, and deterioration in concrete and steel.

Some of the instrumentation is devoted to providing static reports of bridge conditions, such as air temperature, wind speed, load, or displacement. Other bridge-monitoring systems are more dynamic, using a constant feed of data to evaluate the change in a particular risk parameter over time. These parameters can include curvature, vibration, corrosion, cracking, or strain (elongation). Once a preset threshold level is exceeded, the systems are often programmed to trigger communication to a central bridge-monitoring station.

The Humber Bridge in northern England, for example, was built with an extensive set of monitoring mechanisms. Bridge engineers are sent constant information about the condition of the anchorages holding the towers, the hanger cables that suspend the roadway, and any breaks or cracks that could undermine the bridge's structural integrity.

BUILDING A SAFER BRIDGE

Following extensive damage during the Loma Prieta earthquake in 1989, the San Francisco-Oakland Bay Bridge underwent extensive reconstruction. The new eastern span, which opened in 2013 and connects Yerba Buena Island and Oakland, was designed with a variety of unusual safety features to protect it in the event of future earthquakes.

HINGE PIPE BEAM

In areas where seismic activity is likely, special beams known as hinge pipe beams are used in the construction of the bridge deck. In addition to moving within a sleeve-type contraption during expansion (heat) or contraction (cold weather), they are designed to deform in their middle, or fuse, section during an earthquake. This absorbs shock to the structure as a whole.

BATTERED PILES

Although used frequently for the construction of oil rigs in deep water, the "battering" of piles—or insertion into the ground at an angle to the vertical—is only now being applied to bridge structures. Driven into the ground with large hydraulic hammers, batter piles provide more ability to carry lateral, or horizontal, forces.

SHEAR LENGTH BEAM

Because so much of the structure of the new roadway of the San Francisco-Oakland Bay Bridge depends on the central tower, special steel beams connect the four vertical elements incorporated within it. These beams are designed to shear—or separate—under excess load, absorbing damage.

TWIN CITY COLLAPSE

In August 2007, a section of Interstate 35W spanning the Mississippi River in Minneapolis collapsed, resulting in 13 fatalities. The bridge had recently been inspected and found to be structurally deficient, a condition that demands significant maintenance and repair but does not mean that the bridge is unsafe. Normally, bridges in such a condition remain open, with or without weight restrictions, as this one did.

Following the collapse a replacement bridge, known as the St. Anthony Falls Bridge, was quickly put in place; it opened just 13 months later. Today, it is among the "smartest" bridges in the nation with a state-of-the-art monitoring system that continuously measures the bridge's structural integrity and performance. Some 323 sensors record the deck's movement and temperature as well as any stresses put upon it.

MAINTAINING BRIDGES

Bridges today are designed to be robust. However, without constant painting, repair, and maintenance, most bridges won't even be able to reach their design life due to corrosion, cracking, and scaling caused by constant exposure to the elements.

A number of techniques are used to protect or fix bridges and thereby prolong their lives. Painting is one, though paint is not necessarily the best protector against rust. Plating, a process that involves coating the bridge with zinc or a zinc/aluminum mix, protects steel even longer. Protective coatings are also used on concrete bridges, and cracks might be filled with concrete or other material that is inserted or pressure injected.

As with any roadway, deck replacement on bridges must be done from time to time. Often this is done in piecemeal fashion, with prestressed concrete panels replacing the deteriorated deck on its supporting girders. New materials, including prefabricated steel and aluminum, are also being tested as replacements for original concrete decks on many older bridges.

More surgical interventions might also be done to protect other key components of existing bridges. The addition of carbon fiber–reinforced plastic laminates to bridge beams, and a similar wrapping of key bridge columns, can minimize damage in locations where seismic activity is common.

RAPPELLING

Sensors alone are not sufficient to indicate bridge condition. Regular visual inspections will be conducted by maintenance workers walking the cable catwalks and rappelling down structural members.

PAINTING

Painting remains a primary method of protecting bridges from rust and deterioration. Today, layers of new forms of epoxy paint create a strong barrier between bridge steel and the elements. The process is not quick: the recent painting of the Forth Bridge in Scotland took 10 years.

INFRARED THERMOGRAPHY

Engineers may use infrared thermography to assist in bridge-deck maintenance. Radiometers measure thousands of separate temperature points each second, producing thermal maps of concrete surface temperatures that reflect the internal conditions of the concrete, including voids, density changes, or rust expansions.

CRACK SEALING

A variety of surface treatment sealers that reduce concrete's permeability are regularly applied to bridge decks. Cracks are filled rapidly so as not to permit moisture and chloride to corrode steel within the concrete.

HOW TUNNELS ARE BUILT

The Industrial Revolution brought with it more sophisticated technology to build tunnels needed to serve mountainous or highly urbanized areas. Some tools, like pneumatic drills, made it easier to cut through rock; others, like the tunneling shield, made tunneling possible in soft soils that previously proved too tricky for construction.

Today, there are three main approaches to building tunnels. The first, known as cut and cover, is used on land and involves constructing a deep trench along the path of the new tunnel and then constructing the tunnel structure within it in sections.

A second approach to tunneling involves "immersing," or laying, a largely prefabricated tunnel tube on the seabed. Doing so is generally quicker, cheaper, and safer in terms of both workers and underwater seismic activity than tunneling methods that involve boring through rock.

The most complex approach to tunneling involves boring a horizontal tunnel deep under the earth or sea. Specially designed tunnel boring machines (TBMs) construct the tunnel as the hole is bored either in hard rock or in soft earth or clay. The latter requires a pressurized environment to withstand ground pressure and, counterintuitively, can take longer than boring through hard rock.

Among the most common TBMs working in softer soil are earth pressure balance machines, which pressurize the ground ahead of the cutting mechanism to offset external pressure. Unlike earlier machines, which needed workers to go through a decompression chamber to withstand the pressure in the cutting chamber, these TBMs allow workers to stay in a nonpressurized area behind that compartment—entering the pressurized area only for necessary maintenance.

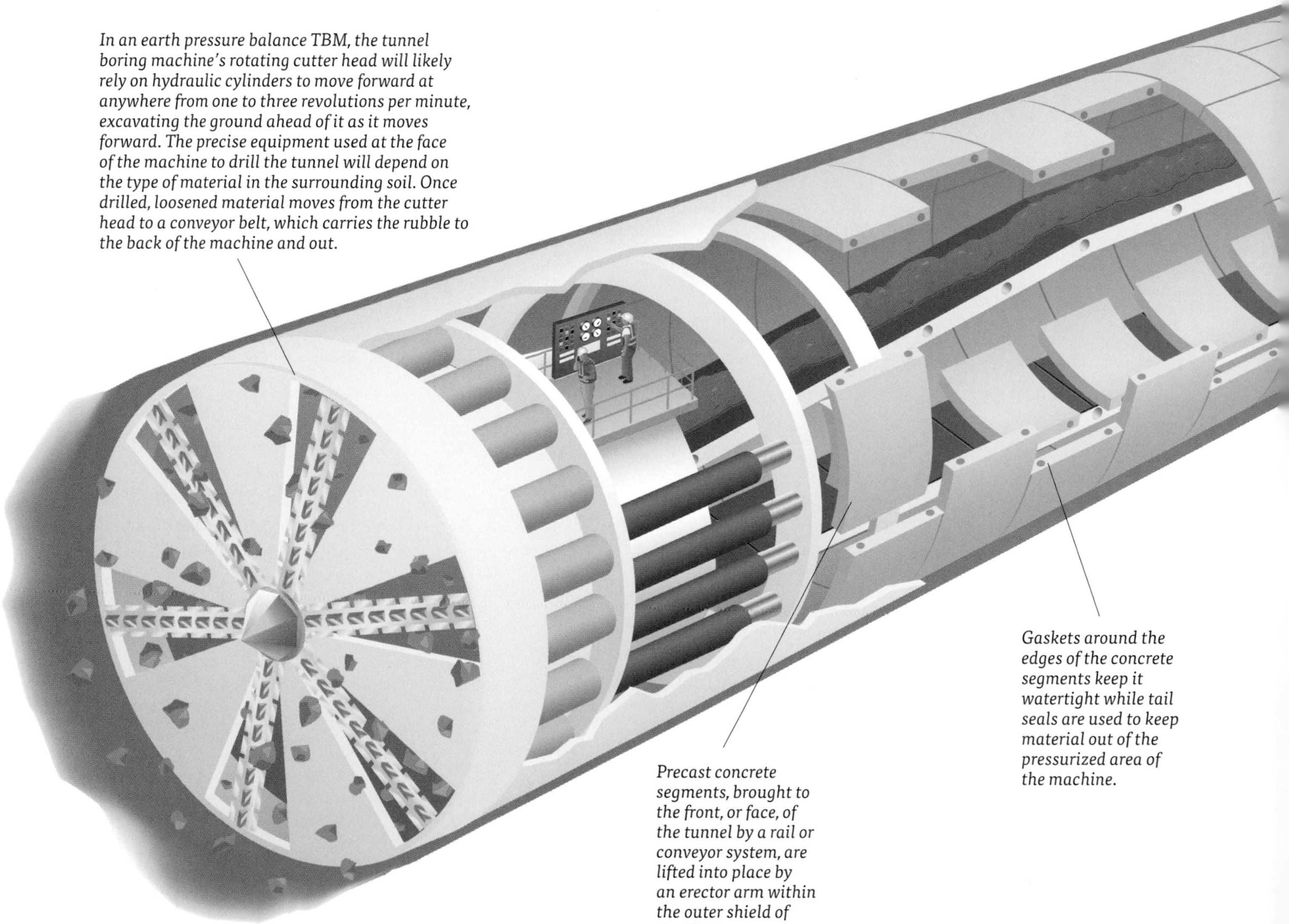

In an earth pressure balance TBM, the tunnel boring machine's rotating cutter head will likely rely on hydraulic cylinders to move forward at anywhere from one to three revolutions per minute, excavating the ground ahead of it as it moves forward. The precise equipment used at the face of the machine to drill the tunnel will depend on the type of material in the surrounding soil. Once drilled, loosened material moves from the cutter head to a conveyor belt, which carries the rubble to the back of the machine and out.

Precast concrete segments, brought to the front, or face, of the tunnel by a rail or conveyor system, are lifted into place by an erector arm within the outer shield of the TBM.

Gaskets around the edges of the concrete segments keep it watertight while tail seals are used to keep material out of the pressurized area of the machine.

Once the key segment completing a ring is inserted, the front of the TBM moves forward.

Waterproof linings consisting of insulated panels might be installed to keep water infiltrating the tunnel from freezing.

Tunnel walls will often be covered with reflective tiles to maximize lighting and visibility within the tunnel.

Ventilation systems are critical to a tunnel's operation, but sufficient casing and space for other utilities and emergency systems must also be installed.

The final lining of the tunnel may be precast concrete panels, cast-in-place concrete with rebar or a sprayed concrete known as shotcrete.

While the largest number of people (up to 20 in some cases) are located on the TBM itself, another half dozen or more people might be involved in completing the walls, roadways, and other components of the finished tunnel.

CUT-AND-COVER

Cut-and-cover construction, used when the tunnel sits fairly close (30 to 40 feet) to the ground and excavation from the surface is possible, can be done either from the top down or from the bottom up. In top-down construction, the tunnel walls are constructed first, followed by the tunnel ceiling. These become integral structural supports for the underground excavation to follow, but can be covered and the ground surface restored immediately. In bottom-up construction, the tunnel is completed before the ground covering is replaced, although on a busy street, decking might be placed above the work site to prevent interruption of traffic.

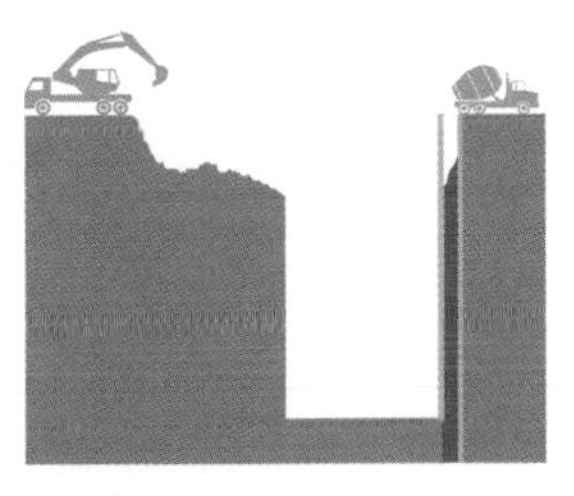

The first steps in bottom-up construction involve the insertion of temporary excavation walls and the excavation of the site.

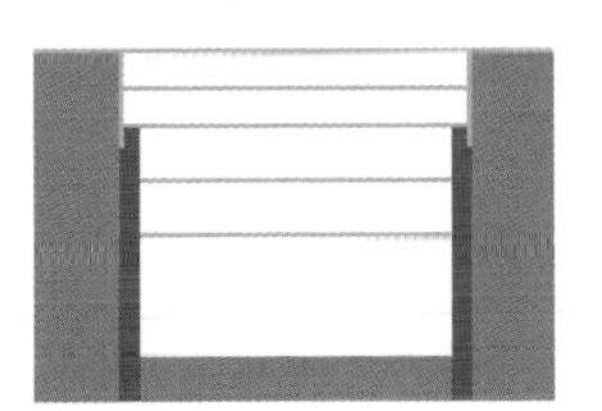

Wall supports are inserted, dewatering takes place, and any permanent drainage system is installed.

Once temporary support walls are in place, the tunnel floor and permanent walls and ceiling are constructed.

Internal finishes are done while waterproofing is applied to the exterior of the tunnel. When the tunnel is completed, the ground is filled to its original level and the roadway or ground restored.

TUNNEL SAFETY

More so than in almost any other form of transportation, safety in tunnels is paramount. The combination of confined spaces and mechanical ventilation means that the risk of a small incident's becoming a large one is very high indeed: rarely is injury or loss of life limited to just one vehicle or rail car involved in a crash or fire.

Although early tunnel travelers feared collapse, asphyxiation, or flooding, the greatest risk to tunnel travelers is none of those—it is fire. As a number of tunnel incidents over time have shown, a fire that begins in one vehicle or as the result of a crash between two vehicles can spread rapidly to other vehicles in the tunnel. Although emergency protocols are in place to respond quickly to such incidents, inevitably response times do not match the speed of a fuel-fed fire. Moreover, smoke and toxic gases will lead to fatalities unless the ventilation system is able to purge the confined space or the occupants are able to egress to fresh air.

For this reason, the types of commodities that are permitted to travel through tunnels are monitored carefully. Combustible materials might be banned or limited to a certain volume. Restrictions address such things as explosives, poisons, and poisonous gases, flammable gases and solids, and radioactive materials. This system of regulation largely relies on self-enforcement or spot checks, although more recently technologies able to detect some of these commodities are being deployed at tunnel entrances.

VENTILATION AND HEAT EXTRACTION

Normal ventilation systems, designed to change and purify the air inside a tunnel, may not be powerful enough to handle fire inside a tunnel. Additional extraction capacity is required to manage both smoke and excess temperatures. Strong longitudinal airflow, produced by jetlike engines and deployed primarily in unidirectional traffic, can be used to assist in these emergency situations.

EMERGENCY LIGHTING/WALKWAY

Most tunnels are required to feature narrow sidewalks, not for pedestrians but for emergency escape. Illuminated signs and strobe lights are used to identify exit routes.

TUNNEL MANAGEMENT CENTER

Most tunnels feature traffic management centers, which provide varying levels of surveillance and traffic control, typically via closed-circuit television. These centers are responsible for providing incident response and reporting as well as for coordinating evacuation of the tunnel if necessary.

MONT BLANC'S TRAGEDY

Among the most tragic tunnel fires ever occurred in Switzerland in March 1999, when 35 people were killed in a fire in the Mont Blanc tunnel connecting Chamonix, France, with Courmayeur, Italy. A Belgian reefer (refrigerated truck) caught fire soon after entering the tunnel, and within minutes the fire spread to adjacent vehicles. With no fireproofing or sprinklers in the tunnel, temperatures at the fire reached 1,000 degrees and it took firefighters more than two days to put out the flames.

Three years later, the tunnel reopened with major safety improvements. These ranged from fire-resistant sheeting on tunnel walls and new traffic lights to additional vents and smoke extractors. Pressurized, concrete-lined emergency shelters were built every 984 feet (300 meters)—fitted with fireproof video links to new, manned command posts. Even the mouths of the tunnels were upgraded with heat sensors to detect overheated trucks as they enter the tunnel.

MAINTAINING TUNNELS

Tunnels are constructed to withstand seismic events and to last a long time—anywhere from 50 to 125 years. Their structures are inspected regularly by tunnel operators, using light hammer sounding, rock cuts, or more sophisticated geotechnical tests. But even the most robust structure won't last long without regular maintenance and repair—including controlling water leaks and moisture buildup, keeping concrete tunnel linings from cracking, and ensuring that tunnel tiles remain reflective (for visual safety).

Ventilation shafts are particularly important. Air in most road and rail tunnels is typically supplied by large fans that push clean air through plenums, or enclosures, located below or on the sides of the tunnel; dirty tunnel air is exhausted or pulled through other openings on the ceilings or tunnel sides. Carbon monoxide levels in the tunnel are monitored either via manned control rooms, where operators adjust fan levels, or in a more automated fashion via sensors connected directly to fans. Opacity levels, which indicate visibility within the tunnel, might be monitored via video cameras.

Cleaning is a regular part of tunnel operations. Some tunnels are designed with black ceilings to avoid the regular cleaning that would be required for white ones. Even with black ceilings, tunnels need to be cleaned regularly—usually at off-hours when at least one lane can be shut down to allow cleaning trucks to operate. Rural tunnels might be cleaned no more than four times a year, or seasonally—so long as the temperature is above freezing. Urban tunnels require considerably more attention; the Holland Tunnel under the Hudson River, for example, is cleaned every week.

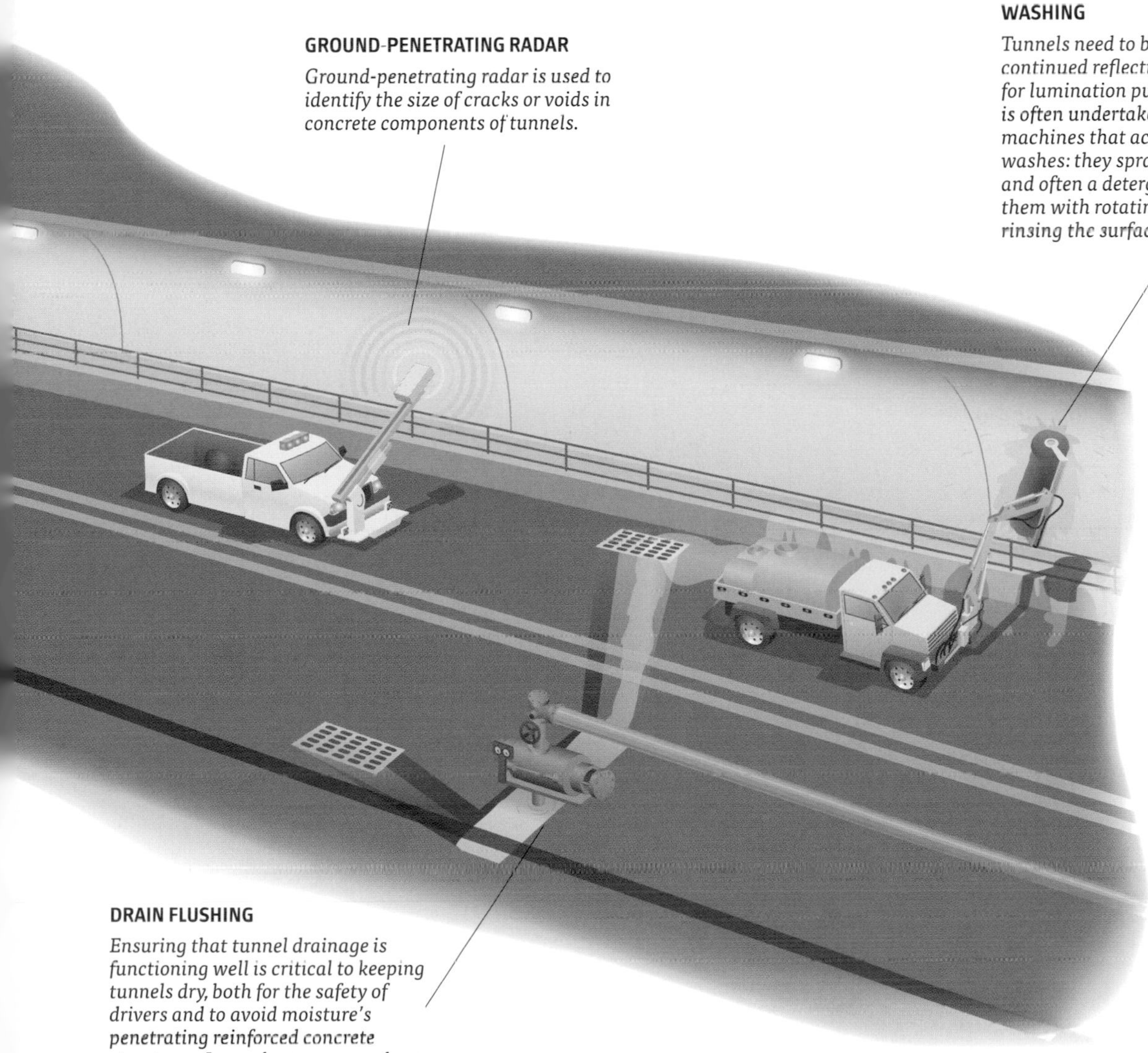

GROUND-PENETRATING RADAR

Ground-penetrating radar is used to identify the size of cracks or voids in concrete components of tunnels.

WASHING

Tunnels need to be washed to ensure continued reflectivity of tunnel surfaces for lumination purposes. Tunnel washing is often undertaken by specialized machines that act like toothbrushes or car washes: they spray the tunnel with water and often a detergent before scrubbing them with rotating brushes and then rinsing the surface with water.

DRAIN FLUSHING

Ensuring that tunnel drainage is functioning well is critical to keeping tunnels dry, both for the safety of drivers and to avoid moisture's penetrating reinforced concrete structures. Ice and snow removal are also key aspects of maintaining the integrity of the tunnel's structure.

AIR

FLIGHT

The human desire to fly can be traced back centuries. Leonardo da Vinci is among those famous for early musings about the mechanics of flight: his fifteenth-century drawings analyze the wingspans of birds and include imaginative ideas for machines that would allow humans to fly. But it was not until the Industrial Revolution allowed powered flight that the skies became a viable means of transport.

Early experiments with hot-air balloons in France by the Montgolfier brothers date back to 1783 and marked the first real liftoff from earth. After reportedly watching ash rise from smoldering embers, they stoked a fire under an inflatable silk bag and tied it to a basket. At first tethered and unmanned, these experiments quickly give rise to untethered manned flights and by 1785 hot-air balloon journeys of some distance—like across the English Channel—were possible.

Most ballooning was devoted to recreational or scientific purposes, such as measuring air pressure. By the midnineteenth century, however, balloons had given rise to "airships"—which relied on a lighter-than-air gas for lift and a steam engine for power. Early airships were slow and cumbersome, but Count Ferdinand von Zeppelin's rigid-framed, internal combustion engine–powered version, which debuted just after 1900, would make airships a viable means of transport.

Work on the mechanics of "heavier-than-air" flight was also under way. Experiments with gliders by George Cayley and others in the first half of the nineteenth century led to new knowledge about the forces on a craft in the air, particularly lift and drag, and to designs for appendages to wings that would allow a mechanical craft to take advantage of these forces to stay in flight.

These experiments informed the innovations of the Wright brothers, Orville and Wilbur, whose liftoff and flight at Kitty Hawk, North Carolina, in December 1903 was the culmination of years of study and experimentation. The Wright plane had no wheels and was launched from a railed catapult (similar in principle to those used on aircraft carriers today) rather than under its own power, but the four-mile flight is nevertheless considered a milestone in modern aviation history.

As airplanes and airships developed, their first customers were the military; the ability to photograph enemy forces and communicate from the air with troops on the ground offered strategic advantages. But the public also became fascinated by airplane technology, flocking to airshows and celebrating aviation pioneers like Amelia Earhart and Charles Lindbergh. The appeal of airships grew too until 1937, when the *Hindenburg* exploded in New Jersey after a scheduled transatlantic crossing—likely as a result of a hydrogen gas leak.

World War II proved a boost to aviation. It offered a stage for air-based weapons and technologies, from radar to fighter-bombers, and for new aircraft, many of which were easily adapted for civilian transportation. Among those technologies was the jet engine, which by the late 1950s appeared on a few commercial planes and would transform the world of flight.

AIR MAIL

No organization played a more formative role in creating the American commercial aviation industry than the U.S. Post Office. For a decade from 1918, the post office established new air routes, trained pilots, and tested aircraft suitable for carrying mail. Roughly 95 percent of airline revenues during this period came from mail—rather than from passengers.

But flying was no party in those days, and transcontinental services were particularly tricky. Without any ground lighting, air mail pilots could fly only by day. Long-haul mail was placed on mail trains at night and collected in the morning to continue its journey by air.

In 1927, the mail contracts were turned over to private companies able to invest in better and safer airplanes. By 1930, legislation had removed direct operating responsibility from the post office—although to this day it has remained in control of awarding air mail contracts.

MILESTONES IN FLIGHT

When the Wright brothers first took flight in 1903, their aircraft traveled only 120 feet (37 m). Through advances in engine efficiencies and aerodynamic design, today's commercial and military aircraft have extended their range to halfway around the globe. The newest Boeing 787 is capable of flying nearly 10,000 miles (16,000 km) on a single tank of gas.

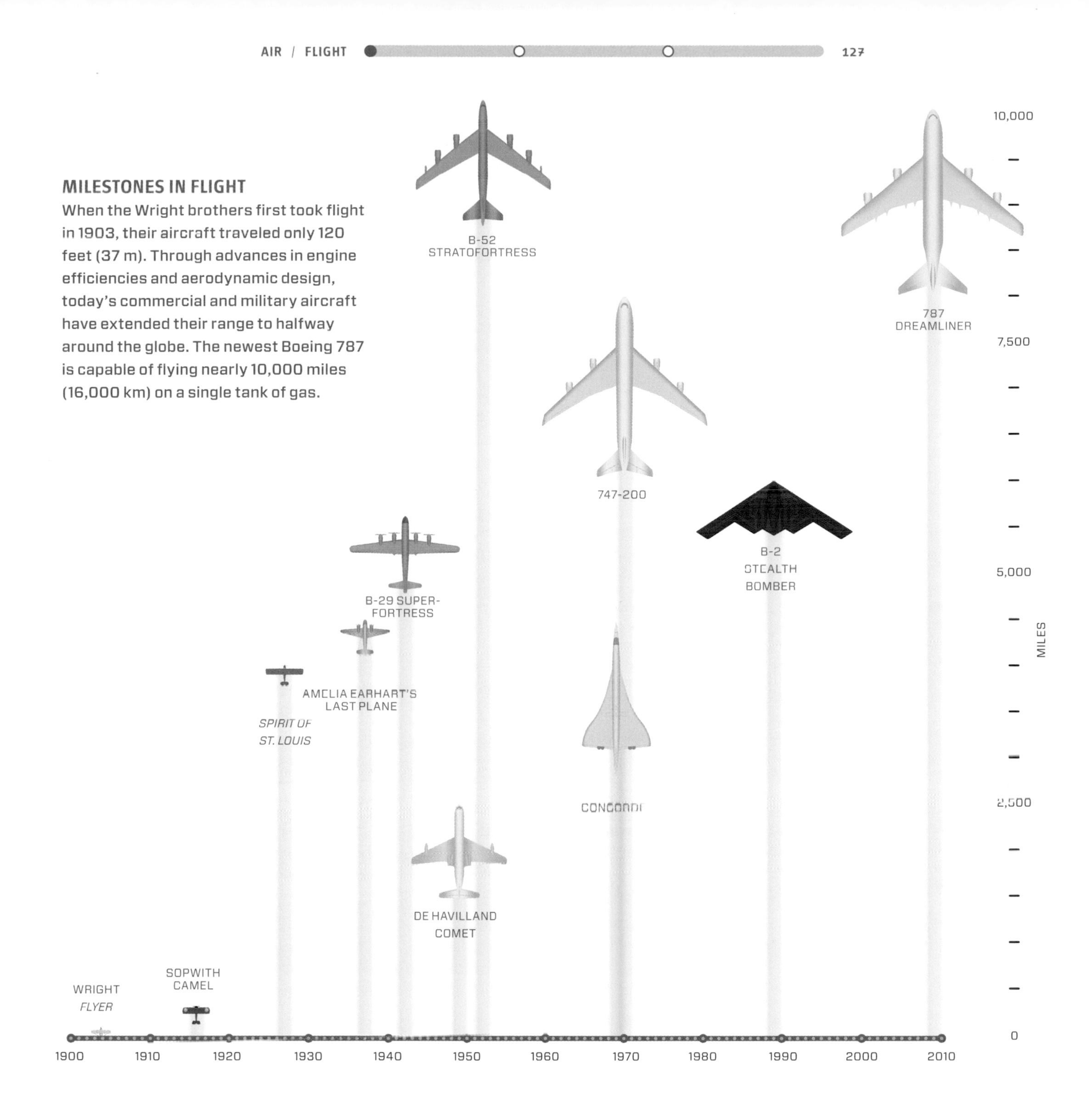

CONCORDE'S PASSING

For those who think the history of transport is all about increasing speed, think again. The fastest plane to fly commercial routes is no longer flying—due simply to lack of demand.

The Concorde was built by a partnership ("concorde") of French and British aerospace firms. The first commercial jet to exceed the speed of sound, it halved flight times from Europe to the United States when it debuted in 1969.

To fly supersonically at altitudes of 60,000 feet, the plane had to be heavier and stronger than the average jet. It featured small windows and a narrow, streamlined body.

While its safety record was good, the Concorde's economics were not. Demand for the service waned, and its operating costs ultimately outweighed its revenues. British Airways removed the last Concorde from service in 2003.

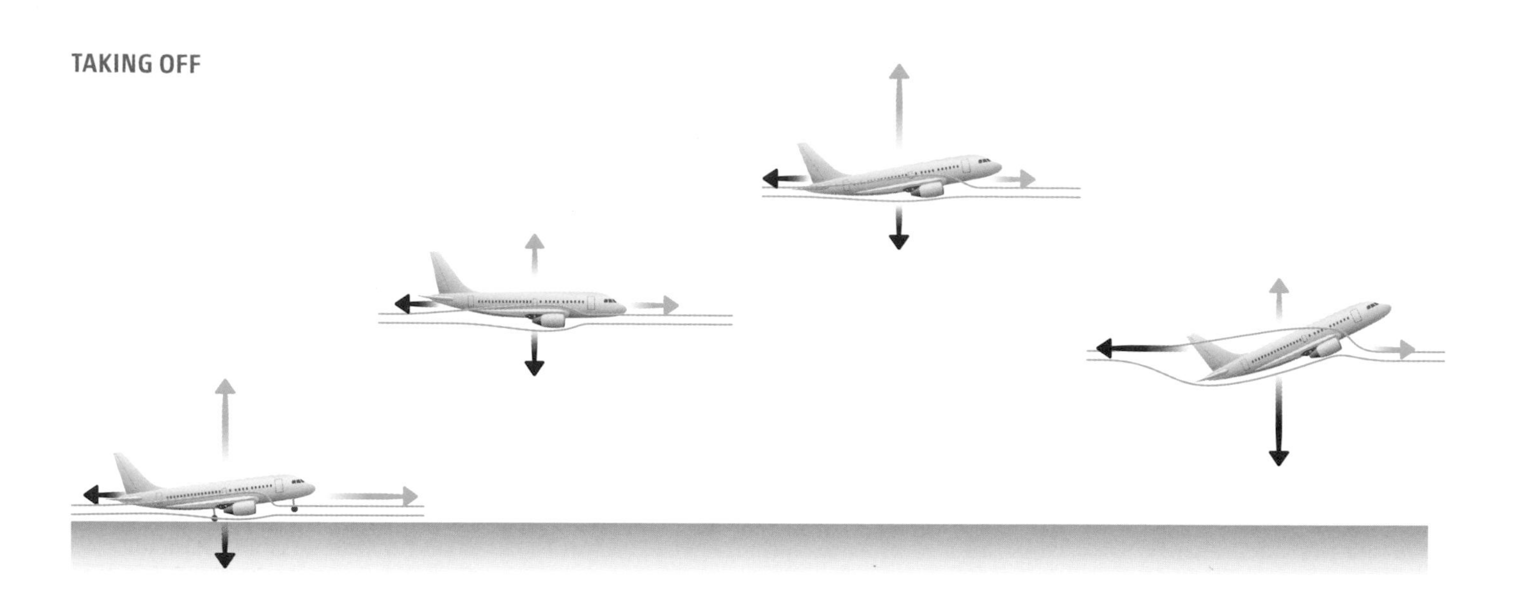

In order to take off, the pilot accelerates the plane with large amounts of thrust to overcome the induced drag of the wing. With enough speed due to the acceleration and a little angle (the angle of attack), the lifting force becomes greater than the weight and the plane takes off.

Once in the air, the plane levels off and all forces are in equilibrium. At cruising altitude, most of the drag is due to the friction on the fuselage, wings, and other appendages.

In order to climb higher, the pilot increases the angle of attack of the wings to generate additional lift. As the plane climbs higher, the air gets less dense and speed over the ground must be increased to maintain the same amount of lift.

However, if the angle of attack is increased too much (usually over 15 percent), the flow of air around the wing will begin to "separate" and lift will decrease precipitously while drag will increase. This is known as a stall (and has nothing to do with the engines' shutting down).

LEARNING TO FLY

What allows an airplane to fly? The answer is a very precise combination of several counteracting forces being exerted on a plane. Lift and weight determine how high it will be off the ground at any point while thrust and drag determine how fast it will be moving through the sky. The balance of all four must be right for successful flight.

Weight is a familiar concept to most people: a function of mass and gravitational pull, airplane weights are large (for example, 1.3 million pounds for an Airbus 380). Less familiar to people is the concept of lift, which counters the weight of the plane to keep it in the sky.

For a wing to generate lift, there must be motion between the wing and the air it is in. The wing is shaped and inclined in a way that changes the direction of the air it is passing through, causing the air to move down and behind the wing; this air reaction gives rise to lift. Lift is proportional to the amount of air diverted by the wing (a function of plane speed and air density) and the vertical velocity of that air (a function of plane speed and angle of the wind through the air).

Said another way, when you stick your hand (wing) out of a moving car and tilt it up a little (angle of attack), you feel your hand lift up. The car's forward movement is pushing a volume of air (that before you came along was stationary) down and behind your hand. Moving all that air down pulls more air from above your hand to fill the void left by the downwash, causing the pressure to lower above the wing. The opposite reaction to your moving all that air down—that is, the adjustment to the pressure differential between the two sides of the wing—is lift. Lift is generated on the top of the wing (it is not a phenomenon of air's acting on the bottom)—hence the location of all the appendages (engines, auxiliary fuel tanks, missiles, et cetera) below it.

Thrust is the force that pushes the airplane forward through the air; generally it comes from an engine. Drag acts against the thrust and refers to the resistance by the flow of air over the wing or other parts of the airplane. When thrust is greater than drag, the plane will speed up; when it is less, it will slow down. When they are equal, the plane will move at a constant speed.

TURNING

An airplane's orientation in the sky is critical to understanding the various pressures upon it at a given point in time. These are often referred to as the angles of rotation—or roll, pitch, and yaw—and operate in three dimensions around the plane's center of mass.

PITCH

Pitch refers to the up and down movement of the nose of a plane—that is, a rotation through a lateral axis. Pitch is controlled by devices called elevators, hinged sections at the rear of the horizontal stabilizer. There is usually an elevator on each side of the vertical stabilizer. The elevators work in pairs; when the right elevator goes up, the left elevator also goes up.

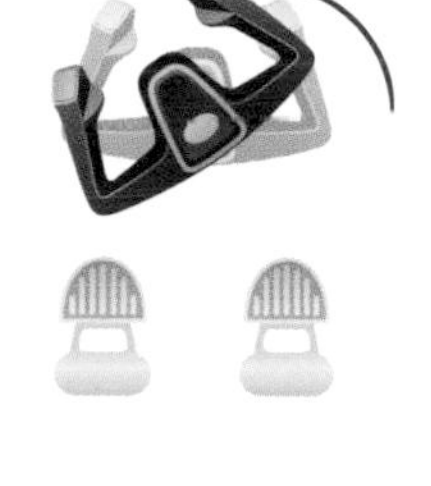

ROLL

Roll refers to the angle of the wings relative to a longitudinal axis that runs through the plane. Roll is often referred to as banking, and is controlled by devices called ailerons located at the rear of each wing. The ailerons work in opposition; when the one goes up (to interrupt lift), the other goes down (to increase it). For example, to turn left, the left aileron goes up to spoil lift on the left side; the right aileron goes down to increase lift. (Ailerons are separate from flaps, which are found at the trailing edge of the wings and are used to increase the lift of an airplane.)

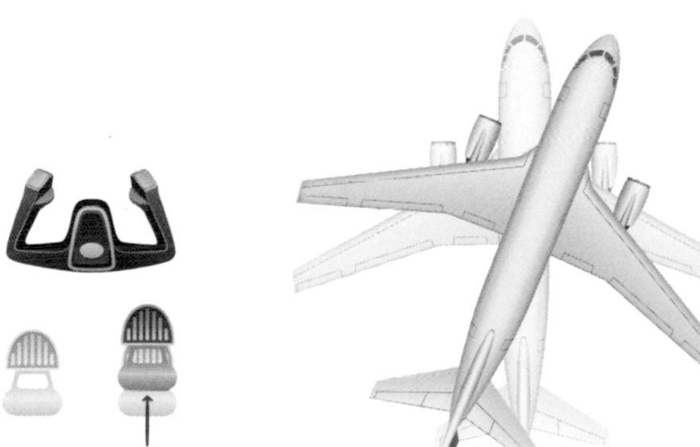

YAW

Yaw is also referred to as heading and indicates the direction along the plane's vertical axis. Specifically, yaw refers to the movement of the aircraft nose from side to side —which is controlled primarily by the rudder, a hinged section at the rear of the vertical stabilizer.

WINGLETS

Although the concept of lift has been understood now for well over a century, engineers continue to find ways to make flight more efficient by improving the way air flows over a plane's wing. In the 1970s, for example, they discovered a way to reduce the vortices, or eddies, of wind created by the differential between pressure on the upper and lower surfaces of the wing. "Winglets," or vertical extensions of the wing that sit at its tip, act as small airfoils that limit the ability of these vortices to form. This can help reduce drag on a plane by up to 20 percent, greatly improving fuel efficiency and thus increasing the plane's cruising range.

FLYING THINGS

Unique among transportation modes, modern aviation holds closely to its past. Old technologies are not relegated to the heap of history, like the steamship or Model T, but tend to survive and get repurposed. At any one time, thousands of balloons, gliders, blimps, helicopters, and a variety of engine-powered planes will be populating the sky above the earth—actively engaged in recreation, transportation, or other more businesslike purposes. These purposes range widely and include sightseeing, advertising, and marketing, and even coverage of sporting events.

Commercial aviation today, however, is exclusively of the heavier-than-air type and relies on either turboprop or jet technology. Most routes lend themselves to one or the other form of propulsion, though some middle-distance

SMALL TURBOPROP

The engines on turboprop planes are similar in operation to those on conventional jets. The energy produced by burning gas moves a series of blades in the compressor section of the engine. However, instead of blasting out of a small nozzle in the rear of the engine, the last of these blades moves the propellers located just in front of the engines.

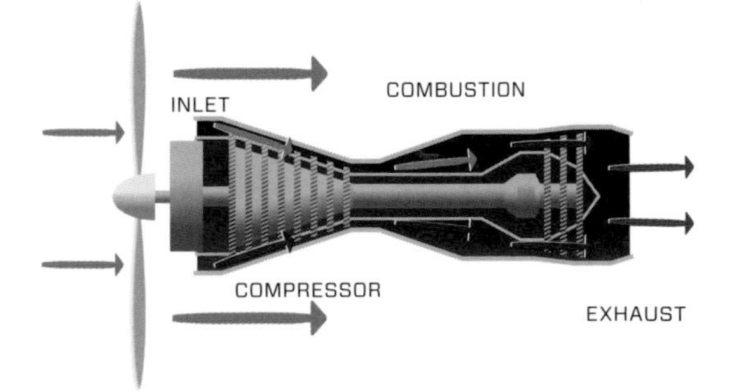

COMMERCIAL JET

In a jet engine, air is sucked into the engine and passes through a compressor to raise its pressure. It is then sprayed with aviation fuel in the combustion chamber and set to burn. The burning gas is directed through a series of turbine blades which spin and drive the compressor before blasting out from the chamber through a nozzle in the rear—creating forward thrust.

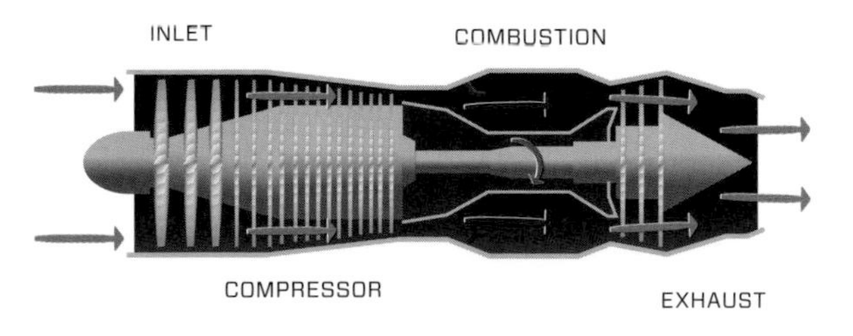

flight routes might be served by both. The choice of craft is the airline's, and a variety of factors may enter into the decision to deploy a particular size and type of airplane on a specific route.

While both propeller and jet-driven aircraft are by now well-accepted fixtures of aviation, aircraft and aerospace manufacturers are always involved in a race to produce newer and better planes. For example, the latest jet to make it into commercial service—the Airbus A380—is bigger and more sophisticated than its competitors. With capacity for 525 people and almost 50 percent more floor area than the next largest jet (the Boeing 747), it has only one problem: it was too big for many airports and terminals, and facilities have had to be retrofitted to accommodate it.

V/STOL

Vertical and/or short takeoff or landing (V/STOL) refers to a type of airplane able to take off either vertically like a helicopter or on very short runways. Its cruising speed, however, is much faster than a helicopter's. Because of these capabilities, V/STOL craft have been designed exclusively for military use or for commercial aircraft operating on very short runways.

BLIMP

Once filled with hydrogen, today's blimps are commonly made of polyester and filled with helium. The rigid internal structure of Count von Zeppelin's day is gone: the only solid parts of modern blimps are the cabin or gondola, which hangs below the blimp, and the tail fins. A blimp relies on inflation or deflation of its bag to travel up or down. Its turboprop engines are located on the sides of the gondola and allow cruising of up to 70 miles per hour (113 km/h). At the rear, the rudder controls direction and an "elevator" adjusts the angle of ascent or descent.

TAKING OFF AT SEA

Takeoff at sea requires special technology as runways are short and lift must be generated quickly. On aircraft carriers in the American and French fleets, a steam-powered catapult featuring a shuttle device moving on tracks below the flight deck is used to produce speed. Other fleets rely on a ski jump configuration at the end of the runway: by facing into the wind, a plane speeding up the sloped runway can generate enough lift.

Takeoffs at sea are not always successful: if sufficient flying speed isn't achieved on liftoff (known in the business as a cold cat), a pilot will activate his ejector seat before the plane falls into the ocean.

HELICOPTERS

Not all lift is created by air flowing over wings. In a helicopter, the spinning of a rotor blade creates the lift needed to stay aloft. This allows helicopters to do a number of things an airplane can't: it can fly backward, it can hover motionless in the air, and it can rotate there. With these abilities, it provides the ideal solution for a range of functions that normal planes cannot do—such as medical evacuation, firefighting, sightseeing, and news coverage, among others.

The idea of a helicopter dates back hundreds of years before it became a reality. Leonardo da Vinci described an "aerial screw machine" or "ornithopter" as far back as 1490. The word "helicopter" can be traced back to 1861, when a French inventor pieced together the word "helicoptere" from the Greek words "heliko," meaning curved, and "pteron," meaning wing. His steam-powered model did not get very far, but the idea stuck.

The year 1907 saw two landmark events in helicopter history: in France, two manned craft—one tethered and one not—lifted off the ground and flew. By the 1920s prototype helicopters were regularly moving half a mile or so at about 50 feet in the air. But a commercial helicopter would not

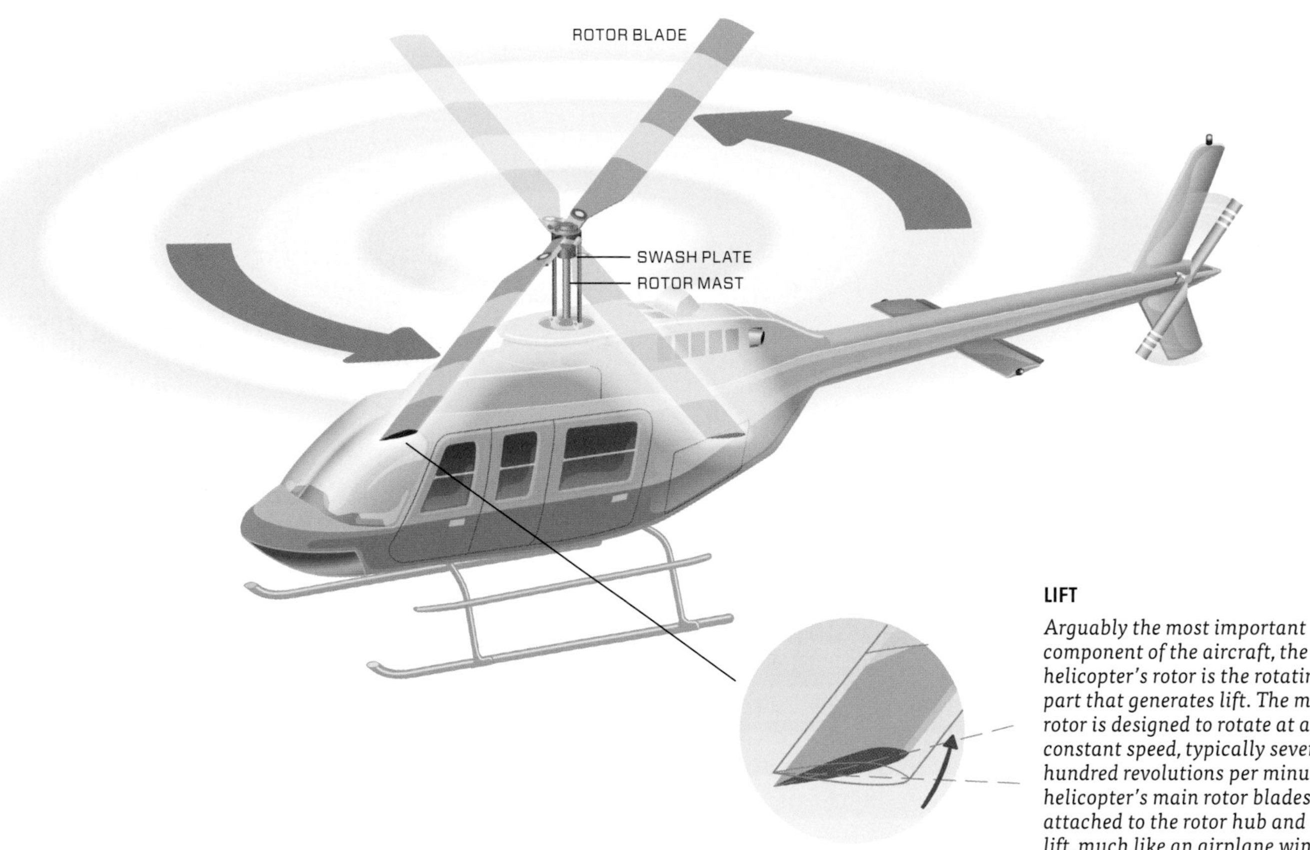

LIFT

Arguably the most important component of the aircraft, the helicopter's rotor is the rotating part that generates lift. The main rotor is designed to rotate at a fairly constant speed, typically several hundred revolutions per minute. The helicopter's main rotor blades are attached to the rotor hub and create lift, much like an airplane wing. As it spins in the air, each blade produces an equal share of the lifting force while hovering.

TORQUE

Torque, a turning or twisting force, is created by the engine's turning the main rotor blade. To compensate, most helicopters use vertically mounted tail rotors to counteract the effects of torque on the aircraft. Other designs may employ a second counterrotating main rotor (for example, the U.S. Army Chinook) or a ducted fan on the tail (no-tail rotor, or NOTAR) to provide countertorque.

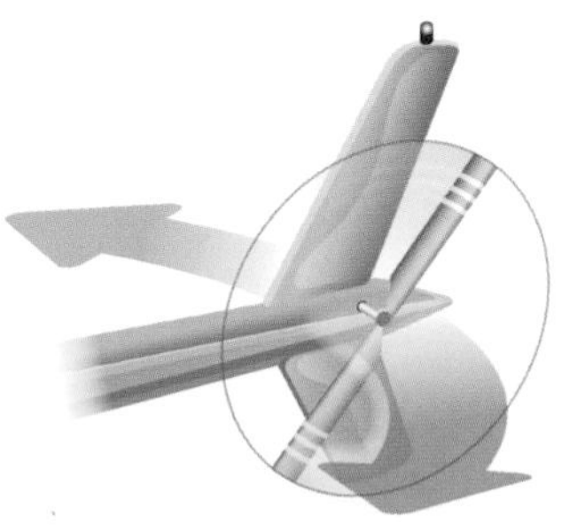

be developed until long after the first fixed-wing aircraft, primarily because helicopters require much greater amounts of engine power than small planes. Once engines had become more powerful and fuel had improved significantly, helicopters served as a viable transportation alternative.

Compared with a jet plane, a helicopter is a relatively simple device mechanically speaking: it contains a metal shaft extending upward from the cabin, a "hub," or attachment point, for the rotor blades at the top of the mast, and the blades themselves—which can be attached in any number of ways (rigid, semirigid, fully articulated). However, it is a much more difficult machine than an airplane to fly—a helicopter pilot must continuously use her arms and legs to keep the craft stable in the air.

The rotor blades provide lift in a vertical direction—that is to say up. The angle of attack of its rotor blades during rotation determines the nature of that lift—both its pitch (tilting forward and back) and its roll (tilting sideways). The pilot must be in control of both, simultaneously counteracting the torque created by the main rotors by relying on the tail rotor or other rotors that are moving countercyclically.

DIRECTIONAL FLIGHT

In addition to moving up and down, helicopters can fly forward, backward, and sideways. Directional flight is achieved by tilting the swash plate assembly at the rotor that alters the pitch of each blade as it rotates. As a result, every blade produces differential lift at a particular point in the rotation.

BLADE FLAPPING

Because of the increased airspeed (and corresponding lift increase) on the advancing blade, it flaps upward; decreasing speed and lift on the retreating blade causes it to flap downward. This imbalance affects the angle of attack on the blades and causes the upward-flapping advance blade to produce less lift and the downward-flapping retreating blade to produce a corresponding lift increase, which in turn balances the lift across the rotor. This phenomenon is known as blade flapping.

IS IT A BIRD? IS IT A PLANE?

When is a helicopter not a helicopter? When it is also a turboprop and can fly quickly over great distances.

The best example of a helicopter-plane is the V-22 Osprey, an American military plane that has the ability to take off and land on short runways (STOL, or short takeoff and landing) or vertically (VTOL, vertical takeoff and landing). Twice as fast as a conventional helicopter, with a range five times greater, it can carry up to a dozen soldiers and is suitable for both combat and rescue missions.

Development and refinement of the Osprey took almost twenty years. The newfangled craft was the product of an unusual collaboration between Boeing and Bell, aircraft and helicopter manufacturers, respectively, under contract to the U.S. Department of Defense; this collaboration began as far back as 1983. Over the next two decades, there were many who doubted the wisdom of such hefty government investment and questioned the viability of the program.

But since its formal introduction in 2007, the Osprey has become an accepted and valuable method of military transport. Both the Marine Corps and the Air Force now rely on their versions of the Osprey in battle zones ranging from Iraq to Afghanistan.

INSIDE THE COCKPIT

The cockpit is the nerve center of a plane. It contains the flight controls and the instrument panel—both critical to maintaining smooth and successful flight. Its windows are protected by a sun shield as well as by an antireflective coating to minimize glare. A heating element keeps them from frosting over.

Steering is primarily controlled by the control wheel, or yoke, which takes the form of either a stick or a wheel. Pulling back on the yoke angles the plane up by moving the elevator up at the back of the plane. Rudder pedals on the floor are used to keep the airplane flying in a coordinated fashion.

A dozen or so instruments help the pilot determine where the plane is in the sky—including direction, altitude, speed, and changes in air pressure (which indicate rates of climb or descent). Other instruments tell the pilot

The turn coordinator is primarily used during instrument flight rules (IFR) to help a pilot gauge the rate of turn with respect to the plane's bank (roll) and yaw. The instrument relies on an inclinometer (a pendulum or ball in fluid) that senses the inertia of the plane and indicates to the pilot if the turn is too steep or not steep enough.

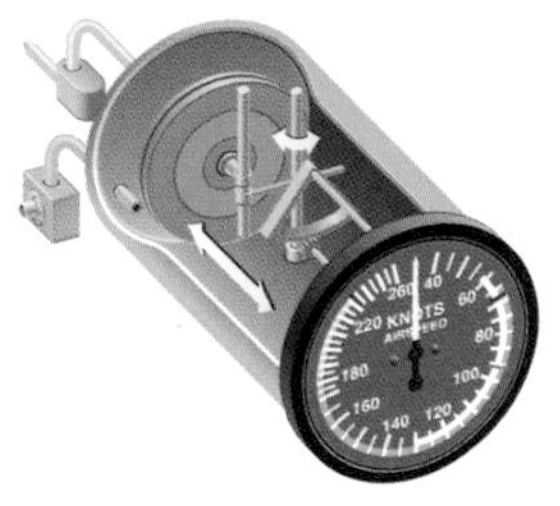

The airspeed indicator measures the plane's speed through the air by comparing the pressure difference between a static air port (atmospheric air pressure) and the "ram air" from the pitot tube that calculates pressure based on the density of fluid flow (dynamic pressure of the high-speed air). The indicated airspeed must be corrected for air density that varies with altitude, temperature, and humidity. The gauge is usually marked with ranges for low-speed flying with flaps, normal operating speeds, and a maximum-rated speed.

While first developed for military fighter pilots, heads-up display (or HUD) is becoming increasingly common on commercial aircraft. At a minimum the display projects airspeed, altitude, a horizon line, and turning indicators onto the cockpit window.

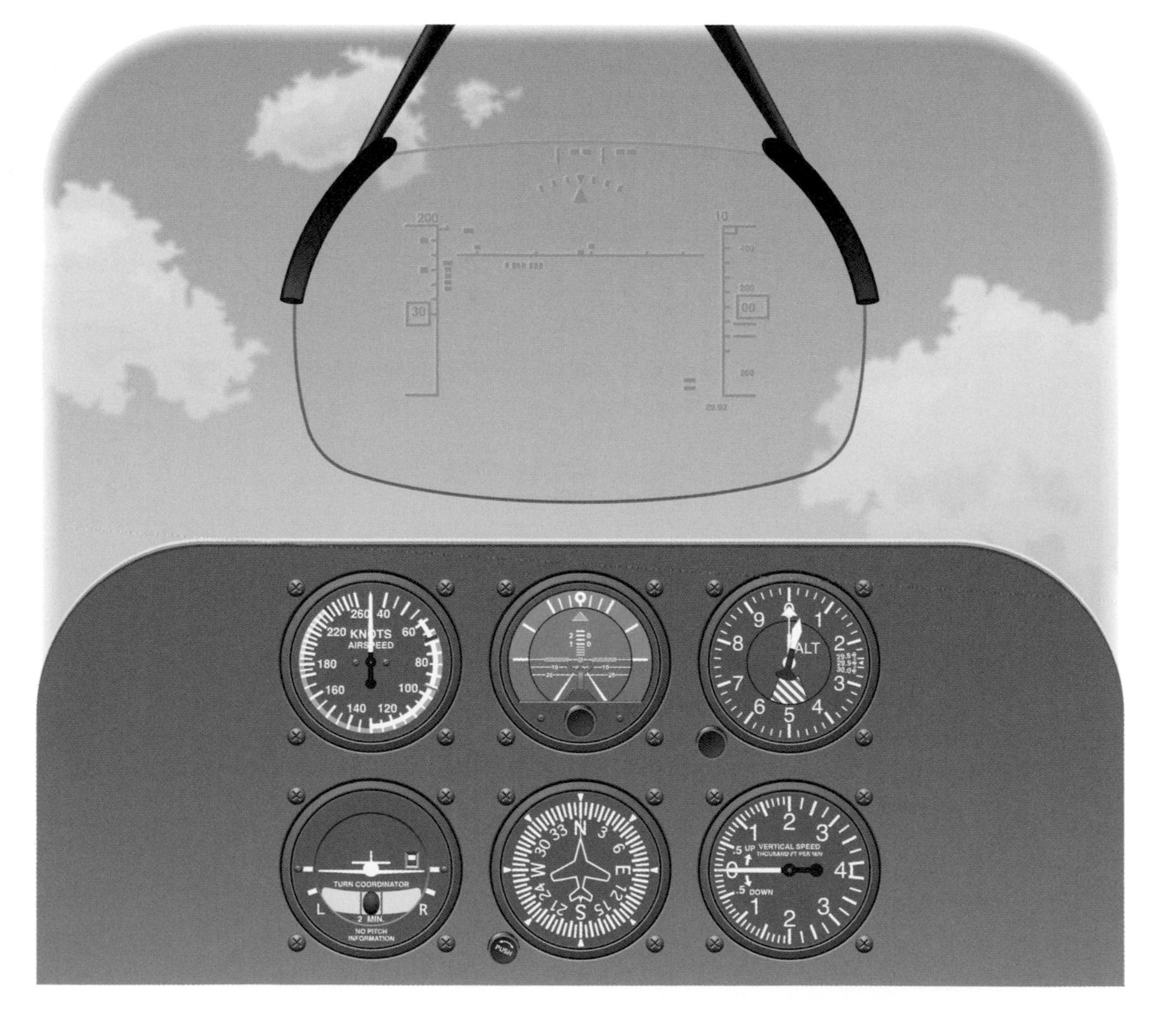

whether the plane's wings are level or banked and what the plane's position is relative to the ground below.

Another set of controls, including the radar transponder and the GPS, determine and show location as well as distance to nearby airports and the time to destination. These controls have become familiar to passengers, who can now watch the progress of their plane on animated maps at their seats.

The attitude indicator, or artificial horizon, is a crucial instrument for flying in difficult weather or with IFR. The instrument shows the aircraft's orientation in pitching (nose up or down) and banking (tilting left or right) and relies on a gyroscope for orientation.

The altimeter measures the aircraft's altitude above sea level by using an aneroid barometer, which measures the atmospheric pressure through a static port in the aircraft's fuselage. Altimeters have several hands, similar to an analog clock, that indicate specific digits in the aircraft's altitude.

All aircraft will have some type of compass and many will feature multiple redundant instruments. A heading indicator is a gyrocompass that is oriented to true north while a magnetic compass points to magnetic north, which can be slightly different. Modern aircraft feature a horizontal situation indicator that integrates heading with navigation information.

A vertical-speed indicator, or variometer, indicates a plane's rate of climb or descent by measuring the rate of change of static air pressure outside the aircraft as altitude changes.

ON AUTOPILOT

The idea of an automatic flight control system for a plane is almost as old as flying itself. Lawrence Sperry, the son of the man who invented the gyrocompass, relied on his father's invention when he flew without hands an airplane at an aviation safety contest in Paris in 1914. Since then, autopilot mechanisms have evolved significantly and become a mandatory feature on planes with more than 20 seats.

The original autopilot connected indicators reflecting direction and position in the sky to hydraulically operated elevators, which controlled the airplane's pitch, and rudders, which controlled its direction. This communication allowed constant and automated correction of the flight course and greatly reduced the burden on the pilot. It also improved fuel efficiency.

A modern autopilot operates in a similar fashion, with a few more bells and whistles. It features a computer with high-speed processors that communicate with sensors located on a variety of control surfaces on the plane. Several times per second, these sensors feed the computer data it compares to control mode—akin to a thermostat in the home. Signals containing necessary adjustments are then transmitted back to the control devices—elevators, rudders, engine throttles, et cetera– to keep the plane on course and ensure optimization of airspeed and fuel efficiency.

FLIGHT RULES

Two sets of regulations and procedures govern flying in many parts of the world: visual flight rules (VFR) and instrument flight rules (IFR). These rules can vary significantly from one another, typically involving different sets of charts and procedures for pilots.

VFR will generally apply when sufficient outside visual cues are present to permit pilots to see certain distances around them—roughly three miles or more. Depending on what type of airspace they're in and whether it's night or day, pilots flying under visual flight rules may not need to file a flight plan, nor will they necessarily need to communicate with the air traffic control system during their journey.

Under instrument flight rules, normally required in bad weather or controlled airspace, the primary means of navigation for a pilot are either a global positioning system or a series of radio beacons located on the ground. In controlled airspace, IFR pilots must have clearance from ground controllers for each segment of their flight. This clearance will specify their route, initial heading, initial altitude, and the frequencies with which they should communicate, and the clearance limit—the farthest they can fly without checking in again and receiving further clearance.

In the United States, pilots need an instrument rating to fly under IFR conditions and the plane must be specially equipped. In other countries, including the UK, so long as the plane is not in controlled airspace the pilot can decide which flight rules he uses based on meeting certain meteorological conditions.

CHARTS AND BEACONS

Just as nautical charts are used by sea captains to navigate across wide open stretches of water, aeronautical charts are used by pilots to navigate safely over land and water. Most will include references to topography, elevations, major land and water features, roads, towns, airports, and radio beacons. But the scale of these charts can vary dramatically: world aeronautical charts scale at 1:1,000,000; operational navigation charts at 1:500,000, and terminal area charts at 1:250,000.

If all else fails, pilots can navigate using these charts and little else—assuming they can see the ground below. Just as early aviation pioneers did, they can use rivers, railroad tracks, and any other visual references noted on the chart to identify their location and direction.

Two types of radio beacons on the ground have historically assisted navigation and are also shown on these charts. Nondirectional beacons (NDBs) emit an omnidirectional low- to medium-frequency radio signal. While they provide no directional info, automatic direction finder (ADF) equipment on the plane detects the NDB transmitter and determines its position relative to the plane. A fairly cheap and reliable method for defining a route for aircraft, NDBs are commonly found in the developing world and in rural areas.

Airways in the United States are no longer based on NDBs; instead, they're based on very-high frequency omnidirectional radio range distance measuring equipment (VORs) or very-high frequency (VHF) omnidirectional radio stations. VOR stations send out both a master signal and a directional high-frequency signal, which together allow the pilot to determine the plane's "line of position," or "radial," in relation to the VOR station. By identifying the intersection of this radial with a similar radial calculated from a second VOR station, the pilot can calculate the plane's exact position in the sky.

The simpler NDB systems can operate at a longer range than the VOR stations, which have a range of roughly 200 miles. However, most commercial and even general aviation around the world now operate on this shorter-range system. The presence of these VOR stations generally identifies the predetermined airways, or roads in the sky, that airplanes will travel and which might be identified on navigational charts.

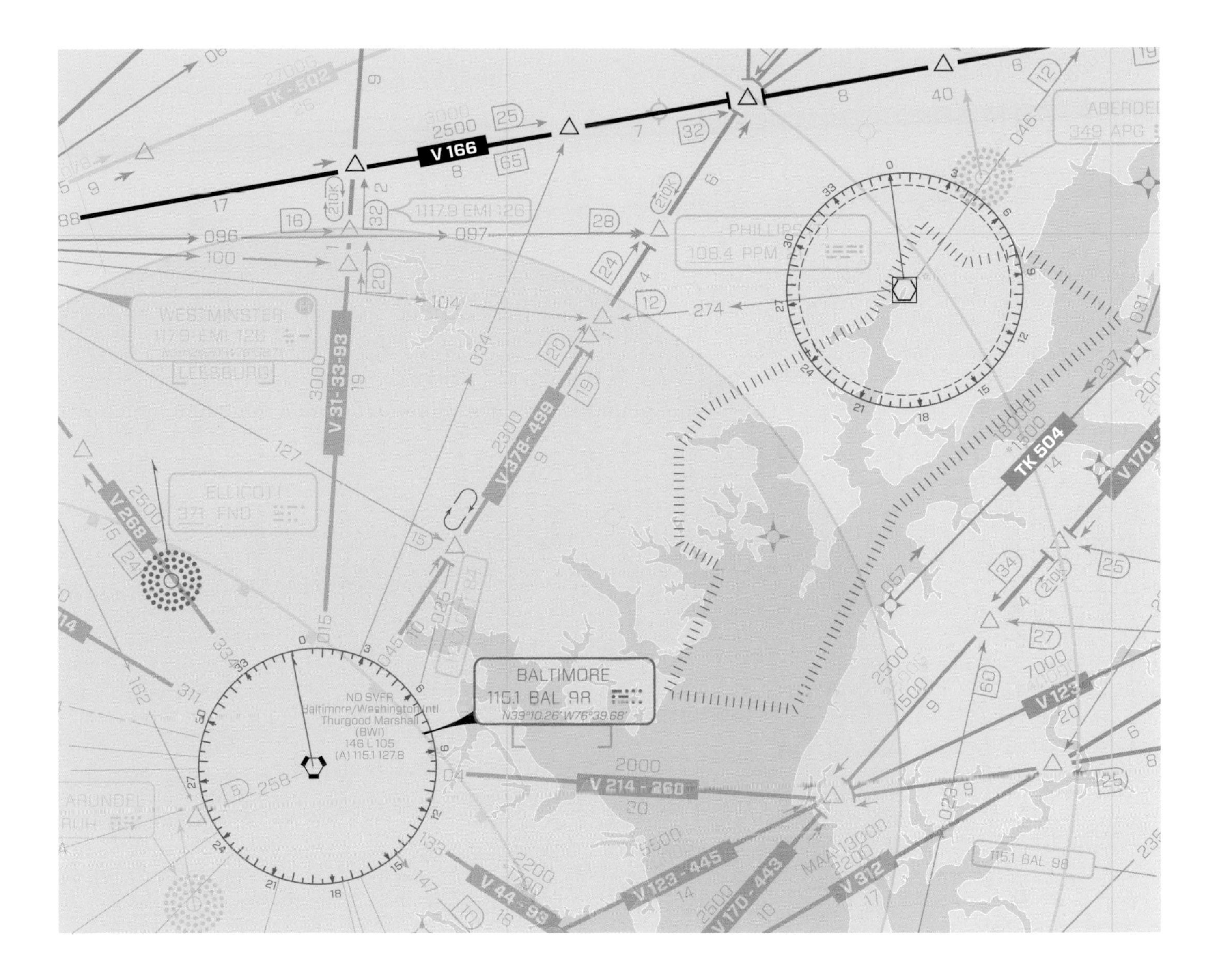

FLYING BY INSTRUMENTS

All commercial aircraft are required to navigate under instrument flight rules. While these charts show less in the way of ground features than their visual flight rule (VFR) counterparts, they do contain valuable information used to guide pilots and control air traffic.

VOR (very-high frequency omnidirectional radio range distance measuring equipment) sends a radio signal and allows a plane to get a single position fix by measuring its bearing and distance to the fixed ground station.

Like VOR, the nondirectional beacon (NDB) sends a radio signal but the NDB signal follows the curvature of the earth and can be sensed at much greater distances and is used to mark long-haul routes.

The military equivalent to VOR/distance measuring equipment (DME) is tactical air navigation (TACAN), a more accurate position system that broadcasts a bearing feature and a distance-measuring signal to military-equipped aircraft.

Low-altitude airways are designated for altitudes from 1,200 feet to 18,000 feet. Also known as victor airways, they are labeled with a V and a number (odd for north-south and even for east-west routes). Jet routes above 18,000 feet are shown on high-altitude en route charts and are preceded by a J and a similar numbering convention to victory airways.

Required navigation (RNAV) routes are designated by a Q prefix for high-altitude routes and T prefix for low-altitude routes and are colored blue. Along these routes, various altitudes are shown above the labels, indicating minimum altitudes for signal reception, obstruction clearance, crossing routes, etcetera.

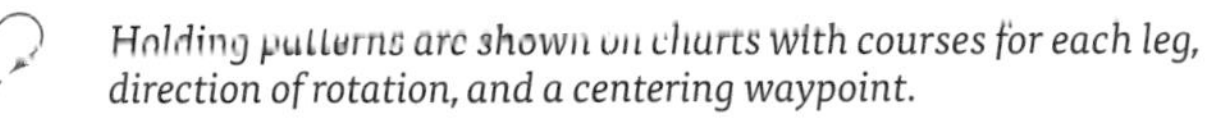

Holding patterns are shown on charts with courses for each leg, direction of rotation, and a centering waypoint.

Airspace that might be controlled with limitations on aircraft operations for security or other purposes is shown on a chart with boundaries and descriptions on the limitation. It is denoted with a prefix of P for prohibited areas, R for restricted, W for warning, A for alert, and MOA for military operations area.

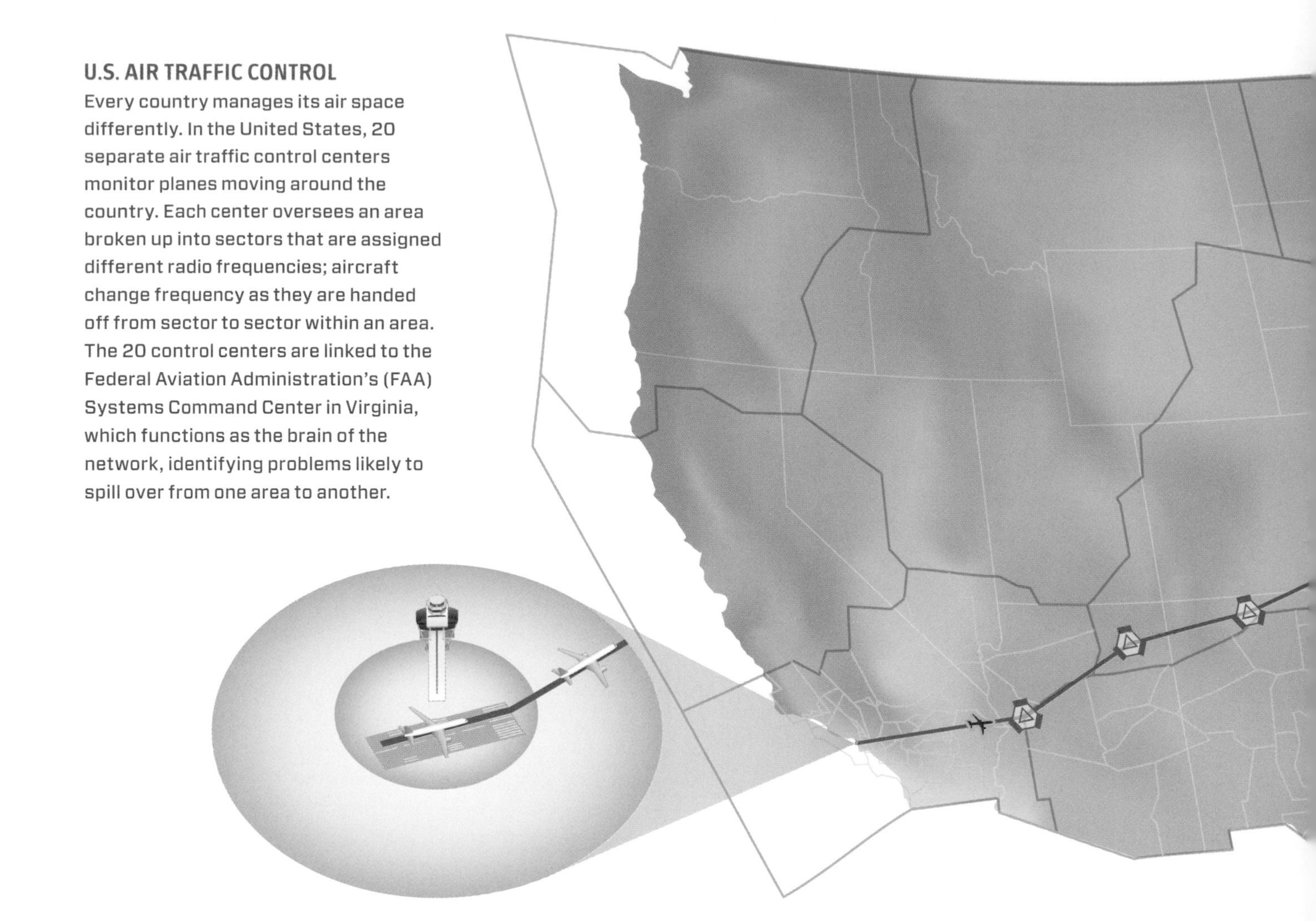

U.S. AIR TRAFFIC CONTROL

Every country manages its air space differently. In the United States, 20 separate air traffic control centers monitor planes moving around the country. Each center oversees an area broken up into sectors that are assigned different radio frequencies; aircraft change frequency as they are handed off from sector to sector within an area. The 20 control centers are linked to the Federal Aviation Administration's (FAA) Systems Command Center in Virginia, which functions as the brain of the network, identifying problems likely to spill over from one area to another.

1. *Flight data/clearance delivery controllers provide flight plan clearance prior to gate departure and can prevent an aircraft from leaving due to weather or air traffic conditions.*
2. *Once an aircraft is clear of the ramp, ground controllers direct aircraft movement on the taxiways and runways.*
3. *The aircraft is handed off to the tower controller, who controls the active runway, authorizes final takeoff, and tracks the aircraft in its ascent to about five to 30 miles (10 to 60 km) from the tower.*

CONTROLLING THE SKIES

Think of the air corridors above 18,000 feet as highways in the sky or jet airways. Like cars on a highway, most of the vehicles traveling along one corridor will be moving at the same altitude, in the same direction, and at the same or similar speed. But unlike highways, these jet airways need close and careful monitoring by third parties.

Those third parties are air traffic controllers. Over 27,000 controllers in the United States alone monitor commercial planes both at airports—where they clear takeoffs and landings—and at en route centers, from which they give aircraft instructions and report weather conditions. In general, air traffic controllers are required to maintain a horizontal separation of between five and 10 nautical miles between aircraft at the same altitude. Vertical separation must be greater than 1,000 feet (305 m) at all altitudes.

Some of these controllers patrol the ocean. Since ground-based radar can't track planes there, historically oceanic controllers have had to use what's known as procedural control to ensure separation by estimating a plane's position from pilot reports (normally via GPS) and computer flight models that project altitude, distance, and speed.

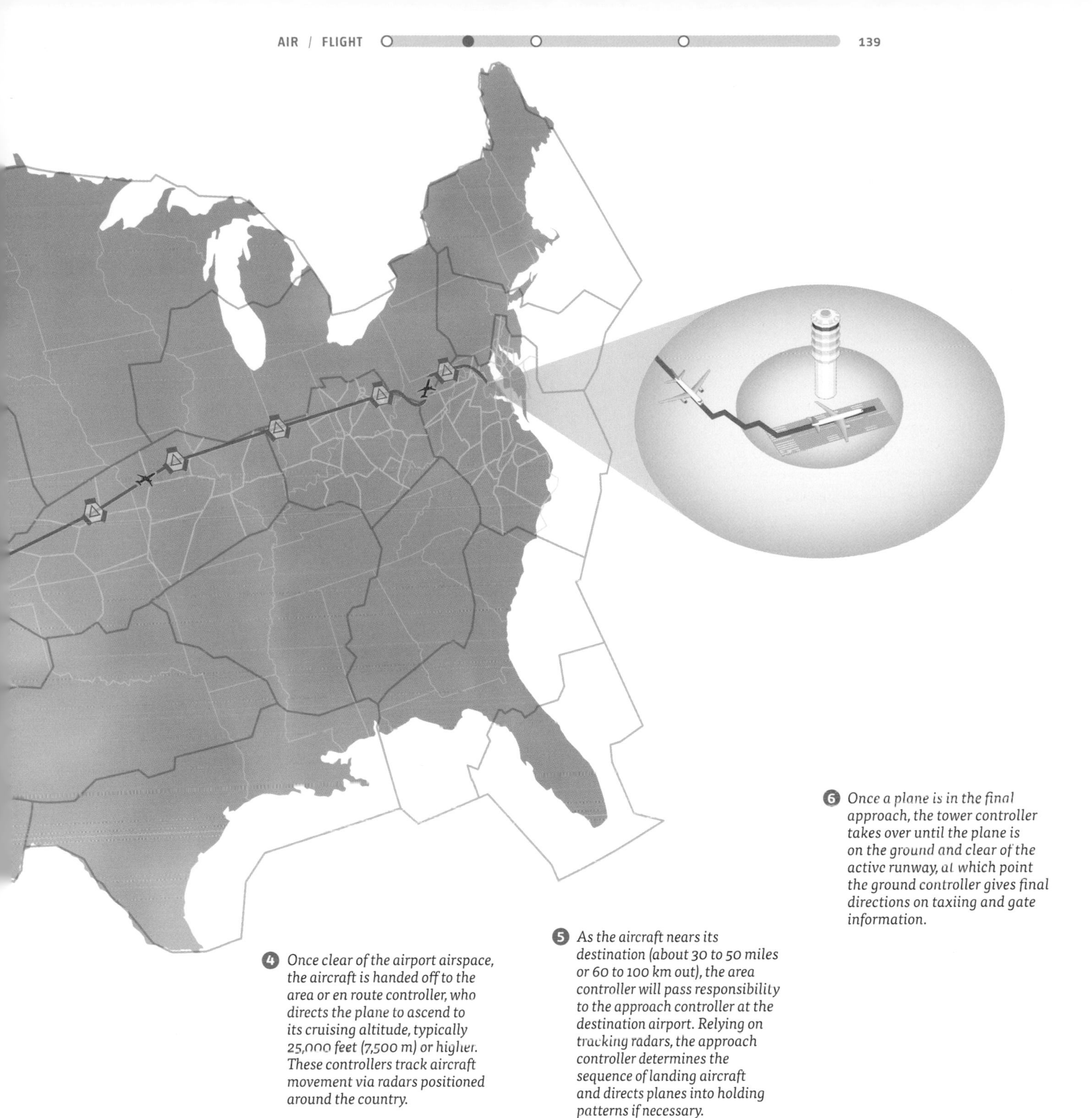

4 *Once clear of the airport airspace, the aircraft is handed off to the area or en route controller, who directs the plane to ascend to its cruising altitude, typically 25,000 feet (7,500 m) or higher. These controllers track aircraft movement via radars positioned around the country.*

5 *As the aircraft nears its destination (about 30 to 50 miles or 60 to 100 km out), the area controller will pass responsibility to the approach controller at the destination airport. Relying on tracking radars, the approach controller determines the sequence of landing aircraft and directs planes into holding patterns if necessary.*

6 *Once a plane is in the final approach, the tower controller takes over until the plane is on the ground and clear of the active runway, at which point the ground controller gives final directions on taxiing and gate information.*

NEXT GEN

As crowded as the skies are today, there is every reason to believe that they will only get more crowded in the future. Technology is making it possible for airplanes to fly closer together in the sky—opening up more slots along routes previously at capacity.

In the United States, airlines are counting on the continued implementation of Next Gen—short for next generation air transportation system. The program involves both the development of new technology and improved connectivity and communications across the existing system. Monitoring of airplanes will move from a system based on ground radar to one based on satellite technology that allows airplanes to "see" one another, even in inclement weather. This new precision will allow for more direct routes and for closer spacing on existing routes.

APPROACHING THE AIRPORT

Landing at an airport involves the work of multiple parties in the air traffic control chain. As a plane nears an airport, generally around 20 to 50 miles out, it is handed off from a controller at an en route center to a terminal radar approach controller, who gives clearance to begin the landing process. Between five and 10 miles out, the plane will be handed off to tower control, a local air traffic controller. This airport traffic controller is responsible for all movement in the immediate airspace surrounding the airport.

The tower controller can use radar or simply rely on pilot reports to determine appropriate approach instructions. Often these directions will include how to enter or leave the circuit—the pattern of aircraft waiting to land at an airport or having just taken off. At very large and busy airports, the circuit is not used and a clearance delivery system allows large commercial aircraft on scheduled flights to take a direct route to the runway and adjusts their departure times from the originating airport accordingly.

In general, the approach circuit (or pattern, as it is called in the United States) is comprised of five legs that form a rectangle around the runway. The rectangle

The airspace around the San Francisco Bay Area is similar to that of other global metropolises, with multiple international airports operating in a small geographic area—so traffic must be managed carefully.

DEPARTING/ ARRIVING

SAN FRANCISCO INTERNATIONAL

OAKLAND INTERNATIONAL

SAN JOSE INTERNATIONAL

MOFFETT FEDERAL AIRFIELD

Plane choreography is handled by San Francisco terminal radar approach control (TRACON), or San Francisco Approach, a facility that controls all airspace around these airports.

Many flights into or out of San Francisco International Airport, depicted by red arrows, cross the takeoff and landing patterns of Oakland International (green) and San Jose International airports (blue). These flights are directed straight through this airspace at different altitudes.

OAK

SFO

NUQ

SJC

CONTROL TOWERS

Airport control towers, though varied in height and architecture, usually rely on windows that encircle their top floors. These windows often tilt outward at 15 degrees so that the controller's computer equipment is not reflected in the windows. Ceilings are often painted black to improve screen visibility.

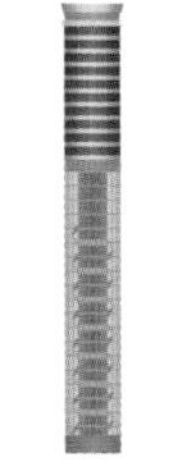

SUVARNABHUMI AIRPORT
Bangkokk, Thailand

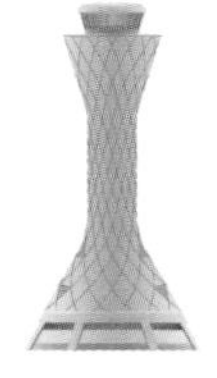

EDINBURGH AIRPORT
Edinburgh, Scotland

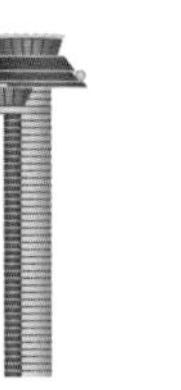

ARLANDA AIRPORT
Stockholm, Sweden

SYDNEY AIRPORT
Sydney, Australia

consists of the two legs parallel to the runway (the upwind and downwind legs) and two legs perpendicular to them (the crosswind and base legs), along with the final approach onto the runway itself.

The circuit is flown at 800 or 1,000 feet in altitude. It is usually done in a counterclockwise direction; that is, all turns in the circuit are to the left—primarily because the pilot's seat is on the left side of the plane and this offers better visibility. In cases of intervening geography or special situations involving noise restrictions, the circuit may be flown in a right-handed, or clockwise, fashion.

Pilots will normally land into the wind to increase the apparent wind speed over the wings and allow a slower approach over land. Because the wind changes direction frequently, airport runways will often be configured as reversible (in terms of where the planes first touch down) and at 90 degree angles to one another so that air traffic control can almost always select a runway and an approach direction that permits landing into the wind (at single-runway airports, the runway is oriented to best take advantage of local wind patterns). Approach directions will be adjusted based on changes in prevailing winds.

HOLDING PATTERNS

Aircraft might be directed into a holding pattern because of weather or traffic. In contrast to a normal approach, these patterns are often completed in a clockwise fashion. The racetrack pattern comprises a series of four legs (that is, a 180 degree turn, a straightaway, a 180 degree turn, and another straightaway) anchored over a "fix," for example, a radio beacon.

When a series of planes are holding, they are said to be "stacked" with each aircraft vertically separated by 1,000 feet (305 m). Planes enter the top of the stack and cycle down to the bottom as each preceding plane is authorized to land.

Depending on the direction of approach to the entry point, the aircraft may have to reverse course to enter the circuit in a "teardrop" turn. New pilots are tested on their ability to enter and exit this pattern.

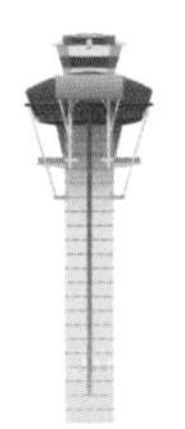

LOS ANGELES INTERNATIONAL AIRPORT
Los Angeles, California

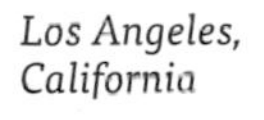

MADRID-BARAJAS AIRPORT
Madrid, Spain

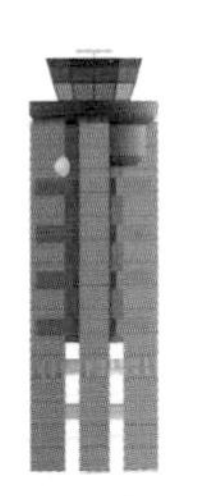

RICHMOND INTERNATIONAL AIRPORT
Richmond, Virginia

HEATHROW AIRPORT
London, England

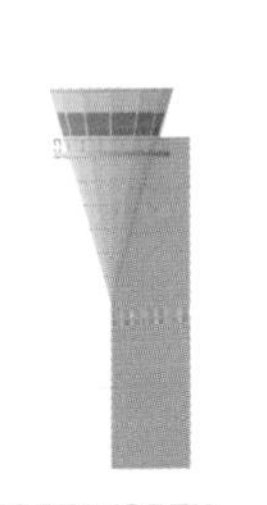

FORT WORTH ALLIANCE AIRPORT
Fort Worth, Texas

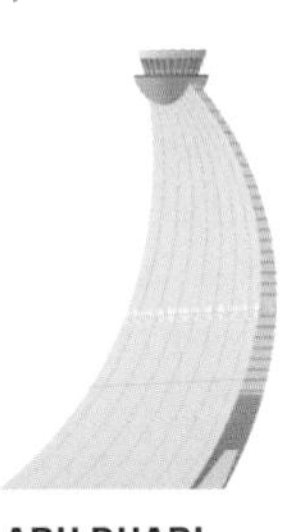

ABU DHABI INTERNATIONAL AIRPORT
Abu Dhabi, United Arab Emirates

UNDER PRESSURE

Maintaining air pressure within a cabin is critical to life in the air. As planes climb higher, air becomes thinner and thinner—to the point where humans can no longer function. Think of a mountain climber in the Himalayas who must take oxygen as she approaches the summit of Mount Everest at 29,000 feet—and then imagine even thinner air several thousand feet higher.

To ensure that passengers continue to function in the sky, airplane cabins are routinely pressurized to an altitude of just under 7,000 feet. This level is sufficient for human comfort but does not create a differential between inside and out so large as to require massive strengthening of the fuselage (which would occur if the cabin was kept at sea-level pressure). Planes that fly higher than conventional altitudes, as the Concorde did, must be reinforced.

Because planes are full of humans inhaling oxygen and exhaling carbon dioxide, the air inside a plane must be constantly changed. Cabin air is a combination of outside and filtered air, which comes in through the valves above the passenger seats and typically exits through grilles in

❹ *As new air is constantly being added, an outflow valve opens to regulate the cabin air pressure (like blowing up a balloon with a pinhole leak).*

❸ *The cool, pressurized air travels to the mixing manifold where it is blended (about 50/50) with recirculated air. This blended air is vented through overhead manifolds.*

❷ *After passing through the engine, the air is very hot and must be passed through a radiator-type system and air conditioners to cool it (some of this hot, pressurized bleed air is not cooled and is sent to auxiliary uses like deicing).*

❶ *Outside air enters the compressor section of the plane's engine. This air is pressurized in the compressor to levels comfortable for crew and passengers.*

THE FIRST COMMERCIAL JET

The world's first commercial jet, introduced by De Havilland in Britain in 1952, was the Comet, an engineering marvel. Relying on a quiet gas-fed engine, it promised a smooth and fast ride—cutting the travel time between London and New York from 18 to 12 hours. But it had a thing or two to learn about cabin pressure.

Within two years of its debut, in 1954, several Comets inexplicably tore apart in flight—killing all aboard. Extensive testing in a water tank, meant to simulate the air pressure on a plane at high altitude, helped identify the problem: metal fatigue occurring at the corners of the square windows.

British aerospace engineers eventually remedied the design flaw, producing a Comet with oval windows. But their efforts were in vain: by then, Boeing had introduced the 707 jet—which would go on to become one of the most successful jet planes of all time.

the floor. Roughly half of this air is exhausted through outflow valves to the rear of the plane; the other half moves through air filters and is then mixed with new air coming in from the engine compressors.

Pressure in the cabins is maintained automatically by an air controller mechanism, which manages the inflow and outflow of new air and exhaust air, respectively. Though some variation in pressurization is often felt at takeoff and descent, in most cases air pressure stays fairly uniform throughout the length of a flight.

PRESSURE VARIATION

Clogged ears are an inevitable fact of life for air passengers—a function of unequal pressure that's built up on either side of an eardrum. Clearing the blockage can sometimes be achieved by swallowing or yawning (not by chewing gum), but the most reliable way is the Valsalva maneuver—pinching one's nose and closing one's mouth while pushing air through the nostrils. Alternatively, earplugs can be used to keep pressure even by restricting the flow of air to the eardrum.

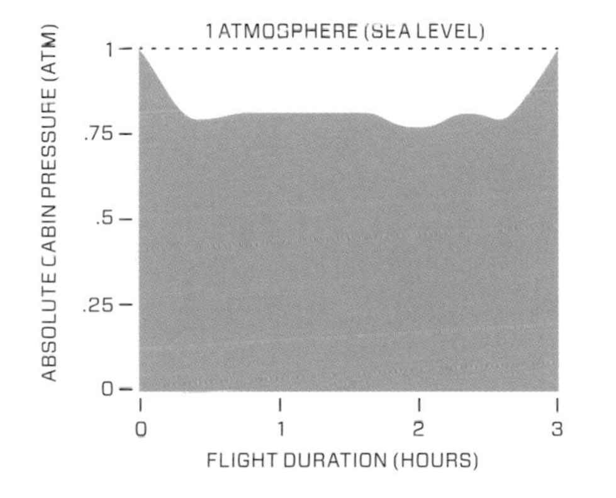

SICK ZONES

While air filters capture almost all of the bacteria or virus-carrying particles in plane air, air circulation systems are often turned off for brief periods while at the gate—leading to a higher than normal risk of getting sick during plane travel. Sick particles do most damage in a radius of two seats around a carrier, with tray tables and seatback pockets serving as conduits to transmit germs from one seat occupier to its successor.

OXYGEN MASKS

Loss of cabin air pressure, whether through leaks or an explosion, is a serious matter. It can quickly lead to loss of consciousness due to hypoxia—including nausea, blurred vision, slurred speech, and mental confusion (similar to the condition that can affect a mountain climber scaling a very high peak). As a result, oxygen masks—though rarely used—are mandatory safety equipment on commercial planes.

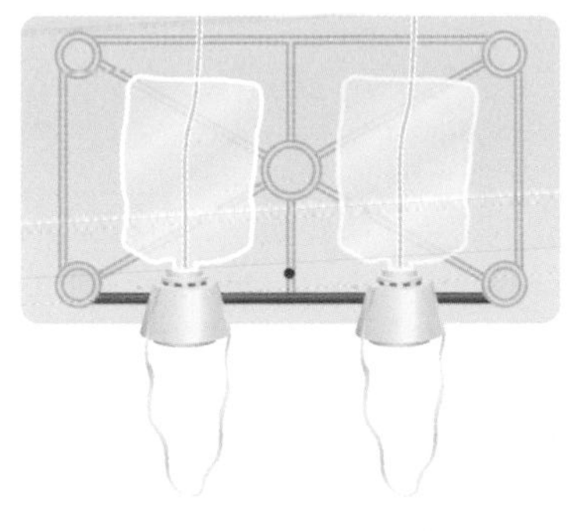

A FLUSH IN TIME

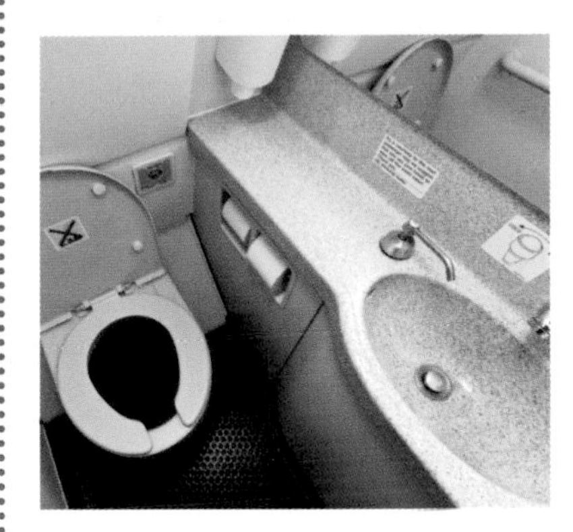

Once upon a time, a flush in the air was just a flush—blue circulating water at the bottom of an airplane toilet simply disappeared at the push of a button. Then vacuum flush systems arrived and made the whole process much noisier—but also more sanitary.

Instead of spilling human waste out of an airplane in midflight, today's vacuum systems move it to a waste port inside the body of the aircraft. Once the plane has landed, a lav agent, or lavatory service vehicle, will arrive to clean the system. For commercial planes, this agent will usually be a truck with a tank configured to easily access the waste storage area. With smaller planes, a lav cart—a smaller vessel towed by a tug—will service the lavatories and dispose of the waste at an airport sewage facility. In both cases, the tank is refilled with a mixture of water and a disinfecting concentrate known as blue juice.

Airplane bathrooms are carefully designed. Perhaps because of a history of nicotine-addicted passengers stealing a smoke in the toilet, these small rooms are designed to offer maximum protection against fire: they come complete with smoke detectors, oxygen-smothering flapper lids on the paper towel bins, and occasionally even fire extinguishers.

BRAVING THE ELEMENTS

Airplanes in flight are exposed to the elements in ways much more extreme than those experienced by trains, cars, or even boats. Foremost among the forces that can jeopardize a plane's safe journey are wind, ice, and lightning.

Turbulence is something that most air travelers experience at one time or another. Though it can be uncomfortable and scary, very rarely will it cause damage to the plane or injury to the passengers inside it. This is particularly true if passengers are wearing their seat belts; of the 80 reported injuries from turbulence in the United States between 1981 and 1997, 73 stemmed from passengers not using their belts.

Rough air can be caused by any number of sources: changes in atmospheric pressure, cold or warm fronts, wind moving over mountains, jet streams, or storms. When the airplane moves through these currents or eddies of winds, it acts as an obstacle to them—similar to the impact a boulder has amid a rushing stream. The solution is straightforward: move the airplane out of the "stream" of wind and into an area of smoother air—as soon as the pilot can find it.

While snow is harmless to planes in the air, ice is not. The buildup of ice on the edges of a plane's wing or

MOUNTAINS

Mountains, particularly tall ones, are one common source of air turbulence. As air hits and rises up the sides of mountains, it moves upward in a sort of wave motion that can affect planes flying above 30,000 feet. The effect can be felt over quite a long distance—up to 100 miles or more downwind of the mountain range.

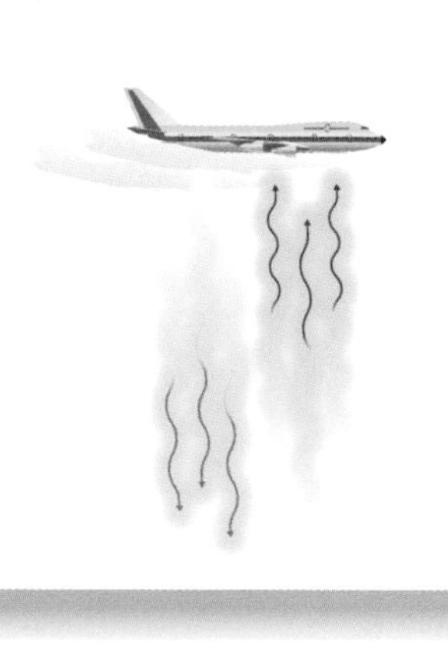

THERMALS

Most people wouldn't expect to experience turbulence on a sunny day. But as any glider or balloon pilot knows, a sunny day can create bubbles of warm air, or thermals, that rise through the atmosphere. These thermals exert an upward force on anything within their path.

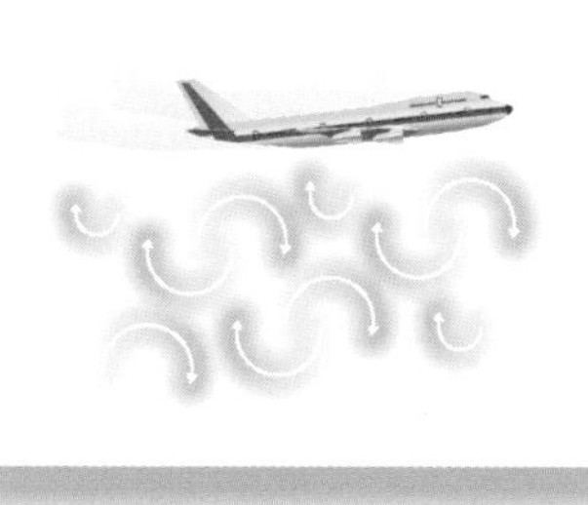

WIND SHEAR

Wind shear involves a dramatic change in wind speed or direction over a short distance. It can move either horizontally or vertically and can occur at almost any altitude. Some of the most dangerous wind shear cases are called microbursts, a term used to refer to sudden wind changes resulting from rain showers or thunderstorms.

TURBULENCE

At higher altitudes, fast-moving air currents in the jet stream can shift and disturb the air around the aircraft, causing the plane to move in the air column.

HURRICANE HUNTERS

Though satellites convey weather information, they can't actually measure weather inside storms. For that you need the bravest pilots in the world—some of whom work for the U.S. Air Force Reserve's 53rd Weather Reconnaissance Squadron.

Their job is to fly into and out of a storm, as low as possible and four or five times in a row, to drop a device known as a dropsonde into its center. As the gadget is falling, it records and sends back information about the storm—wind speed and direction, temperature, barometric pressure, and levels of humidity. The data are sent twice each second to the National Hurricane Center, located at Florida International University in Miami, and are analyzed there by meteorologists to predict the path and likely development of a storm.

tail can disrupt the airflow over them, increasing drag and undermining the ability of the wing to create lift. For this reason, deicing planes before takeoff is a common practice during winter at snow-belt airports around the world.

Ice can also build up at high altitude during flight, from freezing rain or drizzle or even from a cloud. In addition to affecting lift, ice can break off while flying and damage the engines or propellers. To prevent that, a variety of heating appliances have been devised to protect windshield, wings, tail apparatus, and propellers.

Lightning is less dangerous than ice and strikes planes regularly—often triggered by a plane as it moves through an electrically charged storm cloud. In most cases, lightning currents are conducted harmlessly and without interruption along the aluminum exterior of a plane's skin and exit through its tail. Specific features, including lightning diverter strips and shielding and surge suppression devices (similar to those found on tall buildings), work to ensure that it does exactly that. In addition, the area around the fuel tank is designed to prevent sparks from reaching it.

HOT WINGS

Ice buildup on the leading edge of the wings can adversely affect the lifting capacity of the wings and fuel efficiency of the plane. Hot, pressurized air from the engines might be piped along the wing to melt any ice buildup. Some aircraft may also utilize electric resistive heaters embedded in the wing.

DEICING BOOT

To prevent ice from forming during flight, some aircraft feature rubber structures along the edges of the wings or tail that can expand during icing conditions. Known as deicing boots, they operate automatically and inflate periodically to crack and expel the ice before returning to their normal size.

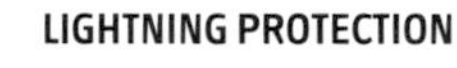

LIGHTNING PROTECTION

When lightning strikes an aircraft, it will normally do so along its perimeter—typically the nose or tip of its wing. As the craft flies through the "circuit" of electricity created by different charges in the clouds, the lightning will be conducted along the plane's aluminum exterior. Planes such as the 787, whose skin is made of composite rather than purely aluminum material, may feature conductive screens within their skins to assist in safely carrying lightning charges.

ST. ELMO'S FIRE

Seamen during the Middle Ages thought it a good omen when static electricity in the form of a flickering blue halo built up on their ship's mast during a thunderstorm and attributed it to St. Elmo, the patron saint of Mediterranean sailors. Static electricity can build up and discharge a similar glow on the wings of a plane flying through an electric storm. What's known as St. Elmo's fire is usually harmless—though it has been known to short-circuit radio equipment or blow a hole in an airplane's skin.

WHEN DISASTER STRIKES

Although many studies show that flying is, on average, safer than driving a car, the impact of accidents in the air is significantly more dramatic. Rapid depressurization of a cabin is one hazard that can be fatal: a 1988 Aloha Airlines accident involving structural failure of the aircraft itself resulted in a stewardess's being blown out of the aircraft.

Fire is another serious hazard on planes, causing nearly one fifth of the 1,153 fatalities on U.S. airlines between 1981 and 1990. A number of systems exist to prevent or limit the spread of fire on planes. In addition to passive fire protection, which involves the use of fireproof or inflammable materials, more active fire protection components are built into today's aircraft: smoke detectors, temperature sensors, air shutoff mechanisms, and a halon system, involving a chemical spray that can be used to put out a fire in a cargo compartment.

When things do go wrong, there are two devices that aviation investigators use to piece together what happened during the final minutes or seconds of a doomed flight. The first is the flight data recorder, which constantly records flight data—such as altitude, speed, rudder position,

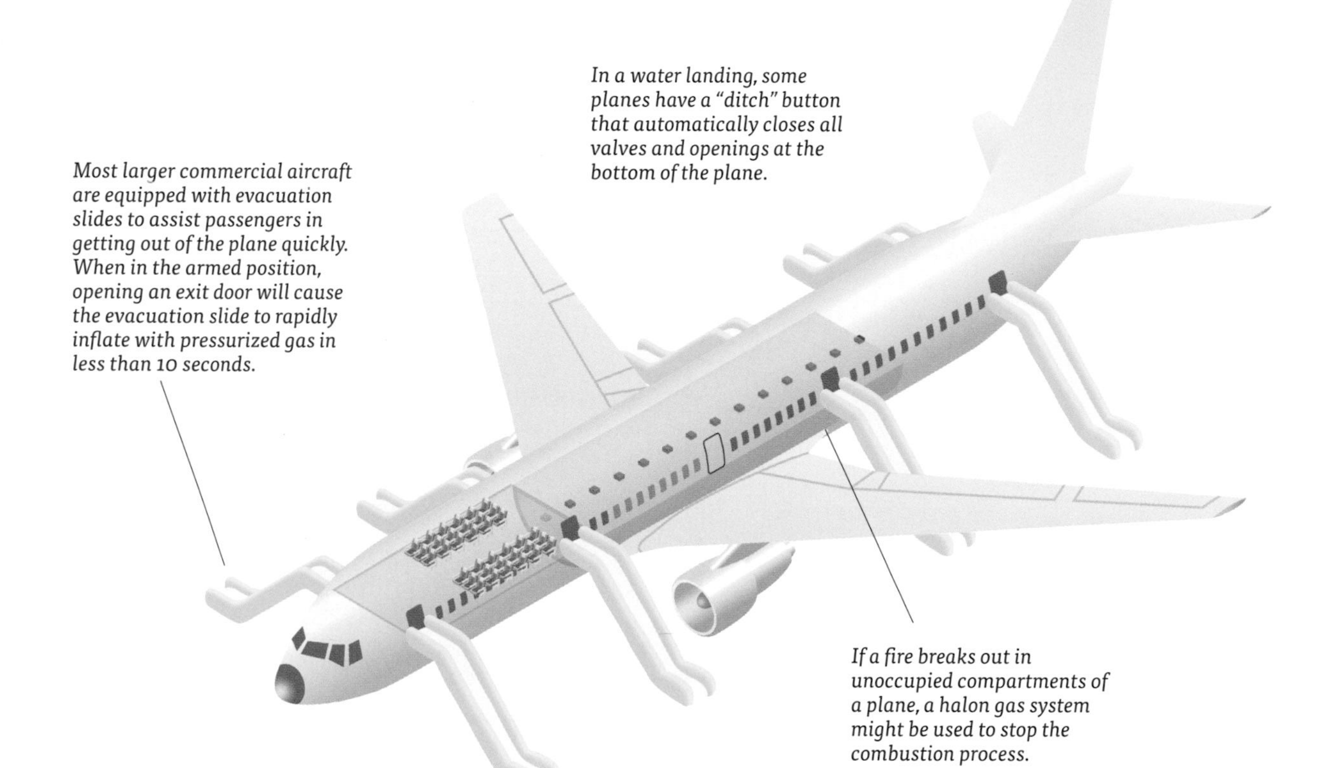

SULLY

Among the most famous would-be disasters in recent history is the one that struck U.S. Airways Flight 1549 en route from New York City to Charlotte, North Carolina, in January of 2009. An apparent bird strike only three minutes after takeoff from LaGuardia Airport in New York caused both engines of the Airbus A320 to lose power as it was climbing just north of the George Washington Bridge, forcing one of the most unusual landings in New York City's aviation history.

Air traffic control directed the plane to make an emergency landing at Teterboro Airport in northern New Jersey. But the captain of the plane, Chesley Sullenberger (or "Sully," as he would become known), instead used his training as a glider pilot to gently drop the plane into the Hudson River west of Midtown Manhattan. Within minutes, local ferries arrived to pull both passengers and crew to safety from the half-submerged and slowly sinking aircraft—giving rise to the incident's "Miracle on the Hudson" moniker.

et cetera—from the plane's own information systems and instruments. The second is the cockpit voice recorder, which tapes the pilots' conversations with each other and with ground traffic control. Both are required on large commercial planes.

Flight data recorders date back to the 1950s, but didn't receive their signature color—not black, but orange or yellow—until the mid-1960s. Powered by the plane's engines, they are generally mounted in an aircraft's tail, where they're more likely to withstand the impact of a crash. Designed to endure extremes of temperature and pressure, flight data recorders must be penetration resistant to meet requirements for fluid and water immersion.

Located alongside the flight data recorder is the cockpit voice recorder, a simpler device that records a minimum of 30 minutes of sound from the cabin. On an average commercial flight, four microphones may record at all times: one in each of the pilot's headsets, another in the headset of the additional crew member, and a fourth near the center of the cockpit. In addition to voices, the voice recorder will pick up other cockpit and engine noise, including things like stall warnings or landing gear extensions.

BLACK BOX

The "black box" (which is usually orange) is really two recording devices, the flight data recorder and cockpit voice recorder. They store all data—including time, airspeed, altitude, vertical acceleration, pilot control positions and voice communications—in solid-state memory cards.

The black box is usually located at the rear of the plane for protection. In order to protect its essential data following a crash, the device must be designed and tested to survive severe impact, high-temperature fires, immersion in salt water, and deep-sea pressures. It even features an underwater locator beacon.

PHONING HOME

Although airlines remind passengers to turn off electronic devices at takeoff and landing, just how dangerous personal electronic devices are on planes is entirely unclear. Mobile phones emit signals stronger than many other devices as they hunt to make connections to a ground-based system. In theory this could affect the plane's avionics, communication, or navigation systems, but most of the evidence that cell phones have affected instrument readings or led to irregularities in the plane's communication systems appears to be anecdotal.

The seat-based phones that some airlines feature are not cell phones; rather, they are part of an airborne satellite system. Their emissions are highly controlled and are electromagnetically compatible with the plane's navigation and other systems.

AIRPORTS

One hundred years ago, you could count the number of airports in the world on two hands. Today, there are an estimated 45,000 of them—covering nearly every nation on the globe. They range from highly automated facilities, with multiple runways and terminals in simultaneous operation, to strictly functional and unadorned concrete landing strips.

Modern airports handle two types of traffic—general aviation, which refers to nonscheduled aircraft movement, and commercial aviation, which includes scheduled passenger and cargo flights. The world's largest airports have historically handled both—though the demand for commercial air travel now relegates general aviation to secondary airports in many busy hubs.

Some of the world's largest airports are also among its oldest, though that title is formally reserved for a handful of small, military airports that no longer handle civilian traffic (if they still operate at all). Amsterdam's Schiphol Airport, Rome's Ciampino and Paris's Le Bourget are among those operating today that began as military fields and opened up to civilian traffic around World War I.

Airports proliferated during the 1920s and 1930s, as technology improved to allow heavier planes, paved aprons and runways, and night flying. Innovations such as approach lighting led to standardized procedures for landing and takeoff, and international navigation standards allowed flights to operate seamlessly on flights between countries.

Many early coastal airports catered to both land-based planes and to "flying boats." North Beach Airport in New York, for instance, catered to Pan American's "clippers," which took off and landed at its sea terminal during the 1930s. Today it is called LaGuardia Airport, and its historic Marine Air Terminal now serves Delta's shuttle flights to Washington D.C., Boston, and Chicago.

Prior to World War II, airport buildings were simple, functional structures if they existed at all. But as commercial aviation grew following the war, airport structures evolved. Terminal buildings became focal

The first airfield is established in College Park, Maryland, in a collaboration between the Wright brothers and the U.S. government.

The first airfield dedicated to commercial air operations opens in Königsburg Devau, Germany.

LaGuardia Airport opens in New York with both a land and marine terminal (the latter for overseas flights).

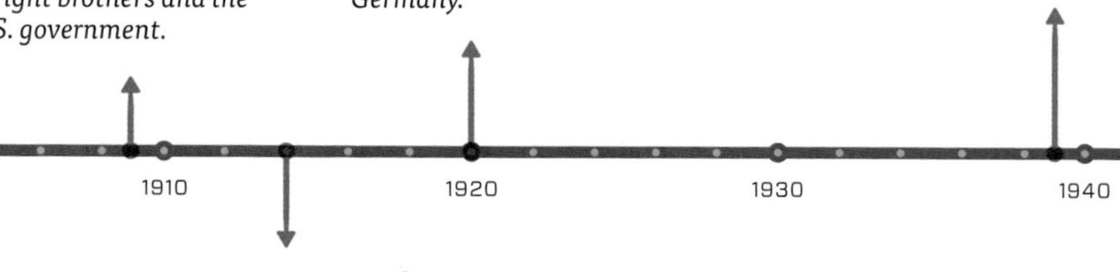

Southampton, England, establishes an "air port" for the Sopwith Schneider flying boat.

points of activity—more rigidly separating land-side activities, such as passenger or cargo processing, from air-side activities, such as fueling and runway operations.

The institution of jet travel in the 1960s made flying even more attractive, and the numbers of air travelers skyrocketed. Airports were designed to accommodate the changes: runways were made longer and stronger, and terminal buildings expanded to handle the greater volumes of people. On or near-airport amenities—retail, parking, and even hotels—grew as well.

Today, the world's largest and biggest airports are cities unto themselves. Sometimes referred to as aerotropolises or airport cities by people in the industry, they offer services well beyond those traditionally associated with flying and serve as huge generators of employment and economic activity for their metropolitan regions. Over 75,000 people work at Heathrow Airport in London, for example; another 8,000 work in airport-related industries just outside its boundaries.

AIRPORT CODES

Most air travelers are familiar with the three-letter airport codes printed on their luggage tags—also known as the location identifier. Most often, the code is made up of prominent consonants in the name of the city served, such as TLV (Tel Aviv), SZG (Salzburg), or CPT (Capetown). But there are many other variations as well.

Those in the United States with an X on the end—such as LAX (Los Angeles) or PDX (Portland)—simply took the previously existing National Weather Service two-letter abbreviation and added an X. Other airport codes represent the first letters of multiple cities served, as in DFW (Dallas-Fort Worth) or MSP (Minneapolis-St. Paul). Airport codes can also be the first letters of a multiple-word destination, as in PAP (Port-au-Prince) or SLC (Salt Lake City). Sometimes airport codes represent the first three letters of the city served, as in CAI (Cairo) or EDI (Edinburgh).

In a number of cases, the airport code reflects an old name of the city, as in PEK (Beijing) or BOM (Mumbai). Occasionally, the code chosen bears no relation to the name of the city. Who would know that MCO, as Orlando is known, was once McCoy Air Force Base? Or that TYS represents Knoxville, Tennessee, because the Tyson family donated the land for the airport?

Some letters never appear in these airport codes. For example, "N" codes are reserved for the military—leaving Norfolk Airport as ORF and Newark as EWR. Other reserved letters include "W" and "K," used for radio stations, "Q," for international communications, and "Y," for Canadian cities.

AIRPORT MILESTONES

The idea of commercial airports is less than a century old. The first airport in the United States was distinctly uncommercial: College Park Airport in Maryland was established in 1909 and was used by Wilbur Wright to test his new flying machine. The first facilities dedicated to commercial aircraft didn't open until 1920, at Sydney in Australia and Schiphol in the Netherlands.

Denver International Airport opens with a Rocky Mountain-like roofline and a troubled baggage handling system.

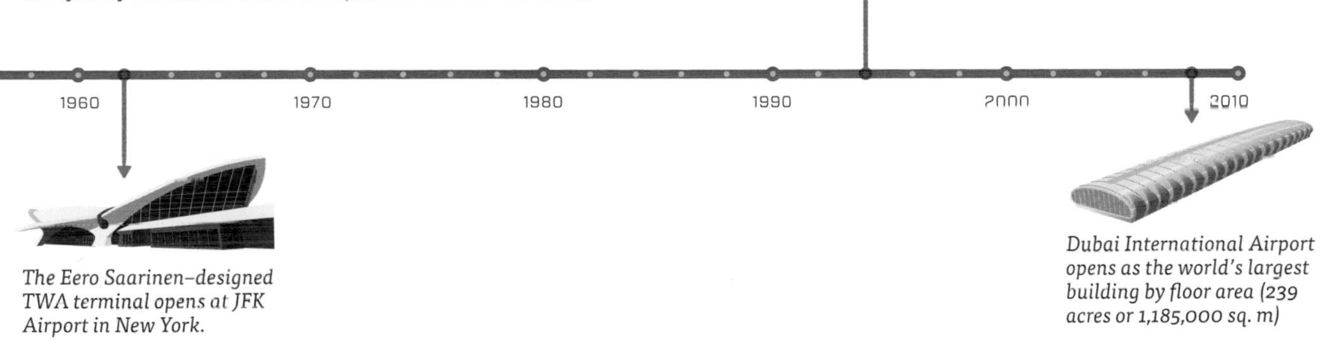

The Eero Saarinen–designed TWA terminal opens at JFK Airport in New York.

Dubai International Airport opens as the world's largest building by floor area (239 acres or 1,185,000 sq. m)

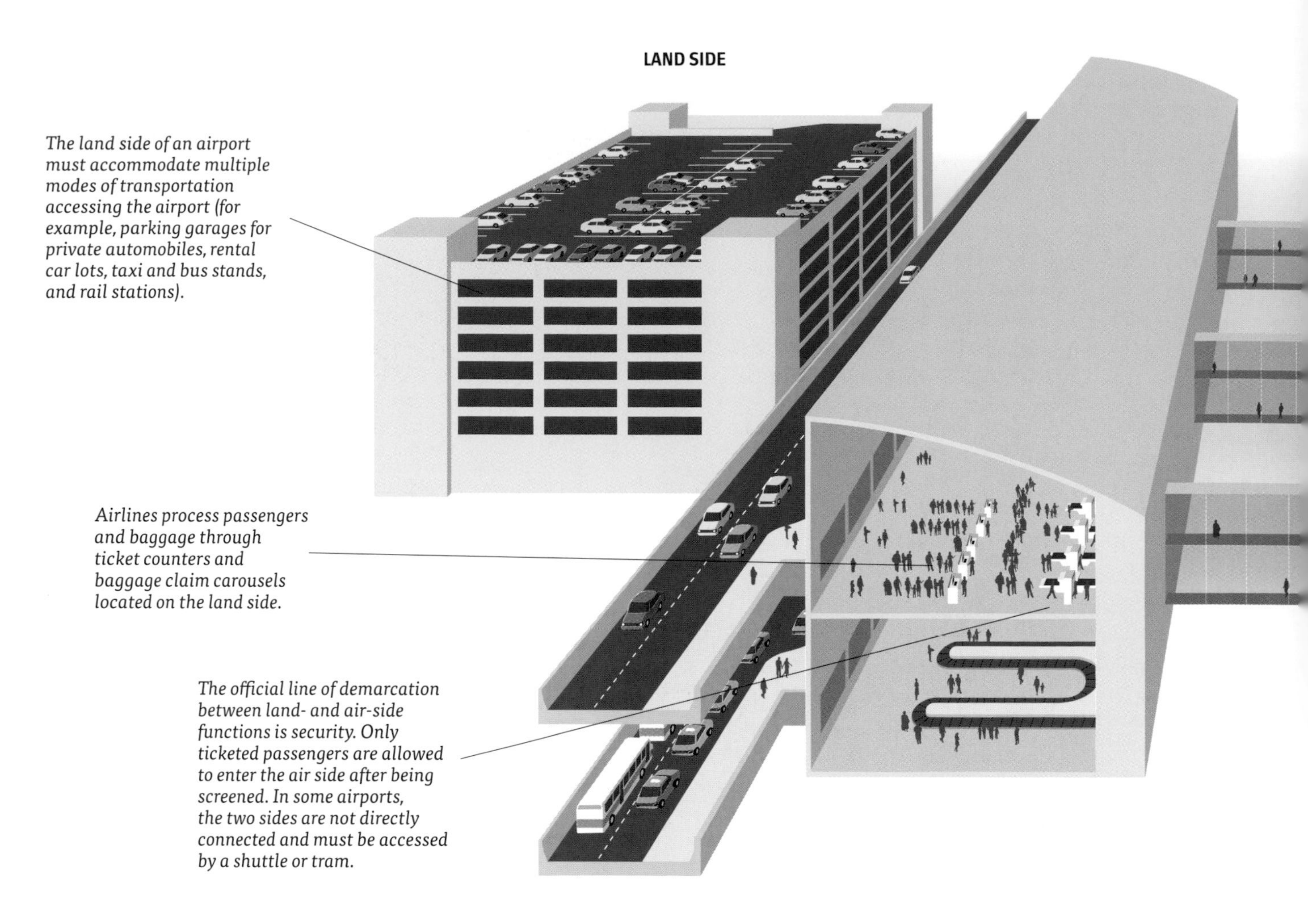

ANATOMY OF AN AIRPORT

Airports around the world are similar in their basic functionality, particularly in the configuration of air-side activities. Ramp areas adjacent to the terminal are known as the tarmac and parking spaces situated remotely from the terminals are known as aprons. Elsewhere on the airfield (though it can also be part of the terminal) sits a control tower as well as a variety of cargo and support facilities, including fuel and catering facilities and maintenance hangars.

Land-side and terminal layouts take more varied forms than those on the air side and vary based on the volume of aircraft and passengers being served. Smaller airports are fairly simple in design. One building will generally provide departure services, including ticketing and waiting, and landing services, such as luggage collection and meter-greeter areas. Arriving or departing planes will be located on the air side of the terminal.

For slightly larger airports serving multiple planes simultaneously, pier structures are often used. A central terminal will funnel its passengers along multiple linear piers, laced with a series of gates that serve individual

TERMINAL LAYOUTS

Airports can be laid out in many different ways, some of them friendlier than others to passengers arriving at one gate and departing from another. Historically, airports relied on the concept of "pier fingers," which stuck out from a central hub terminal like fingers from a hand. But airports today are also making use of long linear terminals or satellite terminals with indoor or outdoor shuttles to move passengers significant distances from the terminal to their gates.

LINEAR

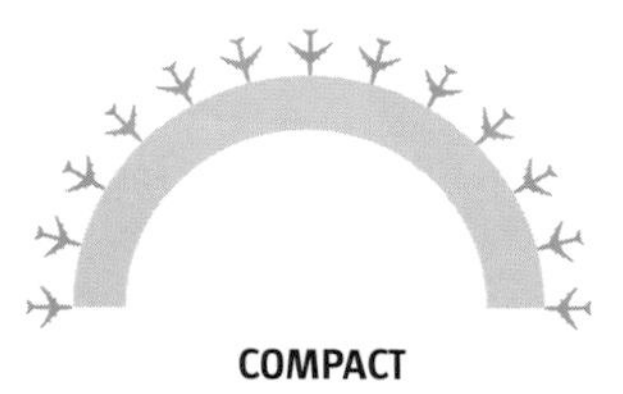

AIR SIDE

Because passengers are encouraged to arrive earlier due to increased security in a post–9/11 world, airport operators have integrated improved retail and dining options on the air side to capture revenue during what they call dwell time.

The focus on the air side is the gates, where airlines strive to turn arriving planes around as quickly as possible after a fast cleaning and crew change.

The air side is home to the ground services needed to service the planes, runways, and taxiways. In addition to flight control towers and maintenance areas, most commercial airports have their own fire departments located here.

aircraft. Jet bridges often connect the pier structure directly to the plane to facilitate boarding and unloading.

Satellite structures can be used to expand terminal capacity. Under this configuration, aircraft gates are clustered around a hub or "satellite" structure that passengers access from a central terminal area. Depending on distance, passengers are asked to walk, travel on a moving walkway or escalator, or take a people mover, train, or monorail.

Increasingly, there are not enough gates to serve all aircraft, and passengers need to be moved from a terminal to a plane parked remotely at a stand on the tarmac. In some cases, this design is selected for its flexibility and aircraft stands are arranged neatly in rows to maximize throughput through the facility.

Cargo facilities, though often owned by the same airlines that serve passengers, are normally remote from the main passenger terminals. They are often purpose designed: for example, perishable centers offering controlled-climate zones or express package facilities fitted out with conveyor belts.

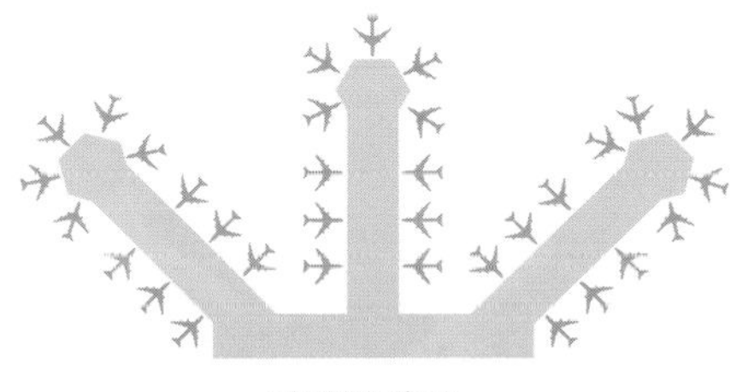

PIER FINGER

SATELLITE

TRANSPORTER

RUNWAYS

Runways lie at the heart of air-side operations. They can be anywhere from 6,000 to 12,000 feet long, though their length varies based on altitude and climate. At higher altitudes and in hotter climates, longer runways are required because air is less dense—which lowers the amount of lift and thrust that can be achieved in a given distance.

Runways are wide—often as wide as a 16-lane highway. And they are thick. At a large commercial airport, they may be up to three or more feet thick—enough to handle the enormous weight of a fully loaded jet (850,000 pounds or 386,000 kilos). Denver airport's new runways, for example, sit on six feet of compacted soil, eight inches of a cement-treated base, and another 17 inches of concrete paving.

Not surprising, runways are one of the most expensive components of an airport. Constructed primarily by a machine that lays an unbroken slab of concrete, they usually feature a series of grooves that run across the runway to drain water from it and allow for thermal expansion and contraction at controlled joints. At a major commercial airport in the United States, for example, runways can cost around $2,000 per linear foot—roughly seven times the construction cost of a typical terminal building.

A series of white lights marks the edge of the runway, turning to yellow over the last 2,000 feet (600 m). Similarly, embedded runway centerline lights are white until the last 3,000 feet (900 m), where they alternate white and red until the last 1,000 feet (300 m), when they turn to red only.

Directional arrows point to intersecting runways and taxiways

Taxiways are denoted by a yellow letter on a black field.

Directions to intersecting runways and taxiways are painted black on a yellow field.

RUNWAY PLANS

Although runways are always linear, multiple runways are not always parallel to one another. Airports may feature runways that intersect in a perpendicular fashion or that together constitute an open V shape. The choice of runway in use at a given time depends on wind speed and direction; the number in use depends on the volume of air traffic and visibility.

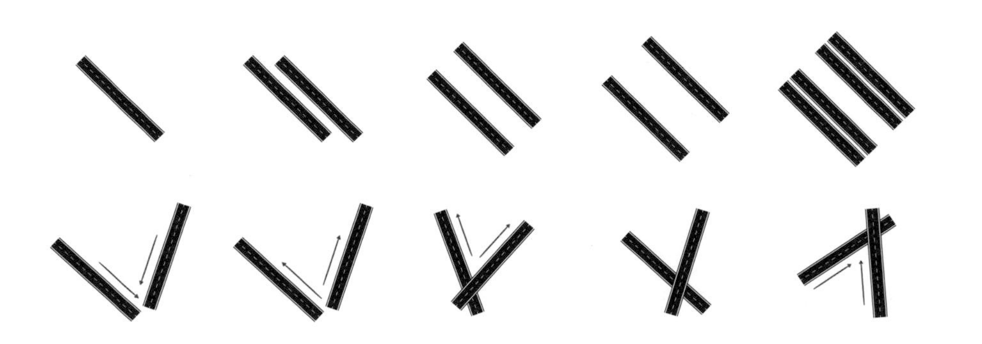

Mounted signs help the pilot navigate crisscrossing runways and taxiways. Current runway location is a one- or two-digit number corresponding to the leading digits of its compass and reciprocal heading.

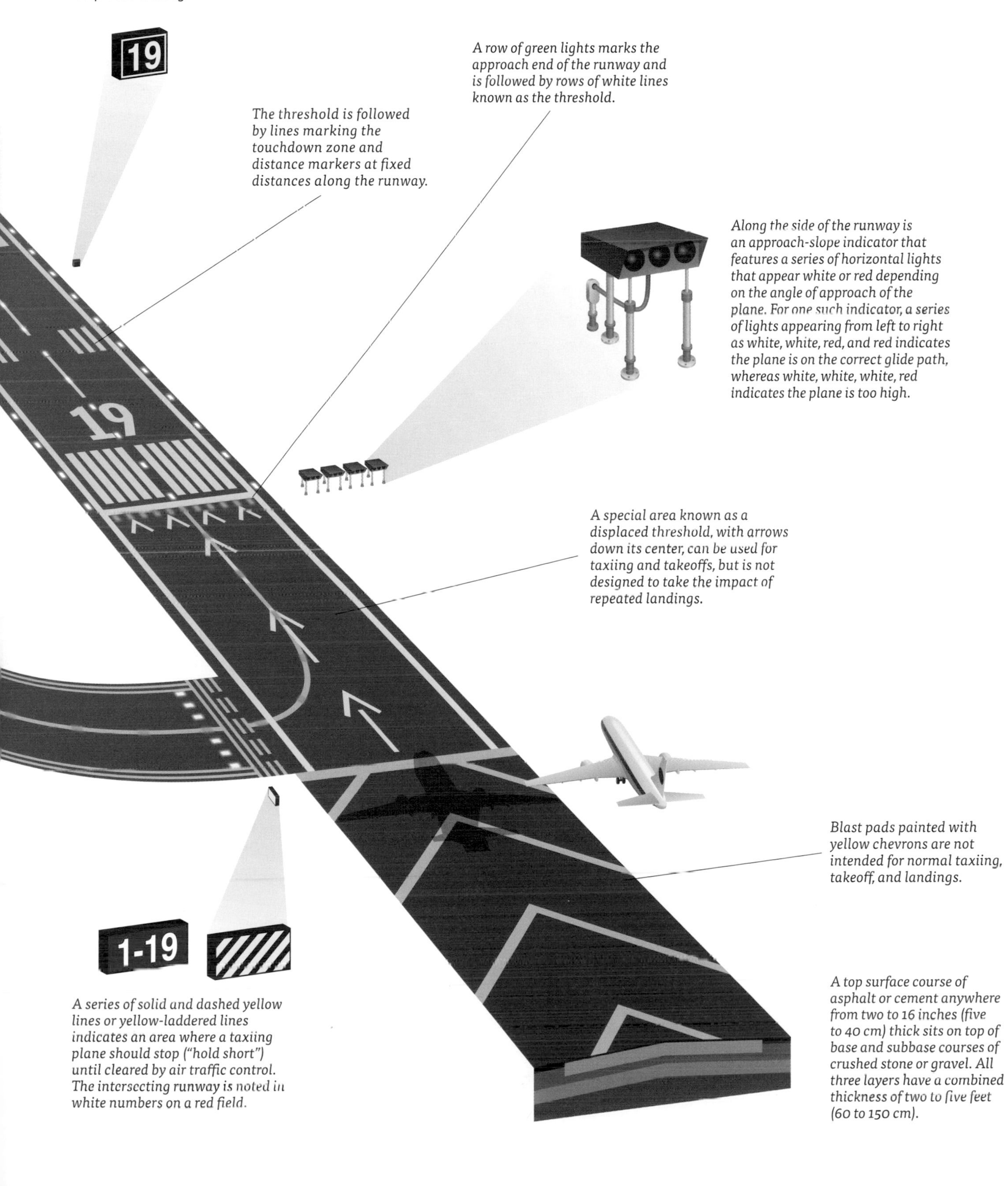

A row of green lights marks the approach end of the runway and is followed by rows of white lines known as the threshold.

The threshold is followed by lines marking the touchdown zone and distance markers at fixed distances along the runway.

Along the side of the runway is an approach-slope indicator that features a series of horizontal lights that appear white or red depending on the angle of approach of the plane. For one such indicator, a series of lights appearing from left to right as white, white, red, and red indicates the plane is on the correct glide path, whereas white, white, white, red indicates the plane is too high.

A special area known as a displaced threshold, with arrows down its center, can be used for taxiing and takeoffs, but is not designed to take the impact of repeated landings.

Blast pads painted with yellow chevrons are not intended for normal taxiing, takeoff, and landings.

A series of solid and dashed yellow lines or yellow-laddered lines indicates an area where a taxiing plane should stop ("hold short") until cleared by air traffic control. The intersecting runway is noted in white numbers on a red field.

A top surface course of asphalt or cement anywhere from two to 16 inches (five to 40 cm) thick sits on top of base and subbase courses of crushed stone or gravel. All three layers have a combined thickness of two to five feet (60 to 150 cm).

LANDING

Landing a plane relies on a number of technologies that have changed little over time. Historically, the undercarriage of most planes was comprised of two front wheels and one smaller wheel, or skid, at the tail. Today, the tricycle undercarriage—with one wheel in front and two in the back—is more common.

Landing equipment on modern general aviation planes can be either fixed or retractable. Retractable gear is

As the plane decreases its airspeed before landing, pilots deploy flaps located at the rear of the wings near the fuselage. The flaps produce additional lift at slower speeds by changing the shape of the wing (they are used during takeoff for the same purpose).

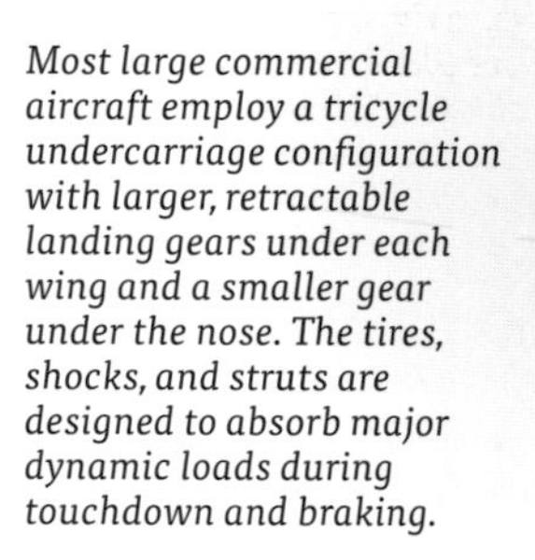

Most large commercial aircraft employ a tricycle undercarriage configuration with larger, retractable landing gears under each wing and a smaller gear under the nose. The tires, shocks, and struts are designed to absorb major dynamic loads during touchdown and braking.

CARRIER LANDINGS

Landing on an aircraft carrier is tricky business. On land, pilots reduce their speed as they approach an airport. At sea, naval pilots approach and upon landing hit full throttle—in case they fail to catch the arresting cables and must take off again.

The technology that helps stop them is known as arresting gear—a series of cables laid across the aircraft landing strip so that they can be caught by a tail hook on the landing aircraft. Spaced about 50 feet (15 m) apart a few inches above the deck, these cables rely on arresting engines on either side of the runway that apply braking force to large reels that hold them. By absorbing impact from the landing plane, the arresting gear can stop a plane within seconds in approximately 300 feet.

Some naval pilots rely on the improved Fresnel lens optical landing system—otherwise known as the meatball, for landing. The meatball is a stack of 12 light cells that together look like a single amber ball. A pilot aligns the meatball with a horizontal row of lights to ensure that the slope of a plane's descent (the glide slope) is correct. If the airplane is too high, the ball will appear above the lights; if it is too low, it will be below them. If it is way too low, red lights will appear.

TAIL HOOK

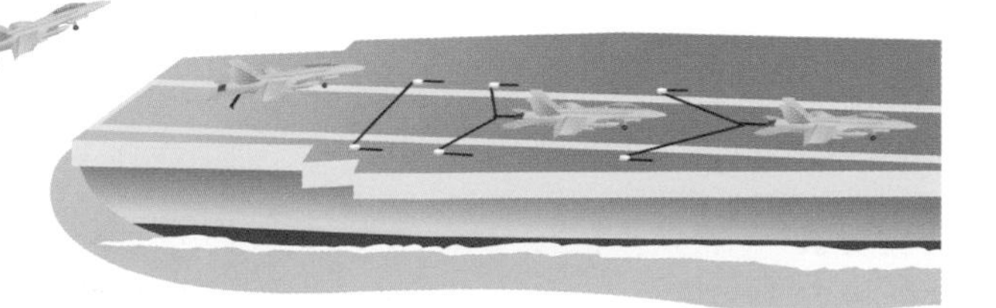

MEATBALL

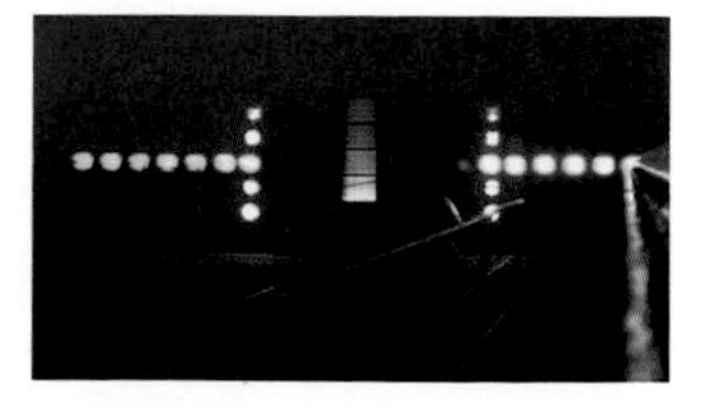

stowed in what are known as wheel wells, and is typically operated hydraulically. Many are designed to lower automatically by gravity if the plane's hydraulics fail.

On large commercial planes, a large number of retractable wheels are present to absorb the plane's weight at landing. Sets of four-wheel bogies can be found at various points along their length. The new Airbus A-380, for example, has a set of four-wheel bogies under each wing as well as two sets of six-wheel bogies under its fuselage.

The plane's landing will be controlled and watched carefully by the tower controllers—both for traffic purposes and in case of mishap. Once the plane has landed, it will exit the runway and be turned over to ground control, which controls all aircraft movements on the air side of the terminal.

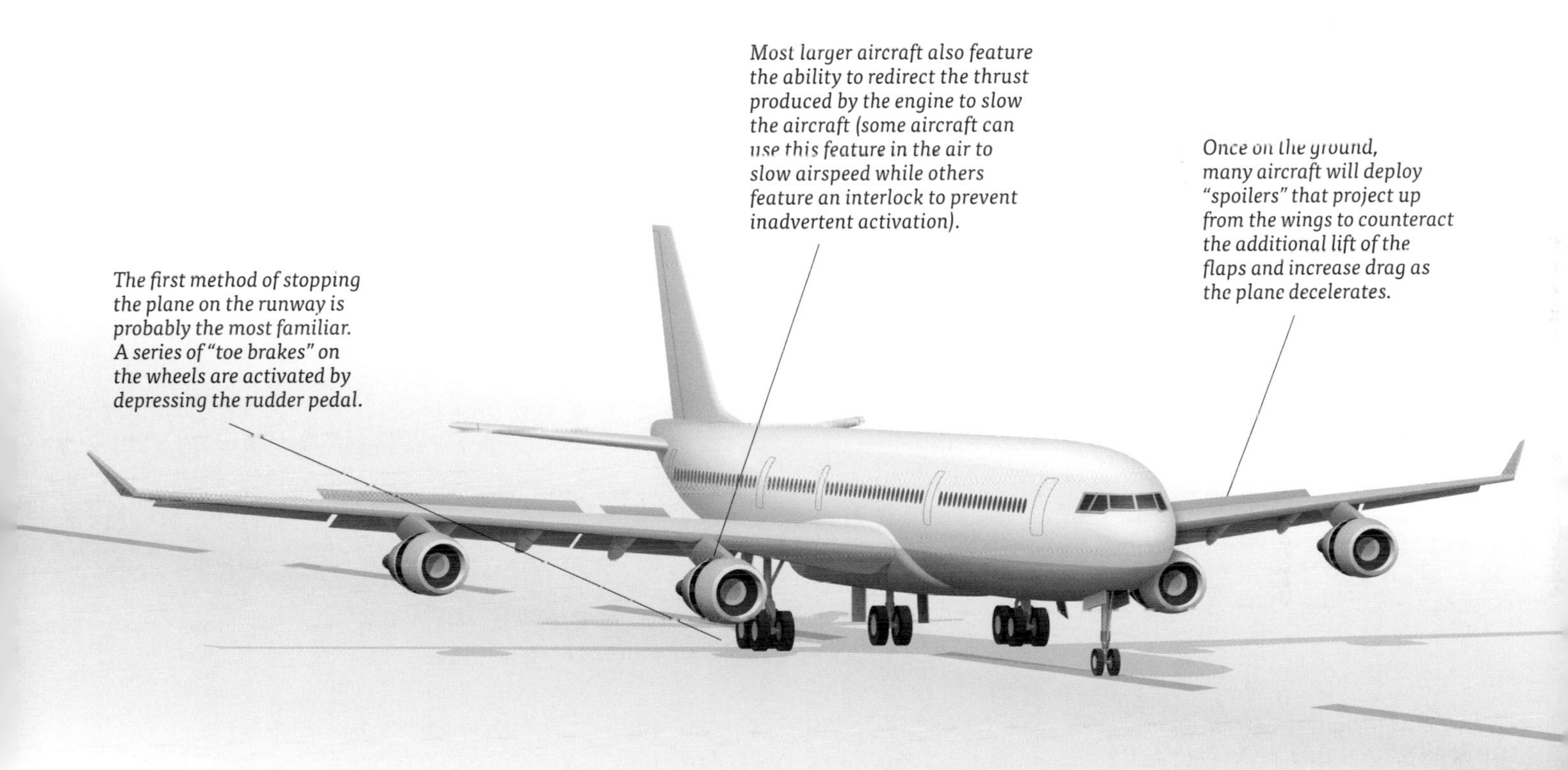

BELLY LANDING

Unfortunately, landing gear doesn't always engage when it's supposed to. In the United States alone, there are on average between 10 and 20 gear-up landings each year—occasionally because pilots simply forget to lower the wheels. Also known as belly landing, these incidents can cause serious damage to the plane but are rarely fatal.

Such incidences are rare, however, on commercial jets. One of the most remarkable landings occurred in 2011, when the captain of a LOT Boeing 767 from Newark executed a near-perfect belly landing at Warsaw's airport. Other than a small fire, which was immediately put out, there was little to show for the incident—other than a very relieved planeload of passengers.

GROUND SERVICES

Once an airplane is emptied of its passengers, baggage from the completed flight needs to be unloaded for passenger pickup. Then, a variety of ground handlers move into action to get the plane ready for its next flight.

First, trolleys with used pallets or containers of food are removed and replaced by the catering company in charge of in-flight meals. Water trucks refill the supply of freshwater on the plane. Lavatory vehicles empty and refill the lavatory tank. Fuel is replenished. If necessary, deicing vehicles stand by to deice the plane before its next flight.

To enable ground service vehicles access to the plane being serviced, a variety of tugs and tractors—often known as transporters—hitch themselves on to the unpowered service vehicles to move them from place to place. When the service vehicles have completed their work and the airplane is ready to take off, vehicles known as push-back tugs will then push it back from the gate or tow it to a safe place to engage the engines. More fuel efficient than the plane's engines, they are also able to provide more precise maneuverability in crowded terminal areas.

CATERING VEHICLE

These vehicles bring prepared food to the aircraft and consist of a lifting system to raise the catering truck body level with the aircraft doors.

LAVATORY VEHICLE

Lavatory service vehicles empty lavatories and then refill them with a mixture of water and a disinfecting concentrate, known as blue juice.

PUSH-BACK TRACTOR / TUGS

Push-back tugs push aircraft away from the gate when it is ready to depart as the plane's engines do not provide the fine control needed in the crowded taxiway and could be damaging in the terminal area.

BAGGAGE LOADERS

Conveyor belt loader vehicles are used to expedite the unloading and loading of baggage and cargo from the cargo hold.

DEICING VEHICLE

These vehicles feature booms with hoses that spray a special mixture of hot water and glycol that melts any ice on the aircraft and prevents ice buildup on the taxiway.

JET BRIDGE

Also known as a Jetway, the bridge is an enclosed, movable connector that extends from an airport terminal gate to an airplane. It can swing radially or extend in length or both.

POWER UNIT

A power unit supplies power to the aircraft parked on the ground while its engines are not running. It may be self-contained or built into the jet bridge.

STORAGE TANKS

Larger, busy airports may have multiple fuel tanks located near the airport, with capacities of well over a million gallons of jet fuel each. These tanks can be connected to the terminal area by a secure underground network of pipes.

FUEL VEHICLE

Aircraft fuel vehicles include both self-contained tankers and hydrant trucks or carts featuring a pump that hooks into a central pipeline network to provide fuel to the plane.

THE ORIGIN OF JET FUEL

The first jet engines didn't need to use aviation fuel—they worked perfectly fine on gasoline. But the dawn of the jet coincided with the beginning of the Second World War—and with rationing of gasoline. As a result, some of the early jet innovators—including Sir Frank Whittle—relied on kerosene, similar to the fuel used in lamps, or on a gasoline-kerosene mix.

Kerosene worked well; indeed, most modern jet fuels continue to be a mix of kerosene and other additives. But as planes demanded more power from their engines, better quality fuels were developed. These included fuels with higher flash points (to avoid accidental ignition) and lower freezing points (to avoid freezing in the upper reaches of the atmosphere).

JET FUEL

Jet fuel, which appears as a clear or straw-colored liquid, is not the same as the fuel you put in your car. It's a form of kerosene that is similar to diesel fuel, but contains a large number of additives that help reduce the risk of icing or explosion at either very low or very high temperatures.

Fuel is loaded onto planes in one of two ways—either directly via a pipeline at a designated fueling area at the airport or by a fuel truck that pulls up to an airplane on the tarmac. On a large jet plane, the fuel enters the aircraft's tanks at high pressure under the wing; on smaller planes, the nozzle is placed above the wing. Often it enters at great speed: a 747 can take on one gallon of fuel per second.

Two types of jet fuel are used globally: Jet A and Jet A-1. They are broadly similar kerosene-based products, with Jet A, widely used in the United States, having a higher freezing point maximum (-53°F or -47°C).
A wide variety of additives can be added to jet fuel, including anti-icing inhibitors to reduce the freezing point of water that separates from the fuel at high altitudes and thus helps prevent the formation of ice crystals that could block various engine valves.

MAINTENANCE

All aircraft require regular maintenance. Depending on the schedule and level of maintenance needed, this work will be done at the gate, at a hangar nearby, or at the airline's maintenance base at a specific airport.

In general, there are four levels of maintenance checks that an airplane will undergo over its lifetime. An A check occurs every 500 to 800 hours at airport gates. Every three to six months, a B-level check will be done in a hangar over the course of one to three days. A more extensive C check, which can take up to two weeks, is done every 15 to 21 months, or less if the airplane has logged an unusually high number of miles. The most extensive check is known as the heavy maintenance visit (HMV), or D check, and involves a total overhaul of the plane. Completed every five or six years, it involves removing the plane from service for up to a month or more and is usually planned years in advance.

Aircraft maintenance is memorialized in the airplane's logbook, which records all data relating to the aircraft's condition, including any unusual events that have occurred during flight, its inspection dates, and the length of time it has been flying. Any damage or repair work is noted here.

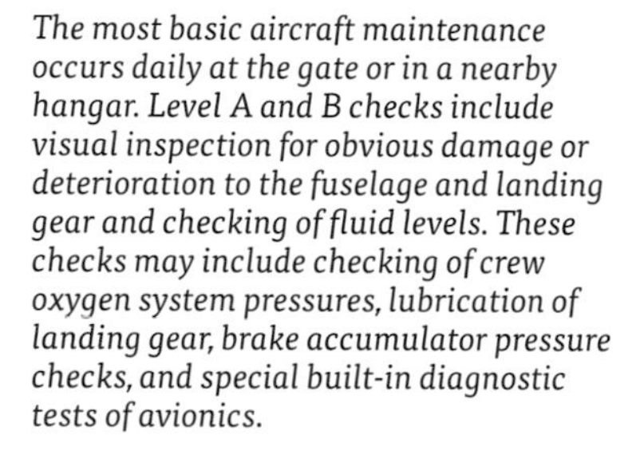

The most basic aircraft maintenance occurs daily at the gate or in a nearby hangar. Level A and B checks include visual inspection for obvious damage or deterioration to the fuselage and landing gear and checking of fluid levels. These checks may include checking of crew oxygen system pressures, lubrication of landing gear, brake accumulator pressure checks, and special built-in diagnostic tests of avionics.

More comprehensive checks (C and D) are accomplished at specialized maintenance facilities and require special tools and equipment. These maintenance periods can take anywhere from days to weeks to complete. During these checks, paint might be removed from the fuselage to test for microscopic cracking of the skin. They may also include structural checks on the wings and landing gear and major engine overhauls and testing.

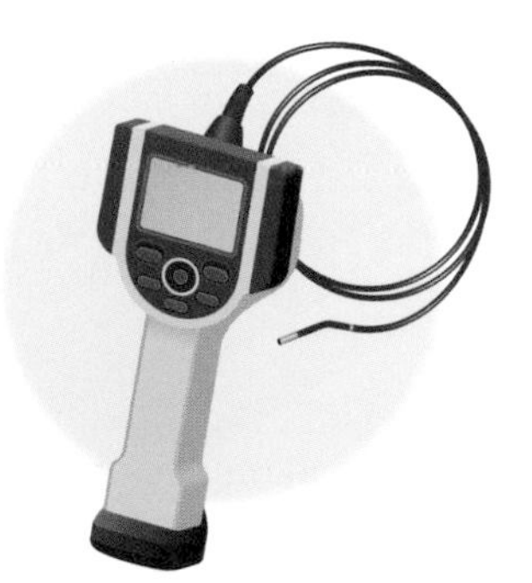

One special tool used for inspecting the internals of aircraft engines is a borescope. The borescope is inserted into the engine components to search for wear, and thermal or mechanical damage. It is typically used to inspect the engine compressor, combustion chamber, and turbine section.

AIR CARGO

Though rarely seen by passengers, air cargo handling at airports is big business. Millions of tons of products move regularly by air—primarily commodities whose travel time needs to be minimized, such as perishable foods and clothing, or those of high value, such as gems and electronics. Many of these products warrant specialized climate-controlled or secure warehouses at airports, where cargo can be sorted, inspected, and stored safely prior to delivery.

The magnitude of this traffic at major international airports is considerable. At JFK Airport in New York, for example, roughly 1,700 acres are devoted to cargo operations. Over four million square feet of built space for handling and storage have been constructed to serve the dozens of cargo carriers that call on the airport.

Cargo flies either in the belly of passenger planes or on special "freighters" designed for the purpose. Many of these are operated by all-cargo airlines, such as FedEx and UPS, although passenger airlines, such as Korean Air and Lufthansa, are also major players in the worldwide cargo market. Large all-cargo airlines generally rely on dedicated, relatively new planes while smaller cargo airlines may refurbish older passenger planes by replacing their windows with solid panels, strengthening their floors, and cutting doors underneath or along the sides of their fuselages.

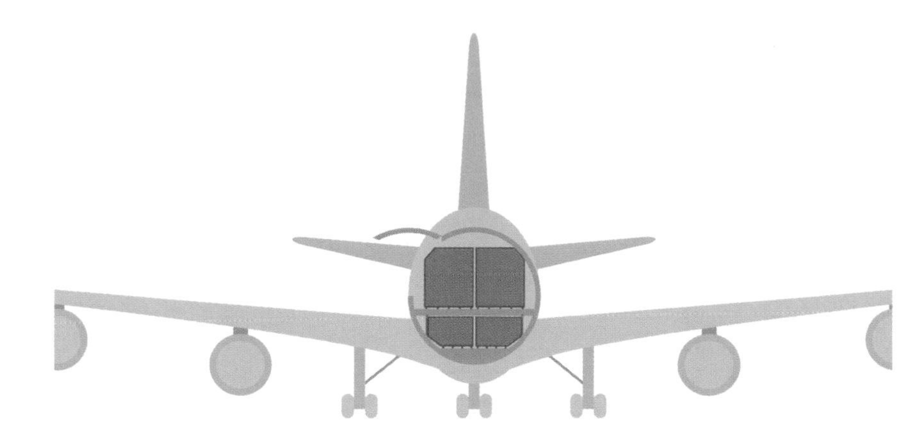

AIR CARGO CONTAINERS

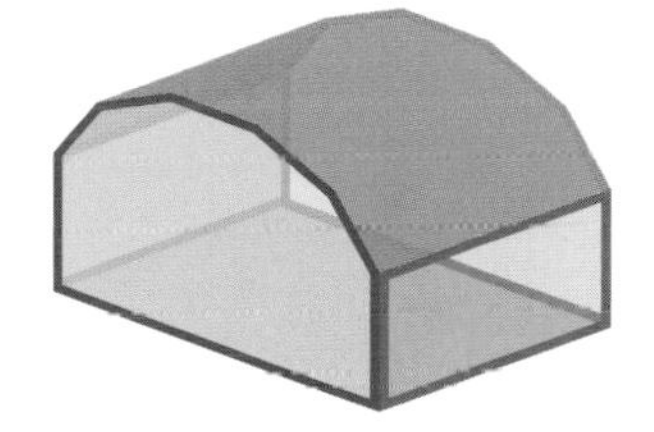

A pallet allows maximum flexibility for bulky containers and can handle up to 420 cubic feet (12 cubic m).

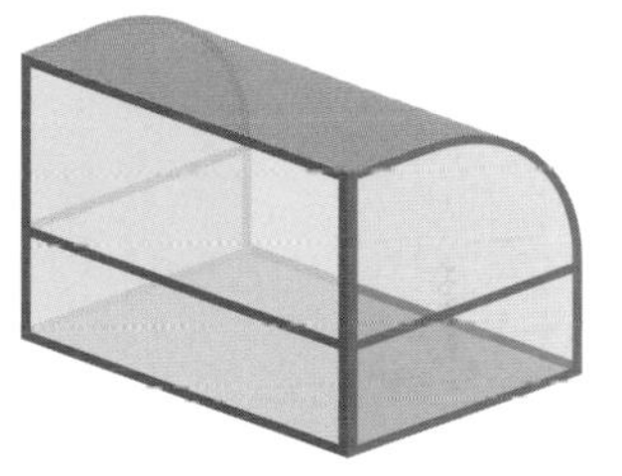

At 96 inches (244 cm) tall, this quarter container is designed to fit inside the turn of the fuselage and can hold 507 cubic feet (14.4 cubic m).

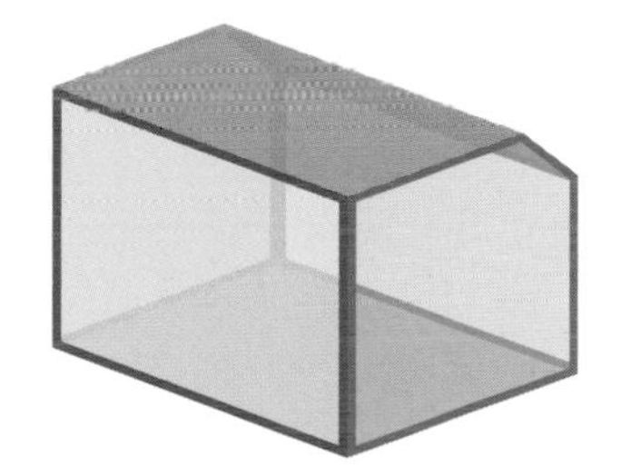

A rectangular container is designed to fit inside the widest parts of the hold and can accommodate over 600 cubic feet (17 cubic m) of cargo.

Designed to fully fit inside the turn of the hold, this container is sized like a pallet and can hold up to 420 cubic feet (12 cubic m) of goods.

HEARTLAND HUB

One of the busiest cargo airports in the world is deep in the American heartland—Memphis, Tennessee. Some 10,000 cargo flights a month operate out of the airport there, serving over 200 countries on six continents. Nearly all of them are operated by one carrier: FedEx.

Within five years of its founding, FedEx opened its hub in Memphis in 1976. Today, the company's 40 miles of conveyor belts on 800 acres operate day and night, 365 days a year, to process 500,000 packages annually. Roughly 11,000 employees undertake the two package sorts each day that provide the backbone of FedEx operations there. So efficient is the hub that the pioneering firm in American aviation—the U.S. Post Office—is now the largest customer of FedEx.

BAGGAGE HANDLING

Billions of pieces of luggage move through the world's airports each year, most of them ending up safely at their intended destinations. Those that don't make up the misplaced luggage that costs the world's airlines $3.8 billion annually.

Baggage handling has three component parts: moving bags from the check-in area to the departure gate, moving bags between gates (for connecting passengers), and sending bags from arrival gates to baggage claim areas. At most airports, this work is done by a combination of people and conveyor belts.

Some of these conveyor systems are small, involving a few handlers loading bags onto an arrivals conveyor belt. Other systems are enormous: Beijing's new airport features 316 check-in counters and relies on a conveyor system that moves bags over a 42-mile (68 km) network to destinations in three separate zones of the airport.

To save on labor costs, certain airports are moving toward more automated baggage handling. Schiphol Airport near Amsterdam is introducing a system that will see 60 percent of all bags being handled by robots unloading luggage containers onto a conveyor belt for sorting and removal to a central processing area. The project is forecast to increase Schiphol's handling capacity from 50 to 70 million baggage items a year.

CHECK-IN

At check-in, the agent matches your itinerary to a baggage tag that contains all the vital information about the journey: origin, destination, and stopover cities, all coded in a unique 10-digit bar code.

SCANNING

After check-in, the bag is scanned by a 360-degree bar code scanner (any bag that isn't read must be manually scanned) and entered into the baggage handling system, which tracks and directs the baggage through to its destination.

TRANSFER UNLOADING

At transfer airports, bags shifting to another plane must repeat the process of being scanned into the baggage handling system and quickly transferred to the next flight.

X-RAY

All checked bags are sent through an X-ray scanner looking for weapons, bombs, and other prohibited items. Computer tomography (CT) scans might also be used for closer inspection on random samples or suspicious baggage. Bomb-sniffing dogs are also used to detect explosives in checked bags. The bag might also be manually inspected by a security agent.

PUSHER

As the bag moves along the conveyor, it may come to a junction where the computer will tell a machine called a pusher to let it pass or push it to another conveyor.

WHAT'S IN A TAG?

What's on a bag tag? Everything but the flight number. The unique bag tag number that travels with a piece of luggage is just an index number that links that bag to a baggage service message (BSM) that contains flight details and passenger information. Reading the code after scanning tells the baggage handling system exactly where that bag should be sent.

The system isn't foolproof, however. The bar code is printed using thermal printers and adhesive paper stock, which helps attach the tag to the luggage but can be hard for laser scanners to interpret. To address misdirected luggage, some airports, including Hong Kong and Las Vegas, are now experimenting with radio-frequency identification chips embedded in the tags.

BUFFERS

During peak handling, buffers are used to store baggage with no immediate priority. Once the peak has subsided, bags are sent from the buffer to their destination aircraft.

Eventually the bag will reach a sorting station for the departing flight. At this point, it will usually be manually loaded onto carts or containers to be transported out to the waiting plane.

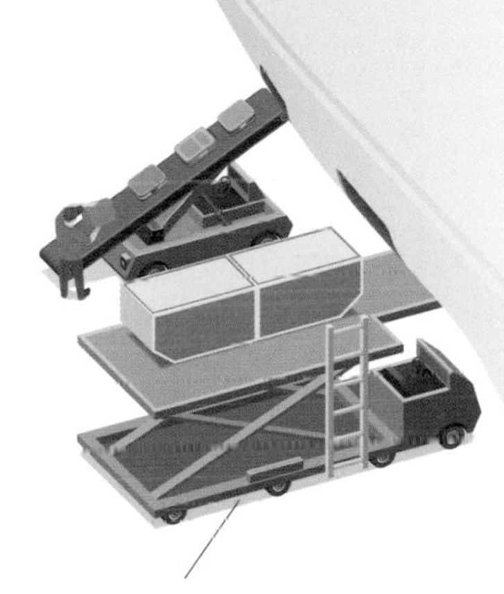

LOADING

While most commercial flights are bulk loaded one by one on a conveyor belt, some cargo may be container loaded on the ground and then placed in the plane.

GARBAGE LINE

In some systems, all baggage that has not gone to the correct belt arrives on the garbage carousel, where it is identified and hand carried to the appropriate aircraft.

CONTROL ROOM

Many airport baggage handling control rooms are manned 24 hours a day and seven days a week. Here, operators monitor the throughput of all departing, transferring, and arriving baggage.

DOLLY & CART

Bags at their destination are placed on a cart and pulled by a tug to a baggage claim conveyor that delivers them to the baggage claim carousel.

MILE-HIGH PROBLEMS

When planners for the new Denver airport announced its integrated automated baggage handling system in 1991, pundits marveled at the idea of thousands of radio-controlled carts (destination-coded vehicles) moving along 26 miles of track without human oversight.

But the system never worked properly in trials: sharp corners made baggage carts tip over and the loader designed to remove bags from the luggage hold sent them flying. The system was still not functional when airport construction was completed in 1993—delaying the airport's opening 16 months.

Things didn't get much better after opening: United, its intended user, tried it for a short period of time then abandoned it entirely in 2005 and reverted to people driving luggage from one gate to another.

SECURITY

Security at U.S. airports dates back to the late 1960s, when airplane hijackings prompted President Richard Nixon to order sky marshals onto planes. Two years later, in 1972, the Federal Aviation Administration in the United States passed the first regulation requiring airlines to search both passengers and their bags. Today, different processing rules apply from one airport and one country to another.

Security screening is now big business across the world. A wide variety of technologies are used to examine passengers boarding planes, from the most basic—which include pat downs and dog sniffing—to the most sophisticated magnetometers and backscatter imaging devices. Very often, new screening technologies are set up in physical spaces not originally designed to accommodate them, leading to crowding and extensive delays.

But airport security is about more than just passenger screening. The primary identification card for international travel remains a passport, though it is an increasingly sophisticated instrument. Today's passports rely on advanced printing technology and special inks and paper to deter counterfeiting and replication. In some countries, they now contain radio frequency identification (RFID), which involves a computer chip embedded in the back cover. The chip contains a digital photograph and other biographical information used in various facial recognition technologies and can be read only by password-enabled chip-readers.

International airports also provide a base for the customs force. Its job is to ensure no contraband materials or weapons—or undeclared goods—are being brought into the country. While technology now allows customs agents to use a machine to "sniff" the air around a package to detect contraband substances, many customs forces still rely on the noses of trained dogs. A familiar sight at luggage carousels, dogs are sometimes brought onto aircraft after passengers deplane to smell for drugs that might have been hiding on a passenger's body.

CHECKED LUGGAGE

Luggage that will be placed in the cargo hold is sent through X-ray machines and chemical sniffers. Suspect baggage might also be sent through CT scanners, similar to a medical CT scan, for a more detailed scan.

IDENTIFICATION

The first line of defense at the airport is identification of travelers. This is most commonly done with a government-issued photo identification such as a driver's license or a passport (if traveling internationally). Travelers are checked against a no-fly database of known or suspected criminals.

CARRY-ON LUGGAGE SCREENING

All carry-on baggage is sent through an X-ray machine for scanning. Different material types absorb the X-rays differently and the machine will show organic, inorganic, and metal objects in different colors to the operator.

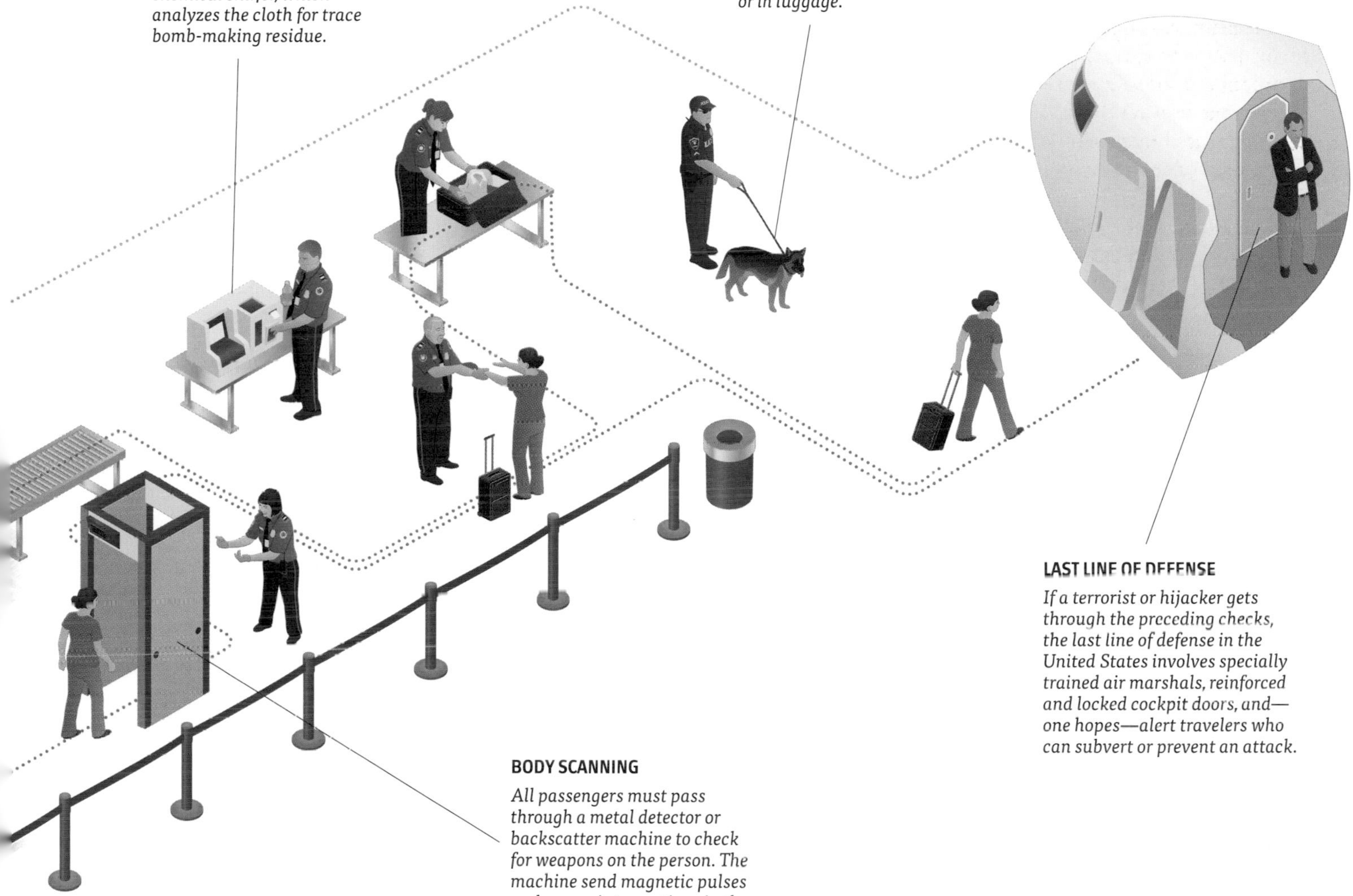

CHEMICAL SNIFFERS

At random intervals or if there is some suspicion, a security officer will swipe a cloth over a person or object and place the cloth on the chemical sniffer, which analyzes the cloth for trace bomb-making residue.

BOMB-SNIFFING DOGS

Specially trained dogs can be used to detect the presence of bomb materials on a person or in luggage.

LAST LINE OF DEFENSE

If a terrorist or hijacker gets through the preceding checks, the last line of defense in the United States involves specially trained air marshals, reinforced and locked cockpit doors, and—one hopes—alert travelers who can subvert or prevent an attack.

BODY SCANNING

All passengers must pass through a metal detector or backscatter machine to check for weapons on the person. The machine send magnetic pulses and senses interruptions in the magnetic field by metal objects.

BACKSCATTER

The latest passenger screening technology used by security officials at airports is also among the most controversial. Backscatter technology relies on a form of ionizing radiation to create a 2-D image of the body—and a fairly graphic one at that. Despite efforts to gray out or distort the images of private body parts, civil liberties groups have protested the violation of privacy that they believe the backscatter image represents.

Unlike traditional X-rays, which penetrate through a target, backscatter detects the radiation that reflects from a target to locate weapons, narcotics, or other contraband items. Used in place of traditional pat-down technologies at airports across the United States, backscatter technology has not been embraced by everyone: many airports in Europe have found the technology ineffective and instead employ more traditional forms of screening.

EMERGENCIES

Airports must be prepared for a wide range of emergencies, not all of them on runways. Risks on the ground range from bombs to spillage of hazardous materials and from natural disasters and weather-related incidents to power failures.

Other serious incidents involving airplanes do not occur on runways but in the air. These incidents might be caused by insufficient lift at takeoff or impaired navigational capability on landing. Often the crash site is outside of the perimeter of the airport and the immediate cause is not always determined.

One cause of aviation emergencies near airports is bird strikes. Between 1990 and 2008, an estimated 100,000 bird strikes were reported by pilots flying over U.S. territory, resulting in 23 fatalities, 209 injuries, and estimated damage of $400 million. Most occur below 300 feet and between July and October, usually on or near an airport.

Two thirds of all bird strikes cause little if any serious damage to aircraft; they may dent a fuselage or shatter a windshield. But those that do can prove serious because of the design of a jet engine and its rotation fan. As a bird strikes a fan blade, it can potentially bend or push it into another blade and cause a series of failures inside that engine.

Other than bird strikes, incidents occurring on airport property are few and far between. Decades ago the biggest

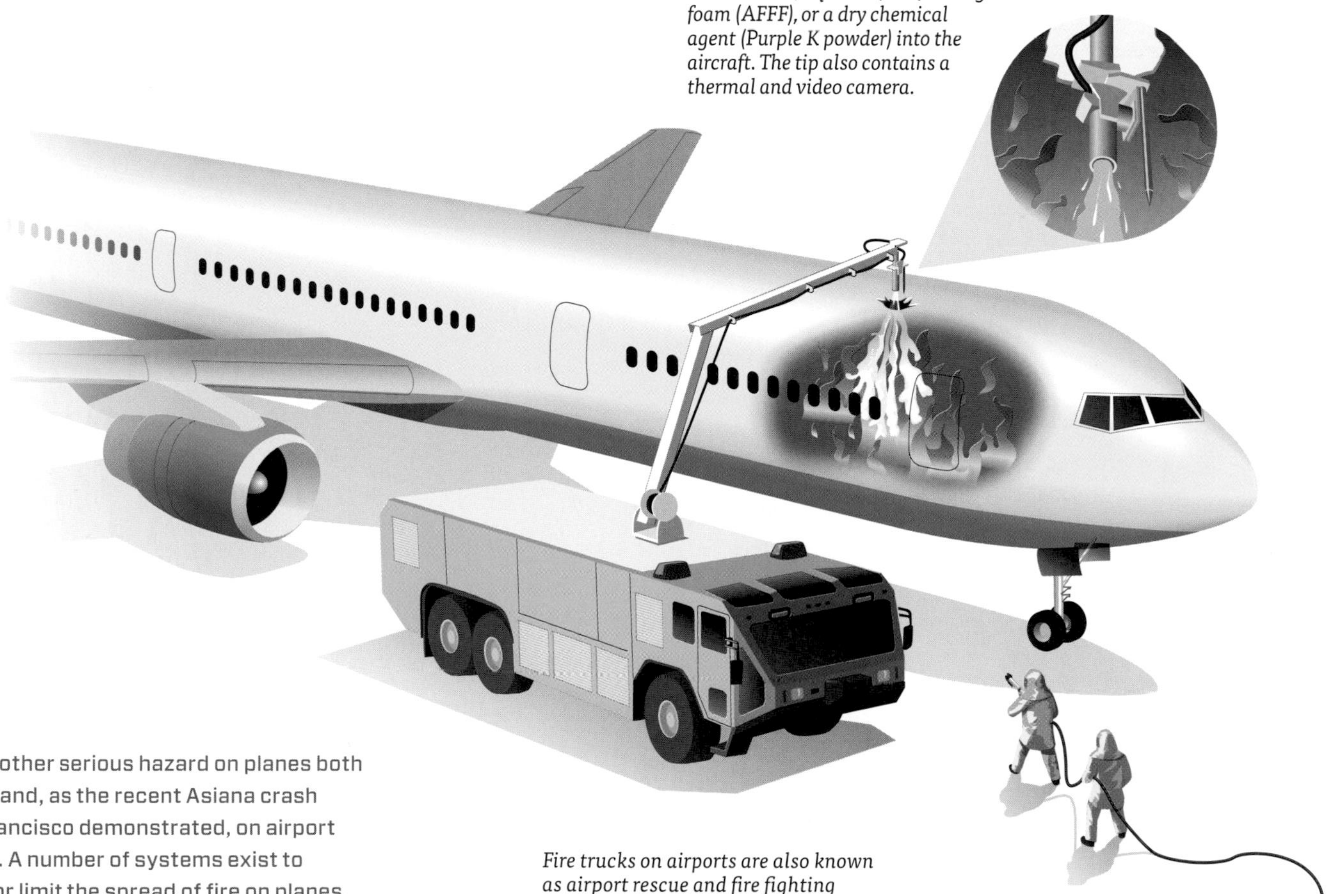

At the end of the extendable turret is the "snozzle," a specially designed nozzle able to penetrate the fuselage of an aircraft and deliver water, aqueous film forming foam (AFFF), or a dry chemical agent (Purple K powder) into the aircraft. The tip also contains a thermal and video camera.

FIRE

Fire is another serious hazard on planes both in the air and, as the recent Asiana crash in San Francisco demonstrated, on airport property. A number of systems exist to prevent or limit the spread of fire on planes. In addition to passive fire protection, which involves the use of fireproof or inflammable materials, more active fire protection components are built into today's aircraft: smoke detectors, temperature sensors, air shutoff mechanisms, and a halon system, involving a chemical spray that can be used to put out a fire in a cargo compartment.

Fire trucks on airports are also known as airport rescue and fire fighting vehicles (ARFF). While they resemble a conventional fire truck, they contain several features unique to airport firefighting, including a forward looking infrared (FLIR) system, a high-reach extendable turret, and the ability to operate at a 30 degree angle.

Firefighters wear proximity suits constructed of aluminized fabric that reflects the heat from high-temperature fires and can protect the wearer in temperatures up to 2,000°F (1,100°C).

fear for airports was hijackings—or rather the landing of planes controlled by hijackers. Security has largely obviated that fear today, and most airport emergency procedures focus primarily on overrun of runways and fires.

Runway overruns are by no means frequent, but do occur. Roughly one airliner a month runs off a runway somewhere in the world—because of excessive speed, late touchdown, or unforeseen conditions on the runway itself. Most are not fatal, though some are: in the United States, roughly one out of every four deaths on airport runways is attributed to a runway miscalculation.

The most frequent emergency at airports involves fire breaking out on a plane on an active runway. The volume of aviation fuel carried by planes means the intensity of aircraft fires can be high, and speed of arrival on the scene by emergency responders is often critical to saving lives. Many of the world's largest airports maintain their own firefighting forces, training regularly on a variety of planelike simulators.

High-expansion foam has long been an important tool for these airport firefighters. Until relatively recently, it was laid on the runway in advance of the arrival of a plane making an emergency landing. Today, due to concerns about the integrity of the airplane braking mechanism, foam is usually reserved for fighting the actual fires themselves once the plane is on the ground.

FLYWAY DANGER

As aviation volumes grow worldwide, the number of bird strikes is increasing. In the United States the numbers of strikes are broadly greatest along traditional bird flyways where, coincidentally, airports are densest.

- *Atlantic flyway*
- *Mississippi flyway*
- *Central flyway*
- *Pacific flyway*

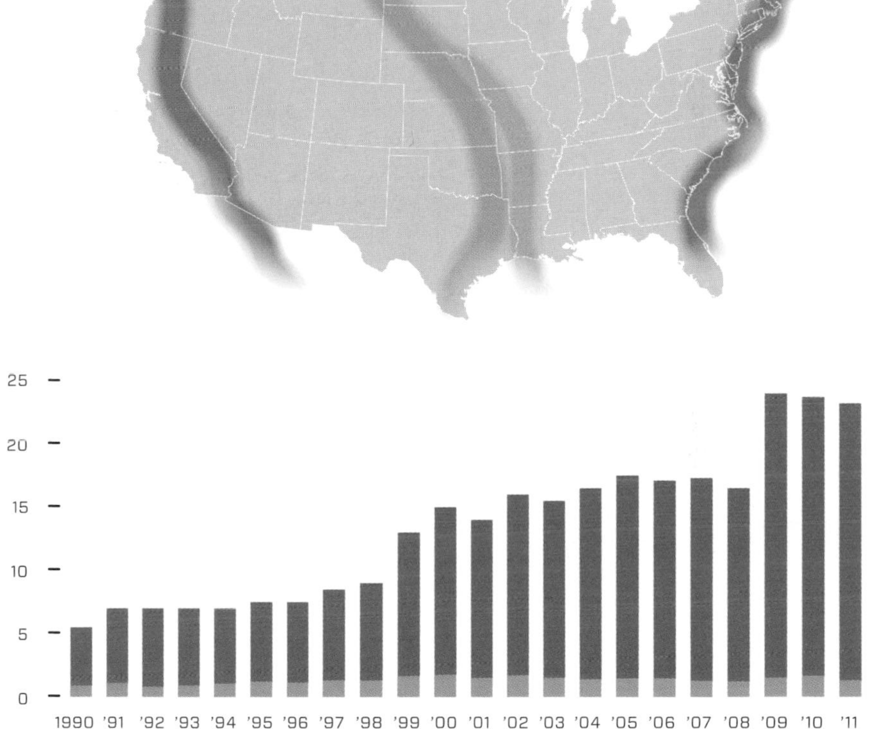

- *Number of bird strikes per 100,000 commercial aircraft movements*
- *Number of damaging strikes per 100,000 commercial aircraft movements*

TAKE 35

Airports deploy highly sophisticated machines to help land planes, screen cargo, and process baggage. Yet even the latest technology can't keep an airport going when nature puts a halt to runway operations.

Normally it is only extreme weather that will stop takeoffs and landings at a major airport. But in the summer of 2011, traffic at New York's JFK was stopped by a different form of nature: six dozen female diamondback terrapins leisurely crossing Runway 4.

Just after 9 a.m. on June 30, 2011, a pilot noticed a large group of turtles heading for a sandy stretch of land sandwiched between the runway and Jamaica Bay—where it appears they intended to lay their eggs. The runway was closed for 35 minutes while the 78 terrapins were removed by truck to a safer destination.

SPACE

No book on transportation would be complete without at least a nod to travel in space. Although much of space travel has not been for the purpose of moving people or goods from one place to another, but rather for purely scientific purposes, transportation innovations have played a critical part in the world's space programs. Indeed, rocket technology can be thought of as an extension of aircraft technology and space travel a branch of aviation.

The earliest rockets were of course not manned: they were born of Chinese gunpowder and used primarily in war. The rockets memorialized in the U.S. national anthem ("the rocket's red glare") were very much of an earlier era: they were signal rockets, carrying a line to shore and signifying a shipwreck.

The father of modern rocketry, Robert H. Goddard, became active in the early twentieth century. He experimented widely with gunpowder, igniting it by electricity and evaluating a variety of nozzles that its exhaust could pass through with force. Working initially with solid fuel, he subsequently focused most of his energies on the development of a liquid-fueled rocket that would rely on a combustion chamber, fed and continually replenished by separate fuel and oxygen lines. His success coupling that with a steam turbine nozzle that efficiently converted the energy of hot gases into forward motion would signify the dawn of modern rocketry.

Even before Goddard's first liquid-fueled rocket appeared in 1926, the Smithsonian Institution's seminal publication *A Method of Reaching Extreme Altitudes* (1919) drew on Goddard's work. But not everyone was convinced: in 1920, the *New York Times* published an editorial poking fun at Goddard's idea of creating and maintaining thrust in a vacuum. Forty-nine years later, a day after Apollo 11 was launched in July 1969, the *Times* published a three-paragraph correction admitting its error.

The world's space programs have largely been limited to nations with sufficient resources to support them—the United States, Russia (previously the Soviet Union), the European Union, and more recently China, Japan, and India. But at the outset there were only two nations involved: the United States and Soviet Union. The paths they chose to pursue their respective ambitions beyond the earth's atmosphere largely defined the nature of space travel over the last half century—and continue to define it today.

At the outset, the two space programs looked very much the same—with individual rockets testing the limits of aerospace technology. A Russian, Yuri Gagarin, and an American, John Glenn, each orbited the earth in a spaceship for the first time within one year of each other in 1961–62. Three years later, in 1965, the first space walk outside of a capsule took place. And by 1969, humans had landed on the moon—meeting the original goal of the American space program.

But the two space programs began to diverge in the 1970s. The Soviet Soyuz program continued to rely

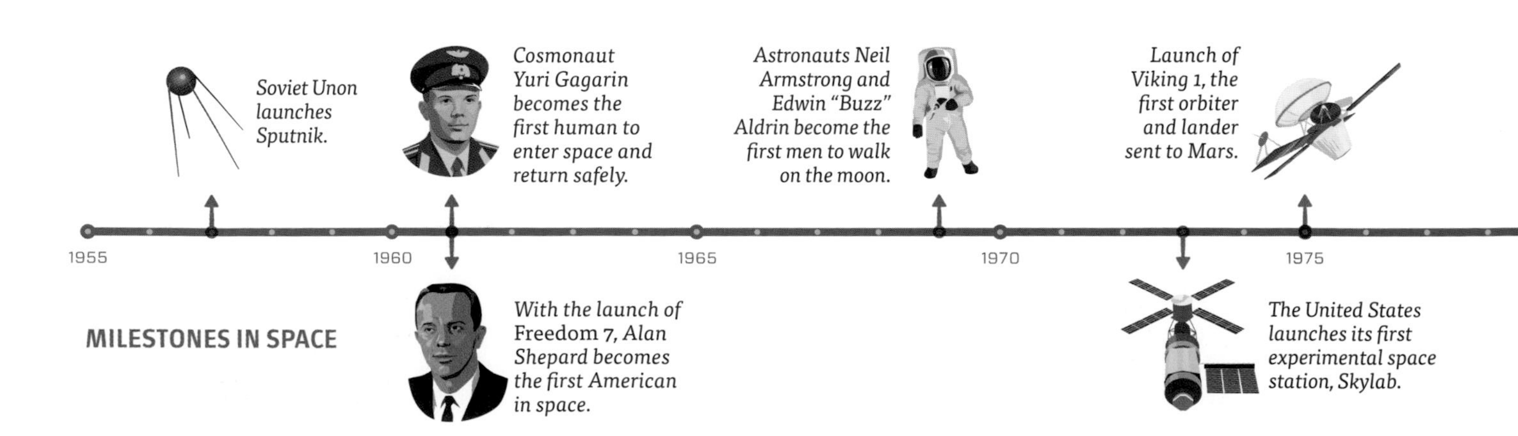

on expendable rockets: after one mission, these rockets were done. The United States instead chose a reusable shuttle system that in aggregate cost roughly twice as much as the Russian system and is now defunct. As one commentator noted, "The Russians won the space race by flying big, dumb boosters."

The United States and Soviet Union would not remain alone in space for long. In 1975, the European Space Agency (ESA) was formed to bring the nations of Europe into space. Its Ariane 5 rockets, first launched in 1988, pushed it to the forefront of the commercial space business—where it stayed for over a decade. Today, the 20 nations of Europe that comprise ESA undertake a variety of forms of scientific research both as an independent group and in conjunction with Russian and American space programs.

The space programs of the United States and Russia, as well as Europe and several other countries, would converge with the historic agreement to create and staff the International Space Station (ISS)—a perennially orbiting space laboratory to be shared by astronauts from different nations. The ninth inhabited space station, it followed in the footsteps of American and Russian efforts such as Skylab and Mir. Since its debut in 1998, dozens of missions have been made to the ISS by astronauts from 15 countries; indeed, it has had astronauts in residence without interruption for the last decade. Notably, it has also played host to the first space tourists—each of whom paid $20 million for a 10-day visit in 2001–2.

ANIMALS IN SPACE

The first animal to achieve success in space was not a dog: it was a fruit fly. Sent 68 miles up into the earth's atmosphere, the fruit flies were ejected from their craft, deployed a fly-sized parachute, and survived the return journey (previous animals, including Albert 11, a rhesus monkey, had died on impact).

The fruit flies did not, however, go into orbit. Laika, the Russian dog, was the first to do that in 1957. Though that brought her worldwide fame, she suffered a rather grim fate—dying somewhere around six hours into the flight from stress and overheating.

For the United States, monkeys turned out to be the right animal for space. Almost three dozen of them flew during the U.S. space program, some multiple times. The launching of Ham, the chimp, in 1961 marked a major milestone. Trained to pull levers to receive rewards, he continued to do so throughout his brief journey—showing that his mental faculties were not at all affected by his time in space.

Since that time, a menagerie of animals has made pilgrimages into space. Monkeys and dogs have been the most frequent travelers there, many of them anesthetized for liftoff. But reptiles and amphibians, including newts, frogs, toads, and snails, have also been launched.

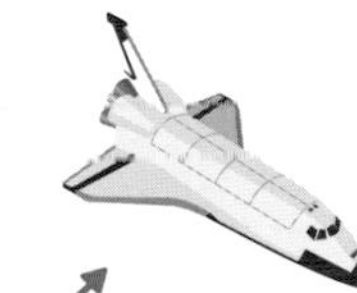
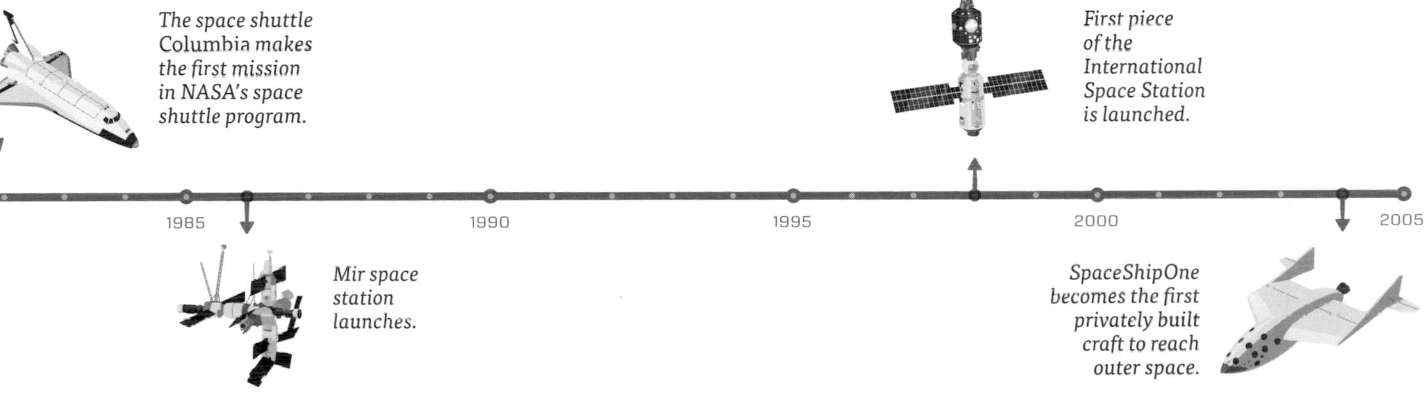

The space shuttle Columbia makes the first mission in NASA's space shuttle program.

Mir space station launches.

First piece of the International Space Station is launched.

SpaceShipOne becomes the first privately built craft to reach outer space.

STAGING A LAUNCH

The earth's rotation flings the rocket into space from west to east—the same direction that the earth rotates. The flinging motion is greatest when the launch site is farthest from the earth's central axis—that is, at or near the equator. Hence launch sites are often found in the tropics (for example, in Florida or Guyana).

However, not all launch sites are in equatorial areas. After the space shuttle's retirement, all manned space missions to the International Space Station were assigned to the Russian Soyuz rocket. Soyuz is launched from Baikonur Cosmodrome in Kazakhstan.

At five minutes into the flight, the second stage completes, separates from the spacecraft, and the third stage takes over.

At approximately two minutes into the flight, the first-stage boosters complete their burn and separate from the spacecraft while the core second stage continues to burn.

At launch, all four first-stage and the central core second-stage boosters of the Soyuz spacecraft ignite using a combination of liquid oxygen and kerosene.

HANS & FRANZ

What travels at a top speed of only two miles per hour and yet burns 125 gallons of fuel per mile? Hans and Franz, the pair of crawler-transporter vehicles used for decades to move spacecraft from their assembly site at Cape Canaveral in Florida to the nearby launch pad.

The largest self-powered land vehicles in the world, each weighs six million pounds and can move 12 million pounds of payload. With a crew of 30, the crawlers lifted the rocket off its assembly pedestal, moved it to the launch site, and lowered it to the launch platform. The journey was slow: it took five hours to cover the five miles between sites.

For 45 years, Hans and Franz serviced almost every U.S. spacecraft—from Apollo and Skylab through the space shuttle program. Just what their future holds now is anyone's guess.

The third stage burns for another four minutes (9 minutes total) before Soyuz separates. At that point, the solar arrays and antenna deploy and spacecraft control is switched to the ground.

LAUNCHING

Two types of rockets are used to launch spacecraft: those propelled by solid fuel and those powered by liquid propellant. Solid-fuel rockets trace their origin back to Chinese, Arab, and Indian warlords who relied on gunpowder to fuel their activities. Though still used in the space program today, the technology is generally found in the solid rocket boosters, used to increase the thrust of a rocket as it lifts off, rather than in the rocket itself.

These solid rocket boosters operate for the first few minutes of flight to give the orbiter an extra push in escaping the earth's atmosphere. Their thrust is enormous—over 3,000,000 pounds (3.3 meganewtons) each at launch. At somewhere around 24 miles (39 km), they separate from the rocket and descend on parachutes to be recovered and reused.

The rocket portion itself contains liquid propellant, which has a higher density than the fuel carried by the boosters and thus requires smaller tanks. Equally important, it can be better controlled, shut down, and tested—so that if a glitch occurs during the preparation for liftoff the launching can be aborted. Liquid fuel propellant can be comprised of a variety of materials, but is typically some sort of fuel (alcohol, gasoline, kerosene, or hydrogen) mixed with liquid oxygen.

The basic principle of rocket launching involves mixing fuel and an oxidizer to facilitate combustion, which produces gas moving at great force through a nozzle. Getting the mix of fuel and oxidizer right determines the amount of thrust available; getting the design of the combustion chamber, nozzles, and ancillary equipment right is critical to ensuring that thrust force is turned successfully into liftoff.

ROCKET SCIENCE

Newton famously explained that for every action there is an equal and opposite reaction. A rocket engine burns fuel and accelerates the hot combustion gases out of the nozzle, producing an opposite reaction known as thrust. It takes a tremendous amount of fuel to lift a rocket to space (the space shuttle orbiter weighed 83 tons empty but required nearly 2,000 tons of fuel for takeoff).

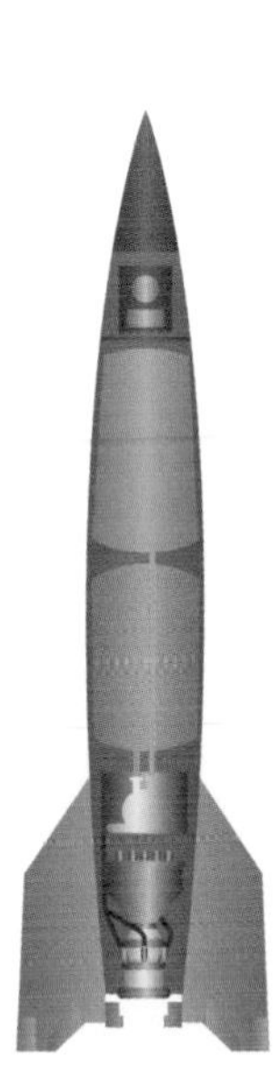

A liquid-propelled rocket combines an oxidizer (usually liquid oxygen) and a fuel (liquid hydrogen, fossil fuels, et cetera) in a combustion chamber to produce the combustion gases that are accelerated out the nozzle. Modern liquid rockets rely upon additional cryogenic cooling systems to keep the combustion chamber and nozzle intact under the extreme temperatures.

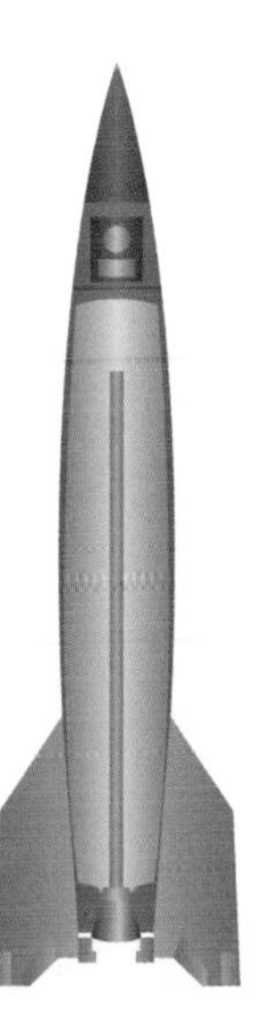

A solid-fuel rocket has a tube down the middle with propellant (a mixture of oxidizer, fuel, and other constituents) loaded around it. At ignition, the fuel burns from the center out and cannot be stopped. Rocket designers will vary the cross section of the center tube from simple circles to multipoint star patterns to affect the burn profile of the propellant after ignition.

The Soyuz TMA spacecraft comprises three elements attached end to end: the orbital module, the descent module and the instrumentation/propulsion module. The crew occupies the central element, the descent module. All but the descent module are jettisoned prior to reentry and they burn up in the atmosphere.

The secondary guidance, navigation, and control system in the descent module allows the crew to maneuver the vehicle using eight hydrogen peroxide thrusters on the vehicle's exterior. At 75 miles (120 km) above the earth, Soyuz's outer surface is heated by the friction caused by the thickening atmosphere.

RETURNING

Bringing a spaceship back into the earth's orbit is a tricky undertaking due to the tremendous heat that plays upon it as it reenters the earth's atmosphere. The same drag, or air resistance, that affects airplanes affects rockets—though to a greater degree because of the enormous speed (30 times the speed of sound) at which they are falling to earth. At those speeds, temperatures on reentry can reach 3,000°F (1,650°C).

In the 1950s, scientists recognized that greater drag on the spacecraft could actually reduce the heat load present during reentry. Positioned correctly, the blunt edge of the spacecraft would compress the air in front of it—effectively creating a cushion of air that pushes the shock wave (and heat) forward and away from the craft. As a result, the Apollo space capsules were designed to rotate so that their blunt end, covered in a heat shield, would lead the fall to earth at a specific angle in order to create this cushion.

To protect spacecraft from a number of risks in space, including heat and debris, space missions have relied on insulating tiles of various materials and depths. Areas likely to be subject to the greatest heat have been covered in tiles made of a composite material that warmed up—and subsequently cooled down—more slowly than others. For additional protection, the most exposed tiles were covered in black glass, which reflects a large amount of the heat back into the atmosphere.

When reentry goes wrong, it doesn't necessarily mean that the spacecraft burns up entirely as it passes into the earth's atmosphere. Over the last 40 years, debris from reentry mishaps has rained down on disparate parts of the world. In 1978, a Soviet craft fell onto the Northwest Territories in Canada; nuclear powered, it left a trail of radioactive debris. Just one year later, Skylab's uncontrolled reentry killed a cow as it fell on a farm in the Australian outback.

REENTRY SHAPES

A significant part of early cold war research on reentry vehicles focused on missiles, particularly the nuclear intercontinental ballistic missiles (ICBMs). As a result, much of this research and technology was highly classified until its adoption for the manned space program.

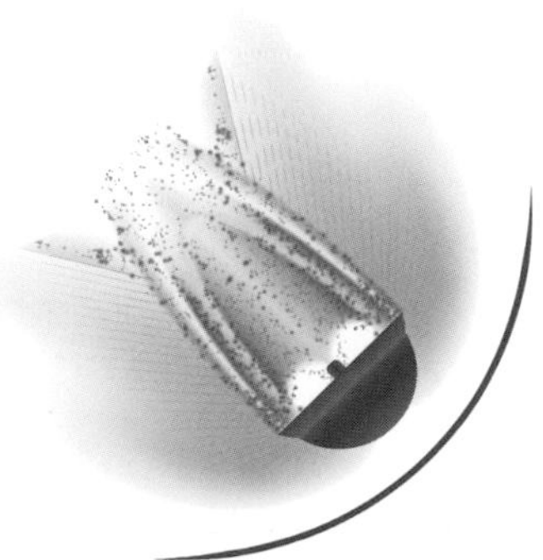

The spherical reentry shape was used in the 1950s and 1960s for reentry shapes on the Soviet Vostok and Voskhod rockets and in Soviet Mars and Venera descent vehicles.

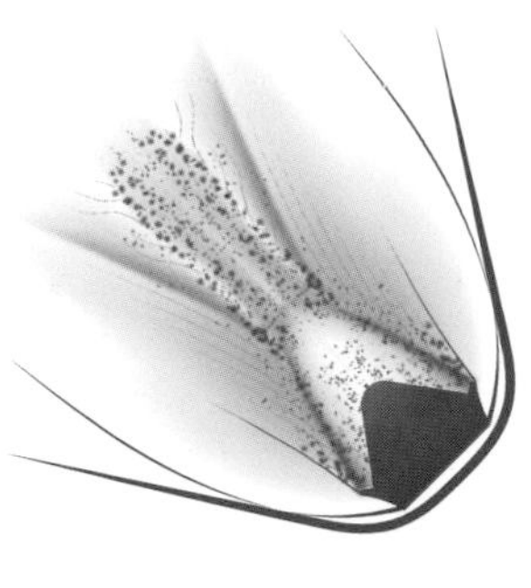

Through the 1950s, missile nose cones under development featured an increasingly blunter forebody heat shield.

The Apollo command module used a spherical section forebody with a cone afterbody.

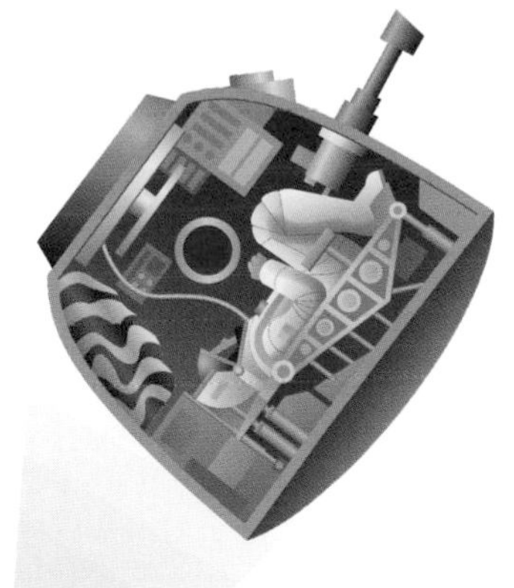

The crew seats are supplied with custom-fitted liners individually molded to fit each person's body. This ensures a tight, comfortable fit when the module lands on the earth.

At 15 minutes before landing, the spacecraft is moving through the sky at over 500 miles per hour (800 km/hr). At this point, four parachutes are deployed to slow the vehicle's rate of descent. Two pilot parachutes are the first to be released, and a drogue chute (measuring 258 square feet/24 square m) slows the rate of descent from 180 miles per hour (285 km/hr).

The main parachute with a surface area of nearly 11,000 square feet (1,000 square m) deploys last. Its harnesses shift the vehicle's attitude to a 30 degree angle relative to the ground, dissipating heat, and then shift it again to a straight vertical descent prior to landing. The main chute slows the vehicle to 16 miles per hour (26 km/hr) and one second before touchdown two sets of three small engines on the bottom of the vehicle fire, slowing the vehicle to soften the landing.

COLUMBIA

On the twentieth anniversary of the first human space flight, in April 1981, the first U.S. space shuttle—*Columbia*—was launched from the Kennedy Space Center in Florida. Its return intact two days later was a cause for celebration. *Columbia*'s success would pave the way for other missions, and other shuttle craft—including *Challenger*, *Discovery*, *Atlantis*, and *Endeavour*.

Columbia would fly 28 missions, 27 of them successfully. On its last mission, however, in February 2003, it broke up on reentry to the earth's atmosphere over Louisiana—just 16 minutes shy of its scheduled landing time. A hole in the shuttle's left wing, apparently formed by a strike from a piece of insulating foam from a fuel tank, was found to be the culprit: the hole had allowed the hot gases of reentry to penetrate the wing and undermine the craft's structural integrity. Seven astronauts were lost in the disaster.

MISSION CONTROL

While only the finest pilots, engineers, and aviators are selected to travel in space, much of the work in guiding manned space travel happens on the ground. And many of those who are ultimately lucky enough to serve as astronauts cut their teeth first on the less glamorous, behind-the-scenes, mission control work necessary to support a successful mission to space.

For the U.S. space program, the nerve center has historically been the Johnson Space Center, located between Galveston and Houston, Texas. Separate control rooms have directed the activities of specific parts of the space program—the International Space Station, the shuttle flights, et cetera. Teams of 50 or so controllers work around the clock to ensure the safety of specific missions, each sitting at a specific desk and responsible for some unique component of the project's mission.

During the early days of the Russian space program, the Russian space agency maintained a mission control center in Kaliningrad, similarly with separate control rooms for the individual space programs (Mir, Soyuz, et cetera). Today, TsUP, the mission control center for the Russian federal space agency, is located in Korolev—just north of Moscow, in Russia. With three separate control rooms and dozens of desks, it is integrated with an international network of mission control centers—in places like Toulouse, France, and Beijing, China.

Though much of the journey of a spacecraft is preprogrammed, mission control specialists are nevertheless critical to the mission. They monitor the flight path of the craft—including liaising with other monitoring entities to ensure that the rockets don't encounter space junk and adjusting the rocket's trajectory if necessary. Mission control specialists also evaluate meteorological conditions on a real-time basis and make adjustments to accommodate changes in landing location resulting from unforeseen weather conditions.

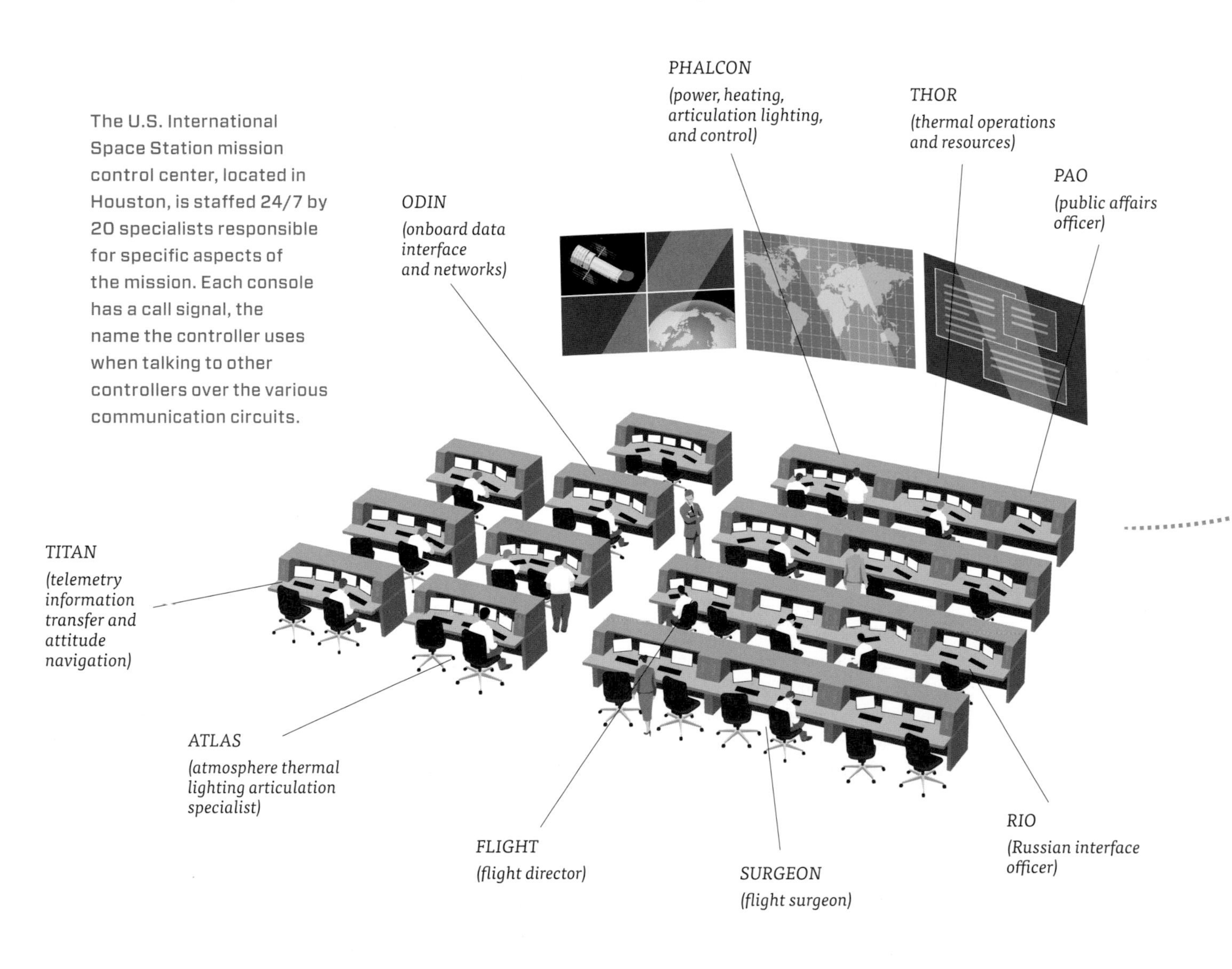

The U.S. International Space Station mission control center, located in Houston, is staffed 24/7 by 20 specialists responsible for specific aspects of the mission. Each console has a call signal, the name the controller uses when talking to other controllers over the various communication circuits.

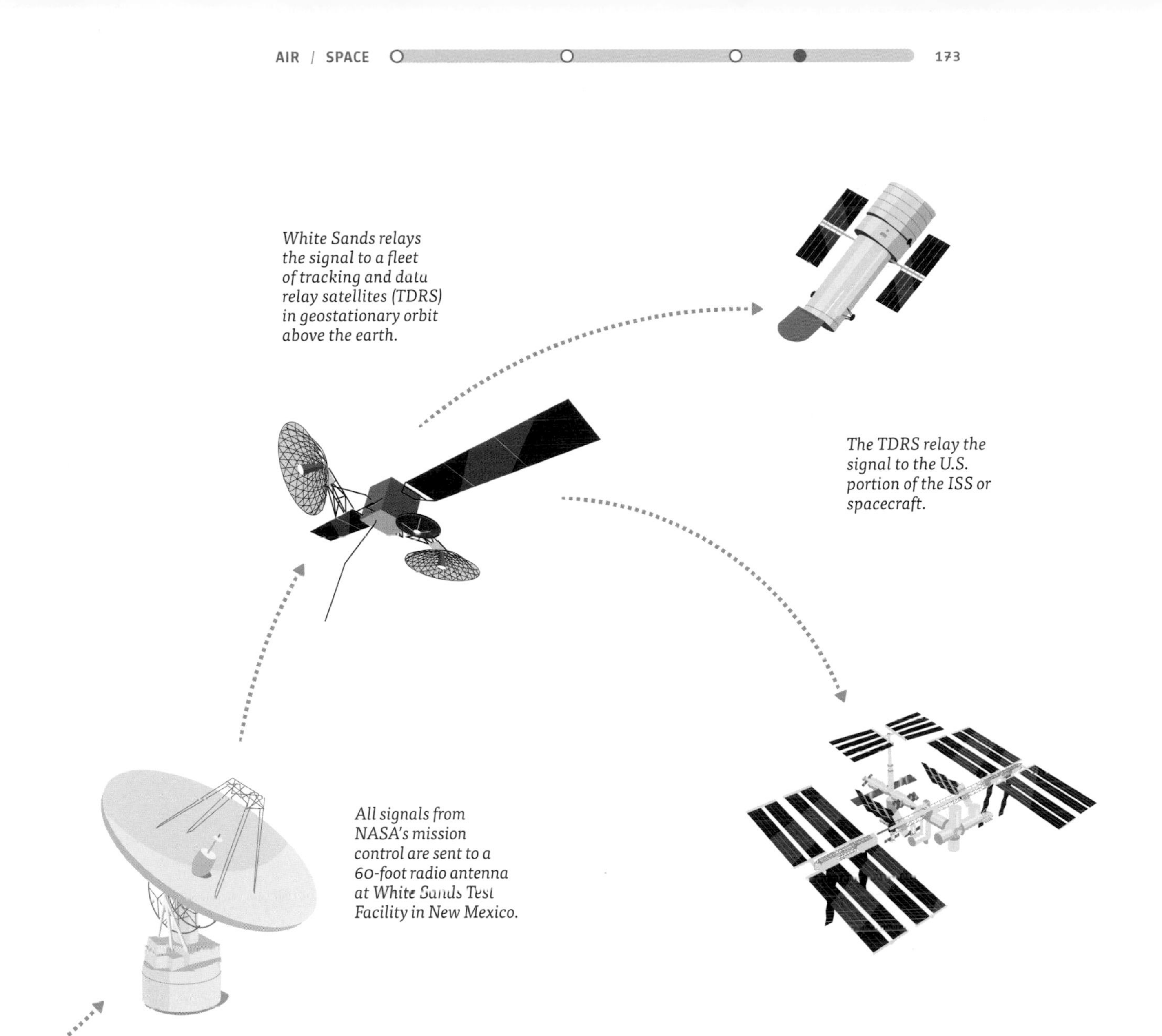

COMMUNICATION

Originally, communication between earth and spacecraft was done by a series of ground tracking stations. For the past 25 years, however, that communication has been done by a satellite system. In the United States, that system is known as the tracking and data relay satellite system (TDRSS). Eight satellites are in service today, providing communication channels for numerous mission control centers tracking a variety of craft aloft—including balloons, aircraft, the Hubble space telescope, and the International Space Station.

Tracking vehicles in low-earth orbit is a lot easier than tracking them in outer space. For that, NASA relies on what's known as the deep space network (DSN)—an international network of giant antennas operated out of three separate locations around the world that covers interplanetary expeditions and the activities of the wider solar system.

Deep-space radio spectrums are reserved for items far from earth—roughly 1.2 million miles (2 million km) and more away. As the earth rotates, distant missions are visible for a longer period of time than satellites in low-earth orbit; as a result, only three large antennas on earth are necessary to cover the entirety of space. They are located roughly 120 degrees apart from one another: one is located in the Mojave Desert of Arizona, another is near Canberra, Australia, and the third is near Madrid, Spain.

These antennas rely on a microwave radio system and are incredibly sophisticated pieces of technology. They feature powerful microwave transmitters, very sensitive receivers, large reflecting areas (which enable them to pick up even weak signals), and a variety of features to screen out background noise radiated naturally by a variety of objects in the atmosphere.

SATELLITES

The transportation industry has used satellites for years, originally in aviation (where they conveyed data between ground control and planes) but also on the roads in the form of GPS systems.Today, there are an estimated 3,500 of them in a variety of locations in space. Only about 1,000 of them are functional; the remainder are retired.

Satellites are launched by rockets at terrific speed to allow them to escape earth's gravitational force. Timing is everything: finding the right spot in the sky largely depends on getting the satellite launch window right. Once in space, the booster rocket is separated from the satellite and the satellite continues in orbit, relying from that point forward on its own onboard rockets to set its course and make minor course corrections.

Satellites rely on fuel for propulsion and orientation and solar panels for energy to operate the electronic devices on board. But neither panels nor fuel supply lasts forever: the average life span of a satellite is less than 15 years. Though the owner is obligated to preserve sufficient fuel to lift the satellite to a quieter part of space or send it into a lower orbit (to burn up in the earth's atmosphere), in practice this doesn't always happen.

As a result, satellite management includes careful monitoring of space trajectories. The crash of two satellites in 2009 highlighted the need to monitor and manage satellite paths via radio signals from earth, where an informal communications network allows owners to pinpoint conflicts and shift satellite locations to avoid collisions.

ANATOMY OF A SATELLITE

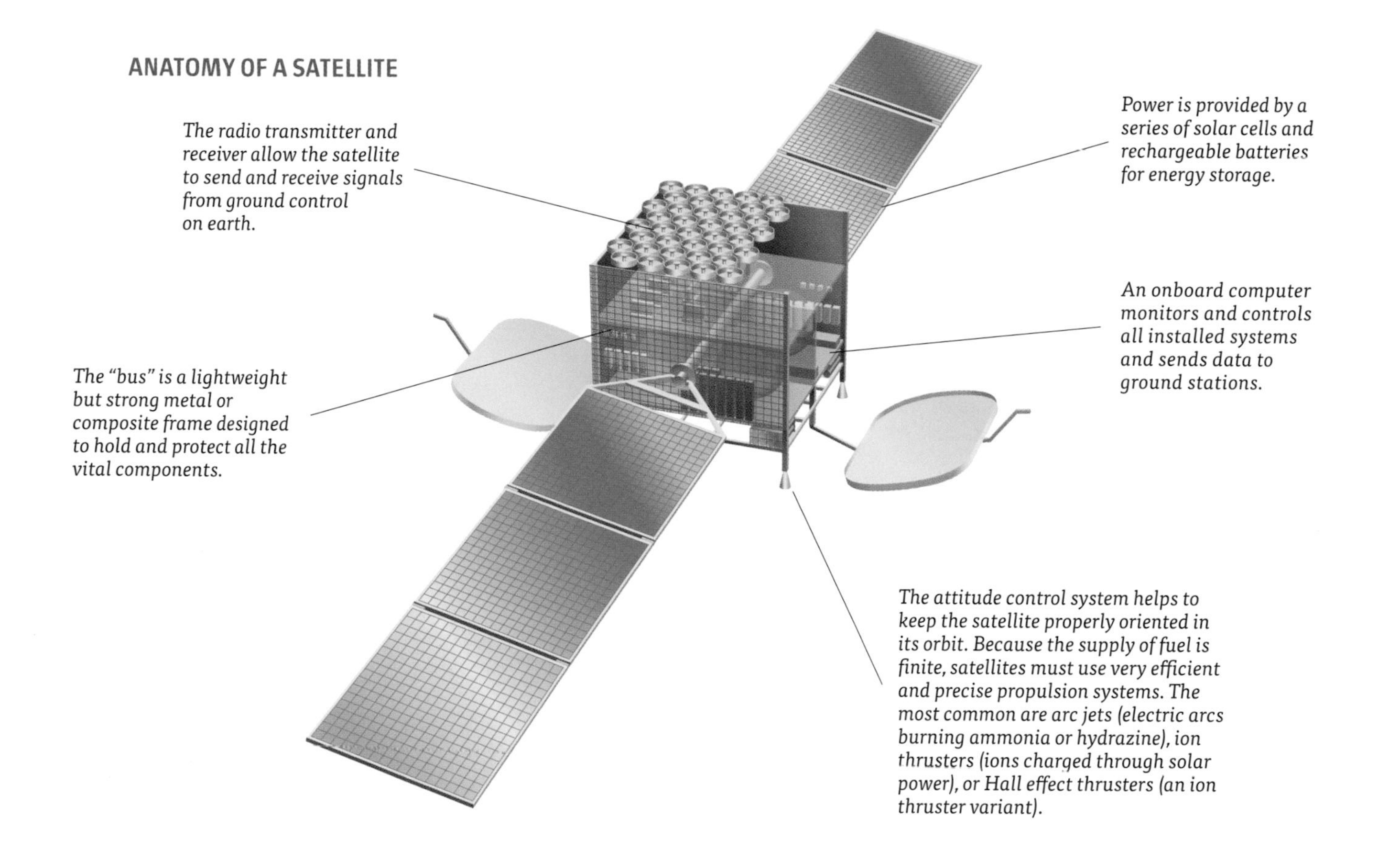

OUT OF THIS WORLD

The closer a satellite is to earth, the faster it has to travel to remain in orbit. Lower orbit satellites provide less coverage of the earth's surface and are asynchronous—passing over the same place on earth at different times of the day. High-earth orbit ones provide more coverage and may be geosynchronous, or able to maintain their position over a fixed point on earth.

GPS satellite
12,650 miles
(20,200 km)

Geosynchronous orbit
22,245 miles
(38,500 km)

International
Space Station
250 miles
(400 km)

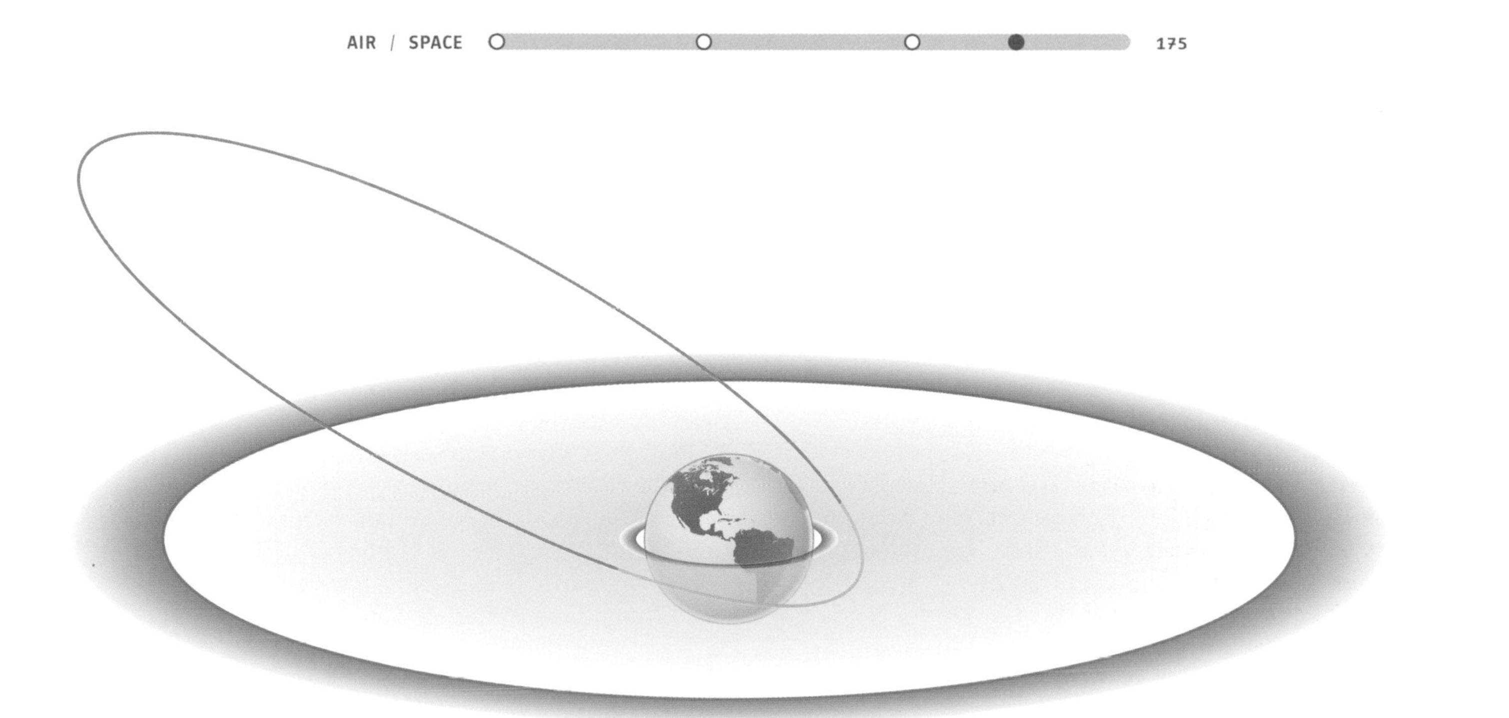

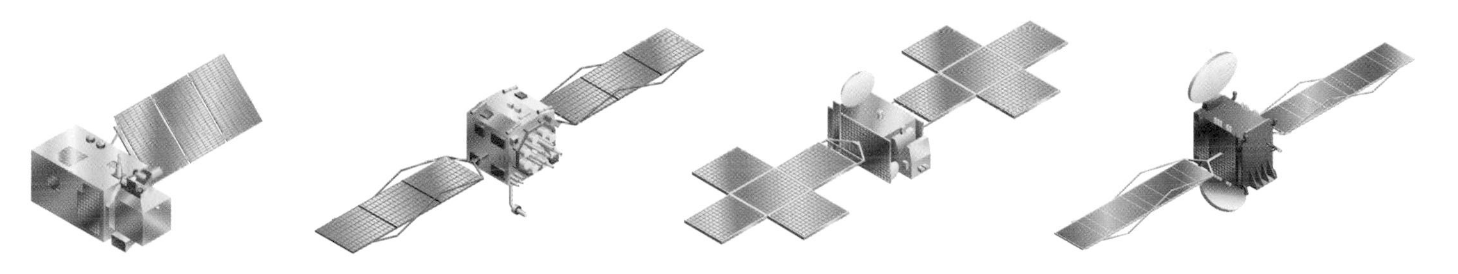

LOW-EARTH ORBIT

Operating at an average altitude of 480 miles (800 km), the European Space Agency's Sentinel-2 pair of satellites will be used for earth observation. Low-earth orbit satellites circle the earth anywhere from 50 to 2,000 miles (80 to 3,200 km) above the earth's surface. In order to stay at this altitude, they must transit at high speeds and require sophisticated ground tracking stations.

MIDDLE-EARTH ORBIT

Twenty-seven GPS satellites fly in medium-earth orbit (MEO) at an altitude of approximately 12,500 miles (20,200 km). Each satellite circles the earth twice a day. The Block IIF is the newest version in orbit. It has a longer life expectancy and is more accurate (atomic clocks accurate to eight billionth of a second per day) than its predecessors.

ELLIPTICAL ORBIT

The Sirius radio networks use a fleet of three satellites. These satellites travel in a 24-hour elliptical orbit around the earth with their closest point of approach to the earth (perigree) at 15,000 miles (24,000 km) and their maximum distance from earth (apogee) at 29,300 miles (47,000 km). Because of their elliptical orbit, only two of the three satellites broadcast at any given time.

HIGH-EARTH ORBIT

The DirectTV network uses a series of Boeing 601 geosynchronous satellites. The geosynchronous orbit is a thin ring above the equator, at an altitude of 22,245 miles (38,500 km). At this height, an orbiting satellite revolves around the earth once in exactly 24 hours, making it appear to hover over a single point on earth. Orbital slots in the geosynchronous ring are highly sought after and are allocated to countries by the International Telecommunications Union, a branch of the United Nations.

Low-earth orbit (up to 2,000 km)

Middle-earth orbit (2,000–38,500 km)

High-earth orbit (>38,500 km)

Moon 240,000 miles (384,000 km)

INTERNATIONAL SPACE STATION

The International Space Station (ISS) orbits the earth today as the focal point of human exploration of space. Launched in 1998 as a partnership between the United States, Europe, Russia, Japan, and Canada, it has been continuously occupied since the year 2000 by one or another of two dozen different crews.

The station is enormous—as long as a football field and four and five times the size of the spacelabs Mir and Skylab, respectively. Composed of two distinct portions, the Russian orbital segment (ROS) and the United States orbital segment (USOS), it weighs approximately one million pounds. Most energy needs are met by solar panels that sit on its skin and are serviced by missions from partner countries.

The station functions as a modern-day science laboratory. Its visitors undertake experiments—on humans, plants, and other materials—to see how they function in space over extended periods of time. It also serves as a data-gathering station for space itself, including the study of meteors and other astrophysical phenomena. Data gleaned from the station will serve as the foundation for the design of future forays into space.

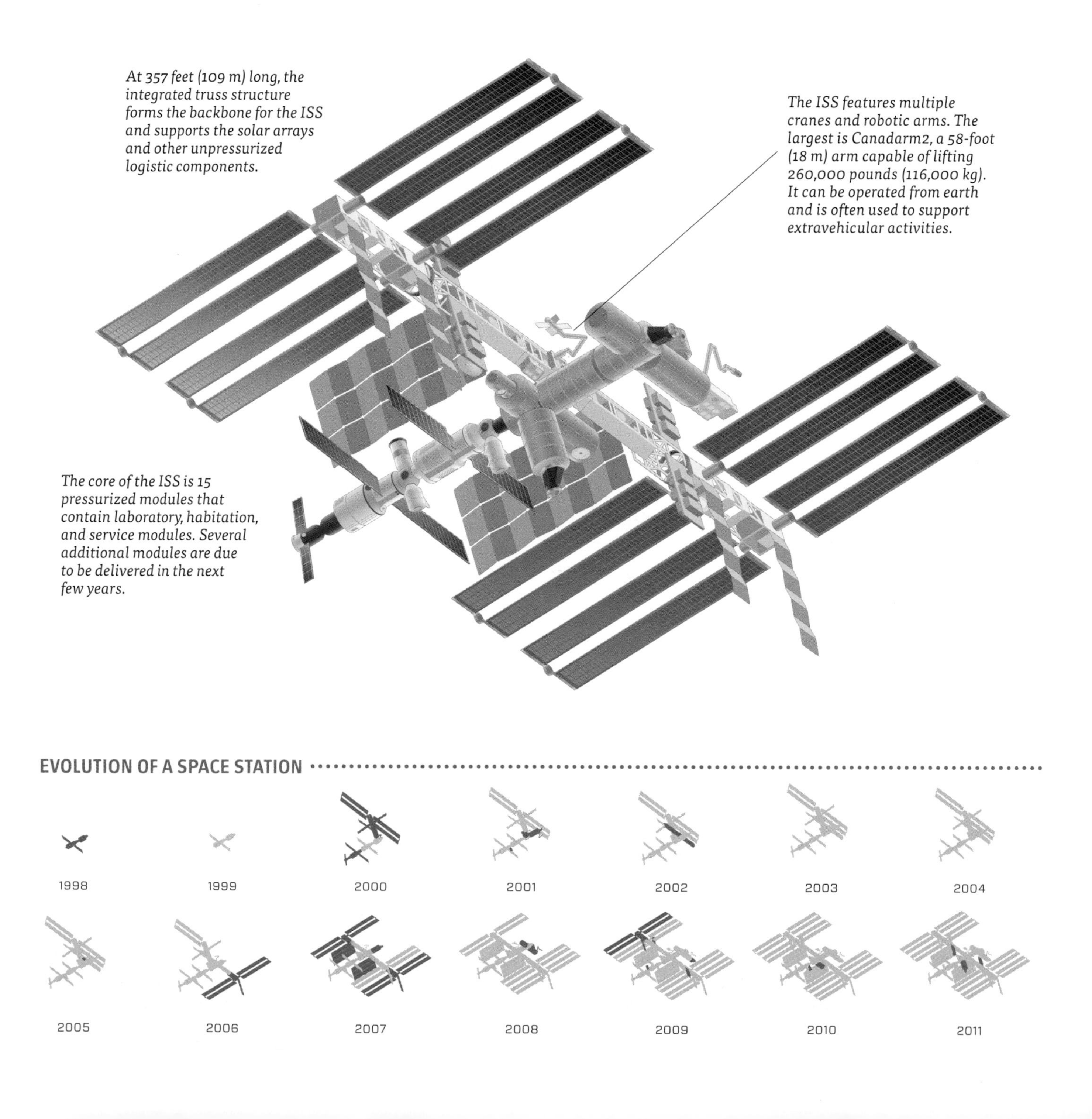

At 357 feet (109 m) long, the integrated truss structure forms the backbone for the ISS and supports the solar arrays and other unpressurized logistic components.

The ISS features multiple cranes and robotic arms. The largest is Canadarm2, a 58-foot (18 m) arm capable of lifting 260,000 pounds (116,000 kg). It can be operated from earth and is often used to support extravehicular activities.

The core of the ISS is 15 pressurized modules that contain laboratory, habitation, and service modules. Several additional modules are due to be delivered in the next few years.

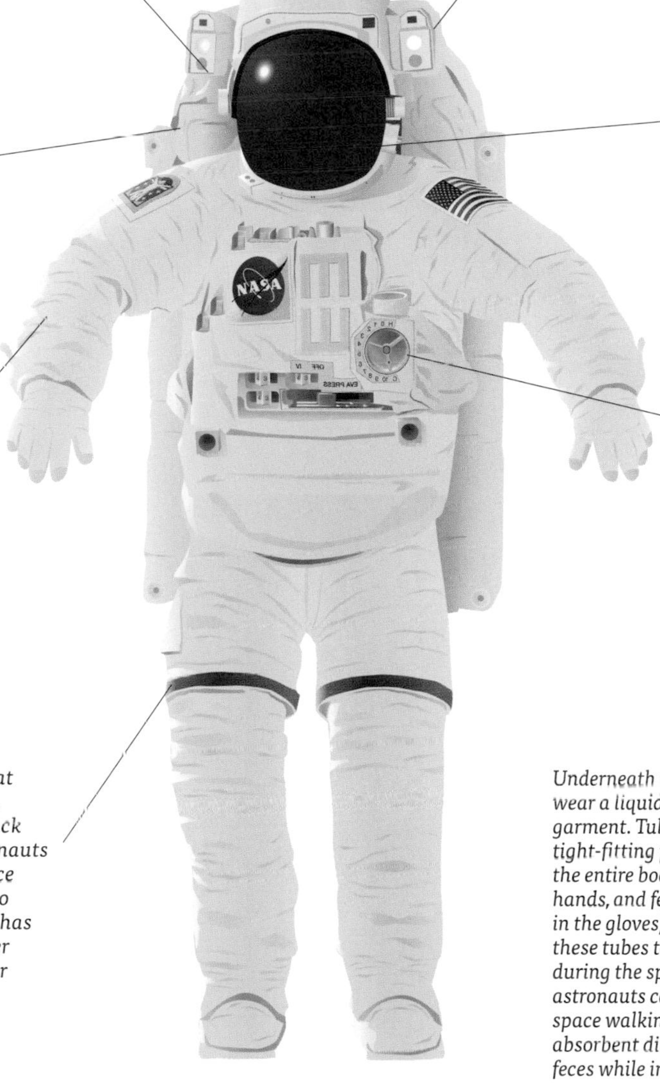

The primary life support system (PLSS) is a backpack worn by the astronaut containing the oxygen tanks, carbon dioxide (CO2) scrubbers/filters, cooling water, radio, electrical power, ventilating fans, and warning systems. Oxygen flows into the suit behind the astronaut's head. Exhaust leaves the suit at the feet and elbows.

While astronauts are usually tethered to the space station, the suits are outfitted with a jet-pack "life jacket," called a simplified aid for EVA rescue (SAFER), to allow an accidentally untethered astronaut to fly back to the station in an emergency.

The PLSS provides up to seven hours of oxygen supply and carbon dioxide removal. The secondary oxygen pack is an emergency oxygen supply that fits below the PLSS on the backpack frame and has oxygen for 30 minutes, sufficient time to get a crew member back inside the spacecraft.

Tinted glass goggles protect the wearer from direct light and glare.

Multiple layers of strong fabric, often finished with a reflective coating, protect the wearer from space debris and radiation. Special joints in the suit facilitate movement at elbows and knees, though movement in a suit so thick remains fairly limited.

On the front of the extravehicular mobility unit (EMU), there is a control module that allows the astronaut to control suit cooling and pressure, the PLSS, and communications with the mother spacecraft and other astronauts. The astronauts cannot see the front of the module while wearing the space suit, so they wear a wrist mirror on their sleeve. All labels are written backward (but backward is forward in a mirror).

White space suits reflect heat so that the astronaut doesn't get too warm. White is also visible against the black background of space, so other astronauts can easily see the space walker. Since space-walking astronauts always go out in pairs, one of the suits always has red stripes in four places so the other astronauts can tell one space walker from the other.

Underneath the space suit, astronauts wear a liquid cooling and ventilation garment. Tubes are woven into this tight-fitting piece of clothing that covers the entire body except for the head, hands, and feet (heaters are provided in the gloves). Water flows through these tubes to keep the astronaut cool during the space walk. Space-walking astronauts can spend up to seven hours space walking, so they wear a large absorbent diaper to collect urine and feces while in the space suit.

SPACE WALKS

Space walks, or what is sometimes termed extravehicular activity, have been a part of space travel since 1965 and today form an essential part of the regular repair work required aboard the permanent space station situated high above earth. The astronaut may remain in an airlock adjacent to the vehicle or float either tethered or untethered outside of it.

But regardless of which form it takes, no space walk is possible without a sophisticated space suit. For a start, there's no oxygen in the vacuum of space and any space walker would quickly fall unconscious. Extremes of temperature could boil or freeze the fluids in the body. At the same time, the astronaut would be exposed to fast-moving dust particles or other orbiting space junk as well as to cosmic radiation.

Space suits protect the astronaut from all of these risks. Temperature controlled and pressurized to protect the human body, they provide oxygen and remove carbon dioxide and sweat from inside of the suit. Both oxygen provision and carbon dioxide removal are handled by canisters either worn on a backpack in space or located within the craft and tied to the astronaut by an umbilical-like tether.

LIFE IN SPACE

Life in space requires that basic human needs be met there—like air, water, and food. The last is perhaps the most straightforward as all the food in space originates on the earth. But air and water must be produced on the vessel itself, and how they are manufactured is one of the lesser-known wonders of space travel.

For life to continue in space, oxygen must be supplied and carbon dioxide removed from the space capsule's atmosphere. Carbon dioxide is removed from spaceship air by a machine and is vented into space. Other gases emitted by people's breath or sweat, like ammonia and acetone, must also be removed from the air.

Oxygen is produced in space by a variety of methods, but most commonly by electrolysis—the process of splitting water molecules into their component atomic parts: hydrogen and oxygen. The energy needed to split the molecules is provided by solar cells on the outside of the spacecraft, and the water is provided either by supply vessels or produced in space.

Because of the length of today's space missions, much of the water used in space today is recycled water. On the International Space Station, water from fuel cells, urine, humidity in the air, and personal hygiene is reclaimed and filtered for reuse.

Unlike air and water, food is produced on the ground and stowed on the vessel. Generally it is arranged in the order of use and packed in clear, disposable containers. Once selected, much of the food needs to be rehydrated in space with water made aboard the vessel. After consumption, the packaging is bagged for disposal. For long stays in space, such as at the ISS, this means being placed in a vehicle to be jettisoned and burned up on reentry to the earth's atmosphere.

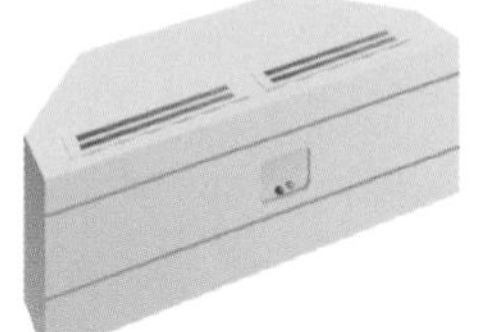

The temperature and humidity control subsystem helps maintain the habitable atmosphere in the ISS by removing heat and humidity and circulating cool, dry air. The moisture is recycled.

CONDENSATE

URINE

Urine recovery is accomplished by a low-pressure rotating (to compensate for zero gravity) vacuum-distillation process. Separated water is mixed with other recovered water and sent through waste processing filtration and purification.

The water recovery system provides clean water by reclaiming wastewater from crew urine, cabin humidity condensate, and waste from space walks. The recovered water is passed through filters that remove gases and solids. Organic contaminants and microorganisms are removed by a high-temperature catalytic reactor before it is tested for purity and sent to storage.

TOILET TALK

While the loss of gravity changes many forces on the body, it does little to change the normal bodily process of producing waste—and getting rid of it remains a part of the astronaut's daily routine. Today's astronauts can belt themselves into an onboard commode and rely on an appliance that fits around the genital area to remove both liquid and solid waste by air suction—rather than water—from their bodies. Liquid waste is recycled on board, but solid waste is compacted, canned, and—on a long mission like the ISS—ultimately jettisoned into space.

Perhaps the trickiest thing relating to bodily functions in space is stomach upsets—particularly nausea or vomiting. One notable incident happened on Apollo 8, when Frank Borman became ill and threw up in the cabin. Without gravity, his vomit floated around the cabin in pieces—forcing the other astronauts to don emergency masks (to prevent themselves from vomiting). After the first initial splats landed on Borman's colleagues, they worked for several hours rounding up the bits still floating or lodged on the cabin walls.

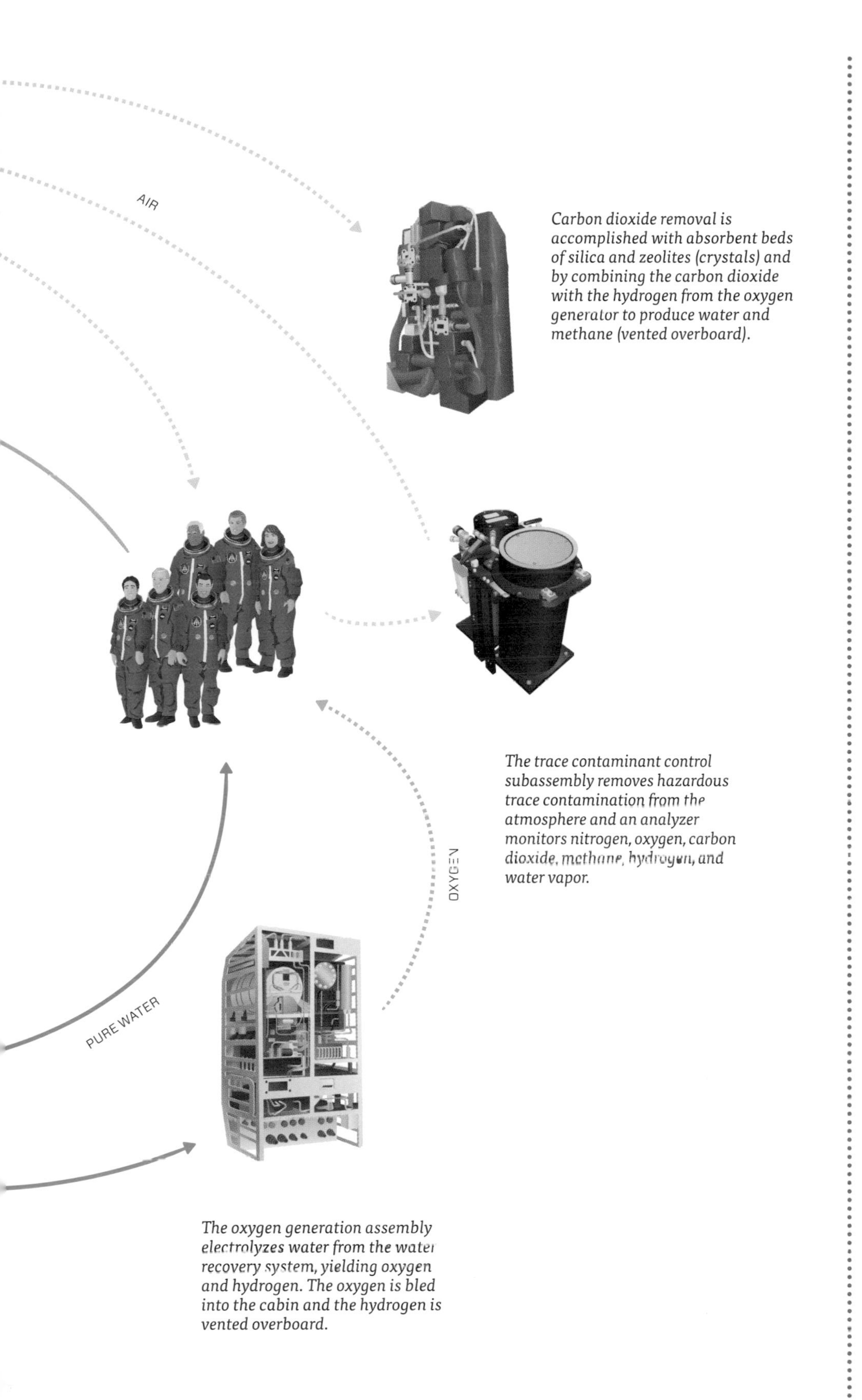

Carbon dioxide removal is accomplished with absorbent beds of silica and zeolites (crystals) and by combining the carbon dioxide with the hydrogen from the oxygen generator to produce water and methane (vented overboard).

The trace contaminant control subassembly removes hazardous trace contamination from the atmosphere and an analyzer monitors nitrogen, oxygen, carbon dioxide, methane, hydrogen, and water vapor.

The oxygen generation assembly electrolyzes water from the water recovery system, yielding oxygen and hydrogen. The oxygen is bled into the cabin and the hydrogen is vented overboard.

FOOD IN SPACE

Space food is no longer freeze-dried powders and liquids in tubes. The International Space Station features food warmers, a refrigerator, and a water dispenser. Yet space food remains largely preprocessed—much of it rehydratable (water is removed to save weight during launch) or thermostabilized and irradiated (typically packed in pouches).

It's not just food that's different in space. Knives and forks stick to food trays with magnets so they don't float away. Salt and pepper, suspended in water and oil, respectively, come in dropper bottles in a liquid form. Water is room temperature or warmer. And bread is banned because of the crumbs it creates: only tortillas will do.

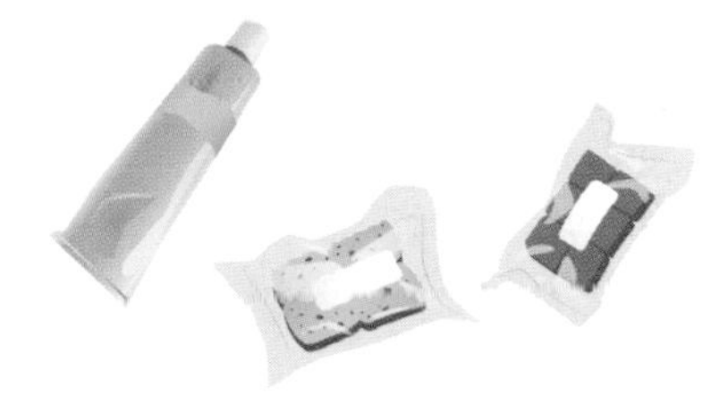

Food from Mercury mission

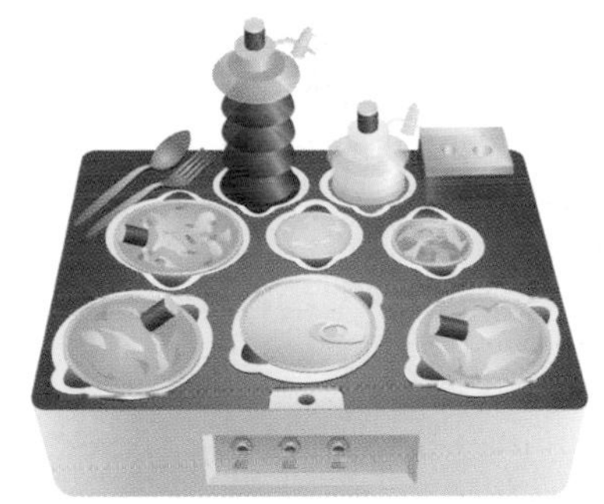

Skylab food tray

Space shuttle food tray

DANGERS IN SPACE

Space travel is not only unnatural for human beings—it's also dangerous. The dangers go far beyond the mechanical hazards inherent in flying at dizzying speeds in a small metal machine tens of thousands of miles above earth. They include the various impacts of prolonged weightlessness (or exposure to what is called microgravity) and the hazards associated with extended exposure to radiation from cosmic rays.

The effects of weightlessness tend to vary by astronaut. Nearly half experience space sickness or space adaptation syndrome: disorientation, vertigo, congestion, headaches, and/or nausea. These conditions stem from the fact that the brain is no longer getting physical signals from the body as to which direction is up. Eventually, the brain adapts to the change and begins to rely on visual, rather than physical, signals to understand position and motion.

The longer-term impacts of weightlessness, however, can be more serious. Space anemia, brittle bones, and a decline in muscle strength can all result from the absence of gravity. Aerobic capacity can also be compromised. As a result, astronauts try to maintain a careful regimen of exercise while in space.

Galactic cosmic rays, mostly in the form of gamma rays, are the most common source of exposure. They are hazardous, fairly continual, and hard to shield against.

Like cosmic rays, solar winds are also continual but less hazardous as most of the radiation comes from low-energy protons and electrons.

Solar flares are intermittent events but are extremely hazardous, with substantial radiation coming from high-energy subatomic particles. With often no more than 30 minutes to react after an event is detected, protecting astronauts is difficult.

RADIATION

Radiation in space comes primarily from cosmic rays and from protons coming off solar flares. While neither represents a danger on earth (they're deflected by the earth's magnetic field and absorbed into its atmosphere), they're far more formidable in space. Radiation from these sources can damage the immune system, leading to mutations in cells and activation of dormant viruses.

The coupling of a weakened immune system and more potent viruses is a dangerous one anywhere, but particularly so within the confines of a space capsule. To protect against it, a variety of shielding mechanisms are used. Most are made of aluminum, although a range of new, lighter materials—able to protect against radiation and deflect what are known as micrometeoroids—are being tested.

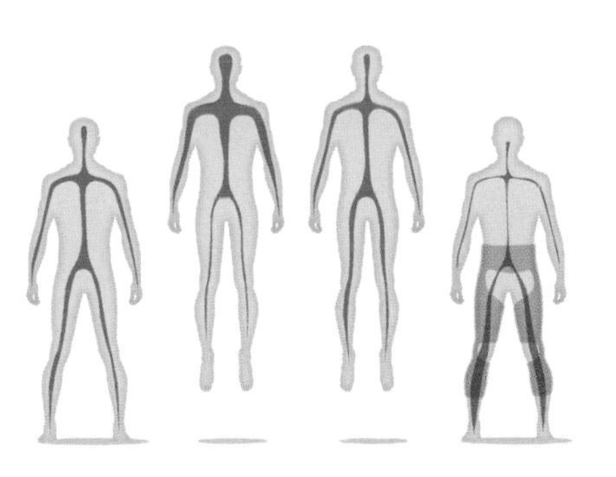

G SUIT

Developed for World War II fighter pilots, G suits, or antigravity suits, offer protection from the loss of blood to the astronaut's brain during takeoff and reentry. They take their name from the G force that measures rates of acceleration; the normal G force for a human being is somewhere between 3Gs and 5Gs.

A space suit can add roughly 1G in tolerance by preventing blood from pooling in the lower part of the body. Form-fitting trousers sport inflatable sacs that fill with water and air, and press against the astronaut's abdomen and legs to increase the blood flow from lower to upper parts of the body.

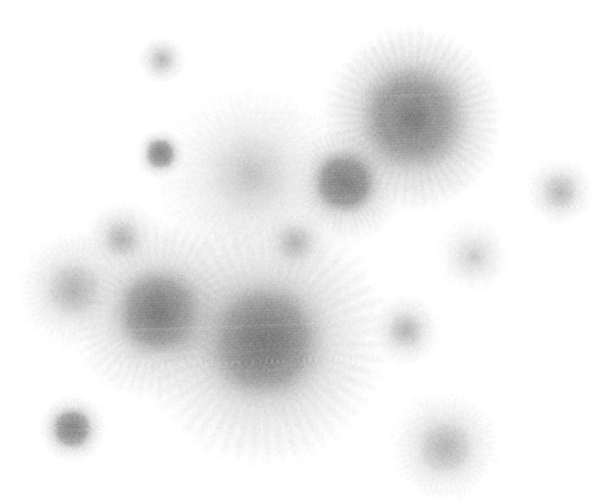

SPACE FUNGUS

Believe it or not, fungus has been found in space—apparently inside of a window on the Mir space station. Though normally found on living or once-living organisms, fungus is now known to successfully proliferate on even never-living objects, including plastic, glass, and metal.

The main concern about fungus in space is the potential impact on the well-being of the astronauts—whose capacity to fight fungal infections is undoubtedly lower in space than it is on earth. But scientists are also concerned about potential mutations in space—and what the impact of these mutations might be if they returned to earth with the astronauts.

FIRE

Astronauts experiencing fire in space are trained to turn off the ventilation system as well as cut power to the affected area of the spacecraft. After that, they do what we do on earth: use a fire extinguisher.

Fire in space, however, is unusual: it burns blue in microgravity and takes on a different shape. On earth, gravity is responsible for pulling colder, denser air to the base of a flame to displace the hotter air. A flame in space with no air flow will not be teardrop in shape, nor will it flicker in the air: its flame is dome shaped and weak.

- *10–15 sec: loss of conciousness*
- *15–90 sec: swelling of body, seizures, and paralysis*
- *30–60 sec: blood pressure drops and blood flow stops*
- *60–90 sec: survival is possible through recompression*

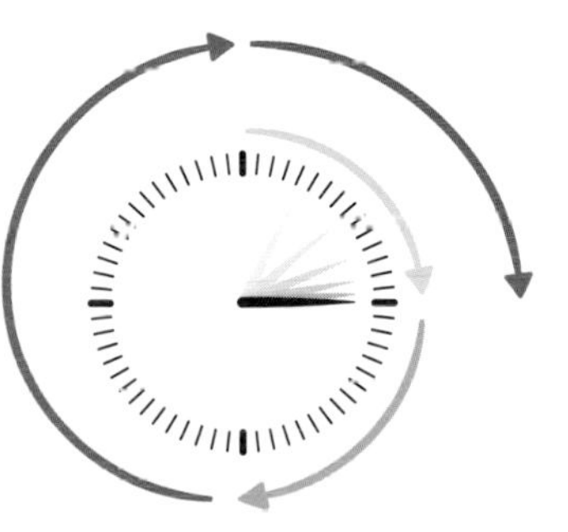

DECOMPRESSION

Decompression is a constant threat in the vacuum of space. Water in the soft tissues of the body vaporizes and causes swelling and bubbles in the veins, resulting in blockages of blood flowing to critical organs. Among other things, this leads to hypoxia, or a lack of oxygen to the body's organs, rendering humans unconscious in short order.

Death from exposure to space is not necessarily immediate and not outwardly violent or explosive like in the movies. Gases might be expelled from the bowels and stomach, leading to projectile vomiting, urination, or defecation; forced breath holding may lead to ruptured lungs. Seizures and swelling may occur (tests on dogs showed that their tongues swelled and were coated in ice).

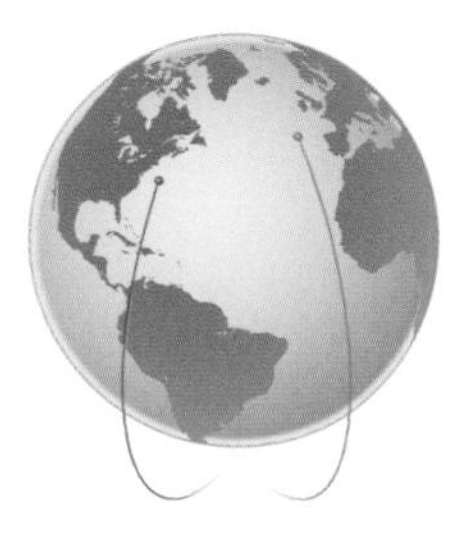

In low-earth orbit, where most space debris is found, pieces travel up to 23,000 miles per hour (36,000 km/hr).

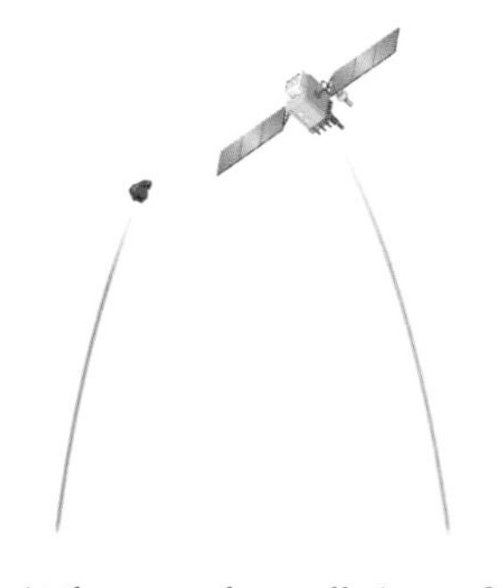

At these speeds, small pieces of debris no bigger than a golf ball can cause significant damage to a satellite or even to the ISS.

A collision between larger objects can release thousands of pieces in orbit.

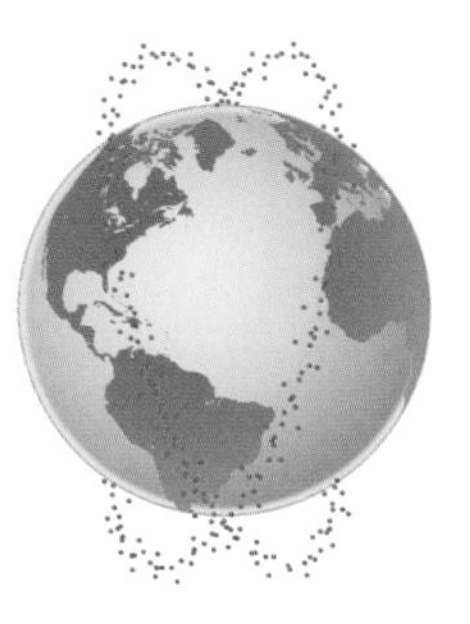

Until they de-orbit and burn up in the atmosphere, these thousands of additional pieces of debris pose threats to all satellites in their path.

SPACE JUNK

On a clear, moonlit night the emptiness and stillness of space is haunting—but deceptive. For on any given night, there are hundreds of thousands of objects of varying sizes orbiting the earth—ranging in size from half an inch (one cm) up. Most of these were never intended to orbit the earth and are serving no function at all.

Space junk or debris is a catchall phrase for things like spent rockets, abandoned satellites, debris from collisions, solid rocket motors, and miscellaneous parts of spacecraft that have simply drifted away from their home. These items range from the very large (for example, solid rocket motors) to the very small (toothbrushes, pliers, and space gloves). Many of these have been orbiting the earth for a very long time: the oldest is believed to be the Vanguard 1 satellite, launched into orbit in 1958 and last heard from in 1964.

An estimated 19,000 of these objects are large and could pose a serious danger to the functional objects in space. Among those at risk is the International Space Station, which operates roughly 250 miles above the earth. To protect it from damage with space junk, a layer of foil, called a meteor bumper, has been placed on its skin. In addition, the U.S. Space Surveillance Network monitors its trajectory and changes its course if a collision is imminent.

In theory, space junk should not be in orbit with operating satellites or other spacecraft. It should be placed in what is sometimes referred to as graveyard orbit, several hundred miles above the normal orbit. However, only one third of the equipment used in orbiting space ever gets there: many satellites run out of fuel first and no longer have the power to be boosted to a safer orbit.

The good news is that space junk doesn't stay in space forever. Due to atmospheric drag, impacts from the moon, and solar radiation, debris circling the earth will ultimately return to it—though at the higher orbits that process could take up to a century.

SATELLITE SLIPUPS

Among the biggest single junk-making events in space was a missile test that China undertook in 2007. As a test of the nation's antisatellite missile technology, the explosion of the weather satellite Fengyun-1C was successful. However, the collision created a massive amount of space junk—roughly 2,500 pieces of trackable debris (roughly the size of a golf ball or larger) and millions more smaller pieces. Nearly all of these pieces remain in orbit today.

Two years later, a U.S. commercial communication satellite collided with a derelict Russian satellite—destroying both. Remnants of this crash likewise remain in low-earth orbit—posing a real danger to other space travelers.

MAPPING SPACE JUNK

NASA uses high-powered radars and optical telescopes to map and monitor tens of thousands of pieces of debris (greater than 4 inches/10 cm) in orbit around the earth. By collecting data on how debris reflects radiation, it is possible to determine material types of man-made orbiting objects in both low-earth orbits and geosynchronous orbits. These data are used in current space environment models and to build better shields for spacecraft.

NASA's main data source for debris in the size range of 4/10 of an inch to one foot is the Haystack radar, operated by MIT Lincoln Laboratory. Haystack statistically samples the debris population by "staring" at selected angles and detecting debris that fly through its field of view.

Observations of geosynchronous orbit debris are taken using the University of Michigan's Curtis-Schmidt telescope located at the Cerro Tololo Inter-American observatory (CTIO) in Chile.

CLEANSPACE ONE

The first satellite designed to clean up space debris is under development. The Swiss space center at École Polytechnique Federale de Lausanne (EPFL) is designing a spacecraft to match the orbit of space junk now circulating the earth. The satellite will attach itself to debris through a grappling mechanism and return it to the earth's orbit—where both components will burn up on reentry into the atmosphere.

Developing the new craft won't come cheaply—an estimated $10 million are required over the next five years to produce the specialized satellite. Even with the money, there's no guarantee that the ambitious undertaking will work: carrying out such tricky maneuvers at speeds of up to 16,800 miles per hour (28,000 (km/h) will require great precision and timing.

THE FUTURE

THE FUTURE

What will transportation look like in the future? Probably not like anything in the past.

For centuries, transportation innovation was devoted to getting places quickly. On the seas, the most prized sailing ships were those that could go fastest. On land, canals were more efficient for moving freight than the primitive roads of the eighteenth and early nineteenth centuries—though their popularity quickly declined once railroads offered a reliable land alternative. In the twentieth century, cars proved faster and more flexible than trains—though they ultimately couldn't compete on longer journeys with the record speeds reached by the airplane.

But the race for speed would halt in the twentieth century—with the world's fastest plane, the Concorde, being retired prematurely due to lack of demand. Tempered too was the drive to go farther: ambitious canals and transcontinental railroads were precursors of the race to the moon that consumed so much American and Soviet resource in the twentieth century—but all three are historical footnotes now.

So what will the future of transportation be about if not going faster or farther? In the short term, probably about more practical things like efficiency, automation, safety, and the environment—in other words, cheaper, more reliable, and less polluting journeys.

The cost of transportation is largely a factor of volume: the more people or cargo, the lower the unit cost of the journey. The same thinking that went into the double-decker bus is today pushing cruise and cargo ships and airplanes to unprecedented sizes—forcing port and airport operators around the world to raise bridges, deepen channels, and rehabilitate docks, runways, and terminals.

Maintaining the reliability of journey times is growing in importance. With the number of trips in all modes increasing rapidly, congestion has undermined many of the advances of the last century. Innovations that help smooth, manage, and speed traffic flow—such as next-gen air traffic control, roadway pricing, and software that allows closer spacing of trains—will become standard across transportation modes.

One big area of innovation for the future relates to the environment. Society's heavy reliance on fossil fuels, and its related impacts on air quality, could hardly have been imagined when the internal combustion engine was invented over a century ago. Minimizing these impacts to meet increasingly stringent regulation of vehicle and vessel emissions around the world will remain at the forefront of the agenda for both government and equipment manufacturers.

One way to reduce emissions is through efficiency improvements. Turbochargers and fuel injection systems, for example, allow traditional cars to perform better on the same amount of fuel and are now being built as standard on more cars. Wingtips on airplanes, which smooth the flow of airplane wings through the air, are now commonly found around the world.

Alternative fuels will play a role in meeting clean air targets, too. Although demand for the fully electric car has yet to materialize, hybrid cars are increasingly popular. Other forms of energy production, including fuel cell technology and compressed natural gas, are—like the electric car—under development but are not yet widely available.

Perhaps the most significant changes to the future of transportation may relate to automation. Taking people out of the driver's seat and replacing them with computers offers opportunities across all modes. It can increase accuracy, capacity, and fuel efficiency by allocating scarce space wisely, by identifying the most efficient engine running speeds, and by selecting the quickest routes. It can also improve safety—by providing real-time condition reporting, by removing room for human error, and—in the case of drones or unmanned spaceships—by going places humans can't or don't want to go.

May 1911

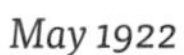

May 1922

July 1925

November 1925

BACK TO THE FUTURE

Imagining our transportation future can take patience. Leonardo da Vinci imagined an ornithopter, a winged structure that would let humans fly, in the fifteenth century. Four hundred years passed before a version of his idea—the fossil fuel–powered (rather than human) airplane—would become a reality.

Other dreams have come true more quickly. Sleek monorails and maglevs operating without engines were the stuff of fiction—until they were built. So too was going to the moon—until rockets powerful and sophisticated enough to take man there made it a reality.

Many have imagined combining modes of travel: cars that fly, amphibious vehicles that move seamlessly from water to land, trucks that ride the rails. A handful of these mode-bending technologies, like Pan Am's flying clipper ships and Norfolk Southern's RoadRailer, have indeed come to pass—but most remain relegated to the covers of enthusiast magazines.

July 1928

August 1931

October 1930

September 1930

AUTOMATION

By the end of the twentieth century, automation had made itself felt across all major modes of transportation. On the seas, computers calculated navigational paths to maximize fuel consumption and avoid weather hazards. On land, automated trains shuttled passengers from airport terminals to gates at remote satellites and cars had learned to park themselves. In the air, some airplane captains complained they were forgetting how to fly due to automatic pilot usage.

More automation is undoubtedly on the way. Today, thanks to leaps forward in computers and telecommunications, data on schedules, congestion, equipment condition, and weather can now be transmitted seamlessly 24/7 to anywhere in the world. So long as increasing reliance on computers brings cost savings, improvements in levels of service, a cleaner environment, and greater safety, the replacement of people with machines will continue.

Among the most exciting examples of what the next century could hold is the driverless car. First shown to the public in the General Motors' Futurama exhibit at the 1939 World's Fair, the "autonomous car" could potentially reduce accidents, mitigate congestion, and lower fuel consumption—not to mention make car travel available to those with limited or no sight. To date, driverless cars have successfully driven hundreds of thousands of miles—climbing winding mountain roads and achieving astonishing speeds on highways without incident—but more testing remains to be done.

THE GOOGLE CAR

Google's driverless car is perhaps the most celebrated, though not the only, prototype of an autonomous car under testing and development today. Several U.S. states, including Nevada, Florida, and California, have passed laws permitting these experimental vehicles to be driven autonomously for test purposes on their roads. Most rely on a range of cameras, lasers, and radar that provide data which help the vehicle keep its distance from cars and other obstructions on the road.

The Google prototype features a laser scanner on its roof. Sixty-four laser beams are continuously sent out to generate a 3D map of the surroundings and determine the distance of objects from the car within a roughly 200-foot radius.

A GPS monitor is located on the top of the car.

Radar is located in four places—on the front and back bumpers and on each of the sides—allowing the car to "see" in all directions.

Wheel encoders keep track of the car's movement, measuring acceleration and rotation around the car's axis.

A camera located near the rearview mirror allows the car to detect the status of traffic lights as well as to read street signs and sense obstructions.

AUTOMATION TIME LINE

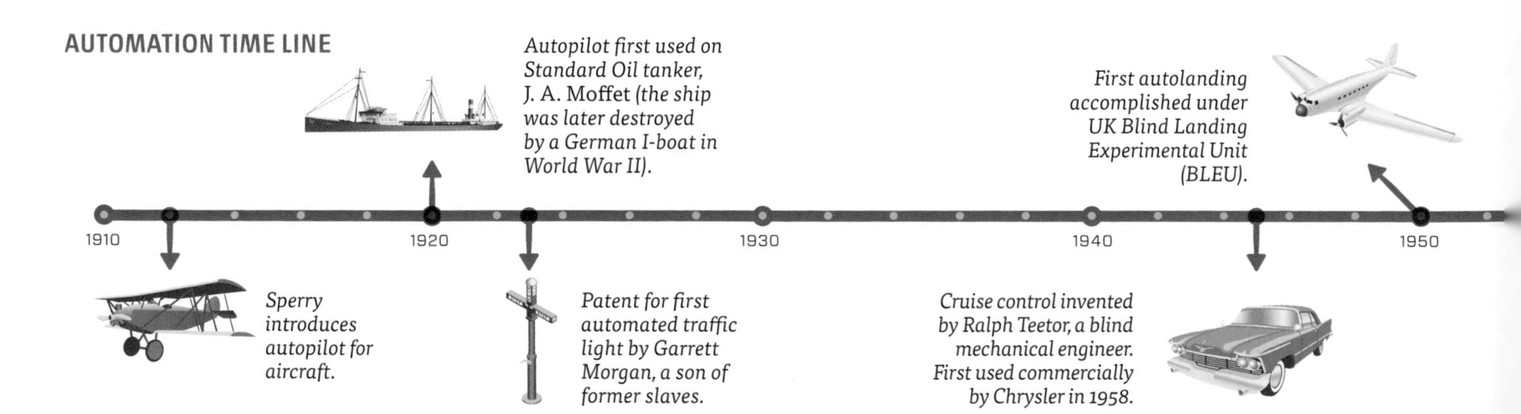

The Google car sees the world as a series of radar-response data points that reflect the presence of objects. A superfast processor is used to interpret data from locational devices and guide the car's operation.

At intersections with lights, the car will move ahead at a green light unless it detects a pedestrian or other object in its path. If the car reaches an intersection without a traffic light, rather than wait indefinitely for another driver to let it go it will inch forward to signal its intentions and—assuming the coast is clear—move through the crossing.

To determine the precise location of the car at any given moment, data provided by inertial measurement technology and wheel encoders are compared with preexisting maps of roads and surrounding terrain and then combined with data from the GPS transponder.

If the car's radar or laser sensors detect a moving object, such as a pedestrian, the car will move to avoid it or—if it is very close—automatically depress the emergency brake. Variations from a preset course will be noted and addressed in much the same way that automatic pilot works on airplanes. These systems can keep a car within .6 of an inch (1½ centimeters) of its predetermined course.

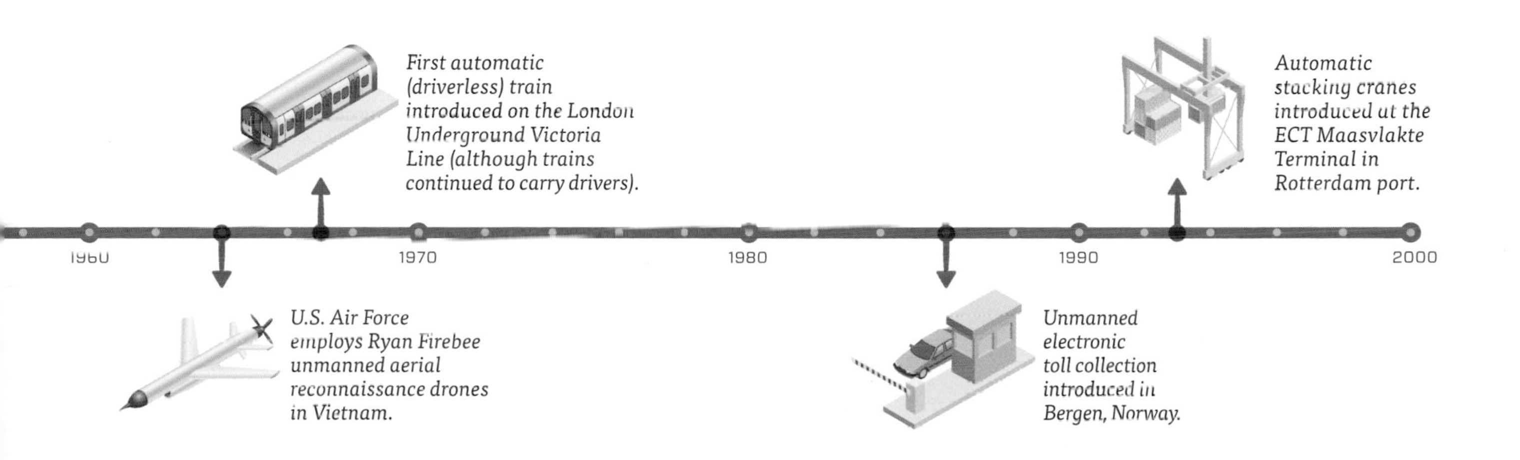

SUSTAINABILITY

The invention of the internal combustion engine revolutionized transport by sea, on land, and in the air—but it has also led to a steady increase in the consumption of fossil fuels. The emissions produced during the burning of these fuels—from carbon monoxide and dioxide to nitrous oxide and particulate matter—are now seen as a major contributor to the deterioration of the ozone layer and the global climate changes now occurring.

Aviation contributes roughly 2 percent of man-made CO_2 emissions. Over the past 40 years, the industry has improved its fuel efficiency and reduced related CO_2 emissions by two thirds. But to achieve its stated goal of a 50 percent net reduction in CO_2 emissions between 2005 and 2045, it will need to do more—including developing more environmentally friendly forms of aviation fuel.

Shipping remains the most energy-efficient method of mass transportation, yet it only accounts for almost 3 percent of all man-made carbon dioxide emissions. Because it can't be attributed to any one country due to its global reach, the promulgation of international standards to increase the energy efficiency of ships has fallen to the body governing world shipping (the International Maritime Organization).

On land, the primary focus of emissions regulation to date has been the motorcar. In addition to hybrid and electric vehicles (EVs), manufacturers are looking at fuel cell technology, which creates power by splitting hydrogen gas into water and electricity, and compressed natural gas (CNG), which produces significantly fewer pollutants than gasoline. Like EVs, however, both require new networks of infrastructure and new engines to supply vehicles.

Biofuels—made from organic substances like corn and sugarcane—can be mixed with gas and may therefore be easier to introduce more widely. Though their emissions are similar to traditional gasoline, the organic plants that produce them fix carbon dioxide as they grow—in many cases more than offsetting the carbon dioxide created when they are burned.

In the United States, small amounts of ethanol (10 percent), produced from corn, have for years been blended into gasoline as a way to support the industries that support its production (primarily in rural areas of the Midwest). Manufacturers now produce vehicles that can handle much larger percentages of ethanol (up to 85 percent), but because the new blended fuel demands dedicated pumps and tanks the product is not widely available.

Brazil is most notable for its commitment to biofuels. Ethanol produced from sugarcane has been an important part of the fuel mix in cars since 1976, when targets were set for the percentage of ethanol that must be included in gas. More recently, 'flex" cars that can run on varying percentages of gas and ethanol have achieved commercial success, causing ethanol consumption to now account for almost half of all gasoline usage in the country.

Like many countries, Denmark has provided incentives for electric vehicle use—lowered tax on car purchases, free parking, and even free battery recharging in some places. But the country is most notable for the lighter touch its electricity has on the environment: nearly a third of the country's power is brought to its grid by wind turbines.

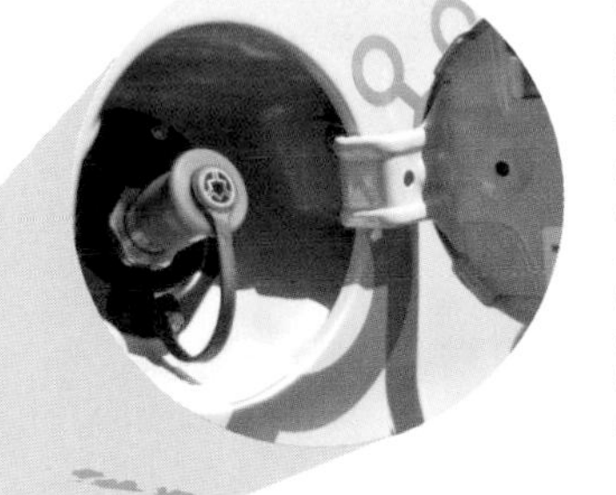

Germany has taken the lead in exploring fuel cell technology. Daimler-Benz's new hydrogen cell-based Sprinter, for example, which becomes available in 2015, can reach a top speed of 105 miles per hour (170 km/hr) and go 240 miles (385 km) before refueling. Four pilot fueling stations operate in the country now, but it is unclear—short of national hydrogen pipelines—how to expand to serve enough of the country to win over a sufficient user base.

Iran relies heavily on cars powered by compressed natural gas, which produces fewer emissions than traditional gasoline. Over two million vehicles in the country use it and are served by an extensive distribution network of 2,000 fueling stations. Other countries relying on CNG include India, Pakistan, Argentina, and Brazil.

Australia is home to a companies looking at more sustainable aviation fuels. A consortium including Airbus and Virgin Australia has been evaluating the potential for mallee trees, a sustainable crop that can help restore land affected by salt inundation to a productive state for local farmers, to serve as a feedstock for aviation fuel through a process known as pyrolysis.

FARTHER

Today's race to space is very tame compared with the one that existed half a century ago. Tempered by cost control and characterized by incremental ambition, it has no central, unifying national focus like putting a man in orbit or better still an astronaut on the moon. And in some places, including the United States, it is no longer even a race that the government enters.

Commercial space flight is now a reality—not just because people other than astronauts can pay enormous sums to experience the feeling of weightlessness. In 2012 and again in 2013, a private company—Space Exploration Technologies (known as SpaceX)—successfully sent its private unmanned spacecraft to dock with the International Space Station in its orbit high above earth. The docking represented the first phase of the $1.6 billion U.S. government contract to ferry supplies to the astronauts in residence there.

SpaceX is not alone in its real quest—which is to take American astronauts, rather than simply cargo, to the space station. Currently, the United States is paying Russia roughly $63 million per astronaut to fly them there on Soyuz

U.S. COMMERCIAL SPACE FLIGHT

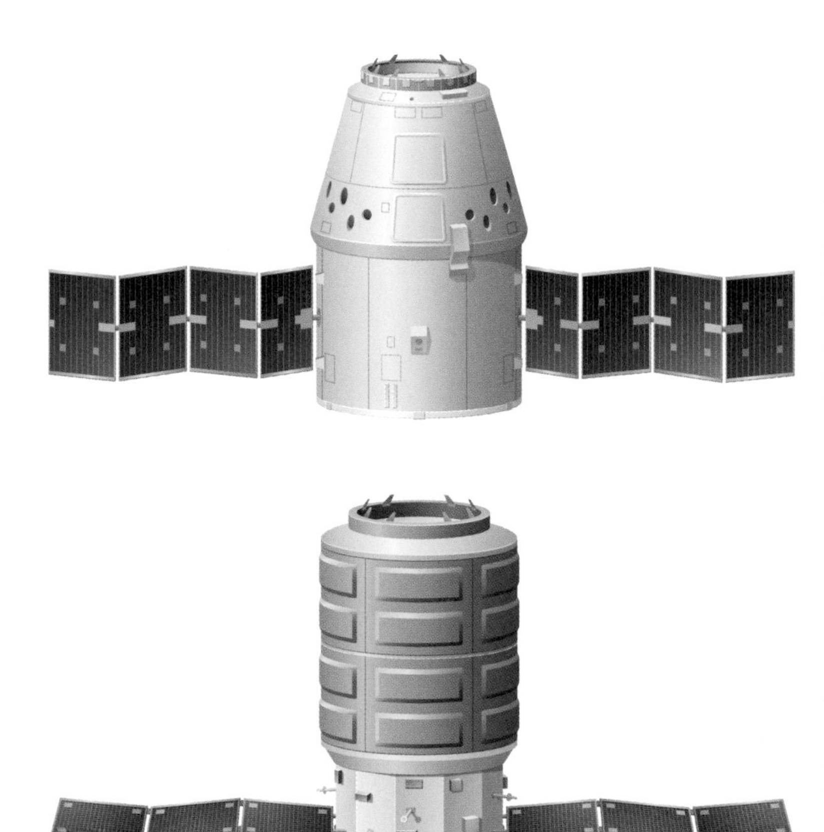

SpaceX's Falcon 9 rocket boosted its Dragon spaceship into orbit for a second time in March 2013 as part of its commercial resupply service contract with NASA. The reusable cargo craft is made up of both pressurized and unpressurized sections and features a heat shield, which protects it on its return journey prior to splashdown in the Pacific.

Orbital Science's Taurus 2 rocket can carry roughly 6,000 pounds (2,700 kilos) of payload in a pressurized chamber. It relies on batteries and solar panels for power and GPS for navigation as it finds its way to the ISS. It is intended to burn up on reentry to the earth's atmosphere, making it ideal for removing trash from the space station.

SPACE TOURISM

If deregulation of airfares opened new avenues of travel to millions who previously couldn't afford it, the same cannot be said of space travel. As governments around the world look to the private sector to assume some of their previous roles in space, a separate industry offering to ferry the very wealthy (or very passionate) to space has mushroomed.

Virgin Galactic was one of the first to offer tickets to the public—at $250,000 apiece, with full payment up front. Those with less cash on hand can reserve a place on a subsequent flight for a mere $20,000 deposit.

rockets—a tripling of the fee charged prior to the end of the U.S. space program. Separately, SpaceX and other companies have been paid nearly $300 million to develop private systems for transporting humans to space; the winner of this race will "capture the flag" left on the space station by the last shuttle crew for the next American mission.

Outside the United States, in places like India, China, and Japan as well as in Europe and Russia, publicly funded space programs continue to support a variety of military, scientific, and economic development ends. The more ambitious programs go beyond research and speak to manned visits to Venus and Mars within time frames that extend to 2050. But the cost and political will to put people in deep space is enormous, and the loose policy pronouncements and uncertain budgets associated with these programs today in no way resemble the focus or commitment involved in putting a man on the moon 50 years ago. For better or worse, the human race has been transported pretty much everywhere it needs to go on earth—and it's not quite sure it wants to go any farther.

MISSION TO MARS

Six unmanned missions have landed on Mars, a planet tens of millions of miles from earth. But putting people on the red planet, and bringing them back, remains an illusive goal. A fanciful alternative to that round trip is Mars One, a one-way journey to the planet offered by a Dutch nonprofit company of the same name (for the year 2023) that now has tens of thousands of applicants.

Spaceships would aim to land on Mars during the window when the two planets are closest—which occurs every 26 months.

Deceleration to land safely on Mars is a tricky business. Unlike on earth, where capsules take advantage of significant air resistance to land naturally, the thinner air on Mars demands using the engines to soften the landing.

Life on Mars would be no picnic: lethal doses of cosmic radiation and extremes of temperature would require living in a highly controlled environment. Absent resupply ships, settlers would be dependent on artificial processes to produce water and any other nutrients necessary for survival.

Wheeled rovers, powered by nuclear fuel, could be the preferred mode of travel around the planet. To date, transportation on the planet has proved tricky, with more than one Martian vehicle getting mired in mud or dust on the planet's surface.

ACKNOWLEDGMENTS

At the close of yet another "how things work" book, I find myself thanking once again the two people who have mastered the art of bringing complex and technical subjects to life: George Kokkinidis and Rob Vroman. And this time I owe them thanks for more than just their hard work: they planted the idea for this book ages ago and lobbied hard to convince me to undertake it.

George, as he did in *The Works* and *The Heights*, has brought his unique blend of intelligence and creativity to the design of each and every page of this book. There is no concept on the preceding pages that he has not considered, digested, and then (literally) envisioned in ways both accessible and compelling. And he has done so with the patience of a saint: his willingness to refine, rework, or redesign a graphic when we felt it lacked clarity or power never flagged over the course of two years and hundreds of illustrations.

Rob Vroman brought to this book, as he had to *The Heights*, his uncanny ability to lay his hands on relevant information within moments of its being requested. As a naval engineer and graduate of Annapolis, he not only understood much of what we have tried to explain here but he taught it, to classes of cadets who followed in his wake. When it comes to things that float, fly, and—thanks to his tenure aboard a submarine—even those that sink, there is little that Rob doesn't know and much that he taught us.

Rob and George were not alone in their interest in transportation. After seven years at the Port Authority of New York and New Jersey, I remain fascinated by the subject—as do many of my former colleagues there. A number of them helped to ensure that this book is, if nothing else, reasonably accurate. I'm grateful to Rick Larrabee and Dennis Lombardi, in the Port Department there, whose maritime experience is measured in decades. And to John Selden, in the Aviation Department, who in one person manages to combine firsthand experience and expertise of airports, flight and the air force. All three were kind enough to read and help steer a half-finished manuscript.

Two other close friends from my days at the Port also helped out. Vicky Kelly's years of managing trains, bridges, and tunnels uniquely qualified her to review more than her share of chapters. Likewise, Jill Lacey's knowledge of rail signals helped clarify how trains really move from here to there—particularly when they get to crossings. I am equally grateful to Jill's husband, John, whose work in the satellite industry makes him the closest thing to a rocket scientist in my immediate world.

As much as people in the transportation industry know about tunnels and airports and containerships, few know as much about the smaller stuff—like cars. For that I relied on Max and Siôn Ellis, the father and son team in North Wales that can take apart and rebuild virtually anything with a crankshaft. Who knew a seat belt mechanism could be so complex?

Once again, I owe thanks to my agent, Sloan Harris, at ICM, and to my editor and publisher at Penguin Press, Ann Godoff. Sloan, since we first met, has nursed a fascination with all things transportation; I hope he will find answers to at least a few long-standing questions here. Ann, while hardly a transportation geek, is no less committed to the product—and as always I am grateful for her continued support and guidance.

On a personal note, I'm especially grateful to my partner, Neal Kamsler—who has watched this book take shape on two continents over the course of the last year, most notably on "work-free" vacations. His patience for my writing hobby is exceeded only by his patience for my kids and for our dog—and I am deeply grateful for all three.

CONTRIBUTING ARTISTS

AARON ASHLEY 43↑, 43↓, 110, 142, 143↓, 144, 145, 146, 154↓, 174, 175, 188–189↑

KEN BATELMAN 12–17, 24–25, 29↑, 53, 60–61, 62, 63, 78–79, 82–83, 84–85, 94–95, 106, 107↑, 128, 129, 152–153, 154–155, 156–157, 168–171↑, 192, 193

DESIGN LANGUAGE 18-19, 21←, 28↗, 29↓, 38–39↑, 42, 48, 50↓, 54, 58–59, 65↓, 74–75, 76–77, 79↗, 82↓, 88←, 90–91, 92, 99, 100↓, 102, 107↓, 121↓, 124–125, 127, 137, 143↑, 150–151↓, 152↓, 159, 165, 169↓, 170↓, 181↓

MICHAEL FORNALSKI 116–117, 132–133, 140↑, 141↑

JIM KOPP 88–89

VIC KULIHIN 22–23, 30, 31, 36–37, 38–39↓, 44–45, 47→, 50, 51↑, 52, 55, 56, 57, 64, 65↑, 68–69, 70, 71, 72–73, 86↓, 86–87, 100–101, 108–109, 118–119, 120–121, 122–123, 138-139, 140–141↓, 162–163, 176, 177, 178–179, 188–189↓

JASON LEE 21→, 26–27, 35, 93, 96–97, 113, 130–131, 148–149, 150–151↑, 164, 166–167, 172–173, 180, 181↑

DAVID PREISS 32, 33, 34, 40–41, 46–47↑, 49, 66↓, 66–67, 80–81, 103, 104, 105, 114–115, 134–135, 147, 158, 160–161, 174↓, 175↑, 182–183

JAMES PROVOST 23↓, 28↙

PHOTOGRAPHY CREDITS

22: U.S. Navy photo by Mass Communication Specialist 3rd Class Christopher B. Janik
28: U.S. Navy photo
29: CCASA Sebastian Brosen
30: U.S. Navy photo by Photographer's Mate 1st Class John S. Lill
32: Engraving by Gustave Doré for an 1876 edition of the *Rime of the Ancient Mariner* by Samuel Coleridge
33: Photo courtesy of Pixabay
34: USS *Francis Scott Key* (www.ssbn657.com)
35: U.S. National Archives photo
37: CCASA Roberto Vhonger
45: CCASA photo courtesy of Equipe C'est N'est Pas une Pipe
46: CCASA Clipper
49: USCG photo by PA2 Mike Hvozda
52: U.S. Navy photo
54: U.S. Navy photo by PH3 (AW) Jenniffer Rivera
57: NOAA National Ice Center
59←: U.S. National Archives photo
59→: U.S. National Archives photo
62: USCG photo by LCDR Steve Wheeler
67: CCASA Dance of Life Image Miner
68: © Time & Life Pictures/Getty Images
72: U.S. government photo
73: CCASA Naquib Hossain
87: © Dan Gottesman
89: © Schenectady Museum; Hall of Electrical History Foundation/CORBIS
90: CCASA Stephanie Irigoyen
91: CCASA Alan Levine
92: U.S. National Archives photo
93: CCASA Mario Roberto Durán Ortiz
95: CCASA Mario Roberto Durán Ortiz
97: U.S. Department of Transportation photo
99: Chicago and North Western Railway
110: NSW State Records Photographic Collection
111: CCASA Gordon Joly
111: CCASA InSapphoWeTrust
111: CCASA Edward Russell
111: CCASA Liz West
112: CCASA Andrew Clarke Parker
113: U.S. government photo
115: CCASA Loz Pycock
118: USCG photo by Kevin Rofidal
122: CCASA Anne Akers
126: Library of Congress—photo by Harris & Ewing, Inc.
127: CCASA Stuart Pearce
129: CCASA Rob Vroman
131: U.S. Navy photo by Mass Communication Specialist 3rd Class Scott Pittman
133: Marine Corps photo by Sgt. Mark Fayloga
135: *Chicago Daily News* negatives collection, Chicago History Museum
136: CCASA Rob Evans
142: British official photographer, Royal Air Force Aircraft 1941–1959: ATP Collection (GSA 325)
143: © Stuart Dee
144: DoD photo by U.S. Air Force Staff Sgt. Manuel J. Martinez
146: © MCT via Getty Images
149: CCASA Julia P (autumn_bliss on Flickr)
154: U.S. Navy photo by PH3 Edwards
155: © AFP/Getty Images
159: CCASA WCM 1111
160: CCASA Jonas N
161: CCASA Thomas23
165: USDA photo by Jenny Mastanuono
167: NASA photo
168: NASA photo
171: NASA photo
183: © EPFL
187: *Popular Science*, May, 1911
187: *Popular Science*, May, 1922
187: *Popular Science*, July, 1925
187: *Popular Science*, November, 1925
187: *Popular Science*, July, 1928
187: *Popular Science*, September, 1930
187: *Popular Science*, October, 1930
187: *Popular Science*, August, 1931
190: CCASA Petr Kratochvil
190: CCASA Sweeter Alternative, sugarcane.org
191: CCASA Andreas Klinke Johannsen
191: CCASA Siemens PLM Software
191: © Glow Images
191: CCASA MrPBPS
192: NASA photo

INDEX

A

abandon ship, 36–37
ABS (automatic block signal), 103
Abu Dhabi International Airport, 141
accelerometer, 97
Adams, John Quincy, 110
aerodynamics, 82
aeronautical charts, 136
aerotropolises, 149
Age of Discovery, 20
Age of Exploration, 32
Age of Sail, 20
air:
  airports, 148–65
  flight, 126–47
  space, 166–83
airbags, 96
air brakes, 83
Airbus A380, 131
aircraft carriers, 20, 25, 28, 131, 154
air mail, 15, 126
air marshals, 163
airplanes, 99
  accidents, 146–47
  air pressure in, 134, 142–43
  air quality in, 142–43
  altimeter, 135
  attitude indicator, 135
  autopilot, 135, 188
  belly landing, 146, 155
  bird strikes, 146, 164, 165
  black box, 147
  brake systems, 155, 165
  and cell phone use, 147
  cockpits, 134–35
  cockpit voice recorders, 147
  depressurization of cabin, 146
  ejector seats, 131
  evacuation slides, 146
  exposure to elements, 144–45
  fire on, 146, 164, 165
  flight data recorders, 146–47
  heads up display, 134
  hijacking, 162, 165
  ice buildup, 144–45, 156
  instrument flight, 134, 136, 137
  jet fuel, 157
  jet planes, 16, 126, 130, 142
  landing, 146, 154–55
  lift, 128, 129, 132, 133
  lightning protection, 145
  logbooks, 158
  maintenance, 158
  navigation, 135, 136–37
  oxygen masks in, 143
  pitch, 129
  roll, 129, 134
  sick zones in, 143
  taking off, 128, 131
  thrust, 128
  toilets, 143, 156
  transatlantic, 15
  turboprops, 130, 133
  turning, 129, 134
  V/STOL, 131, 133
  water landings, 146
  wheel wells, 155
  winglets, 129, 186
  Wright Flyer, 15
  yaw, 129, 134
airports, 148–65
  anatomy of, 150–51
  approaching, 140–41
  approach lighting, 148, 153
  baggage handling, 160–61, 162
  baggage tags, 160
  cargo facilities, 151, 159
  codes, 149
  control towers, 138–39, 140–41, 150
  dwell time in, 151
  emergencies, 164–65
  ground services, 156
  holding patterns, 141
  hubs, 159
  international navigation standards, 148
  and jet fuel, 157
  landing at, 154–55
  land vs. air side, 150–51
  milestones, 15, 148–49
  runway overruns, 165
  runways, 141, 152–53
  security, 150, 151, 162–63, 165
  terminal layouts, 150–51
  turtles at, 165
air pressure, 134, 142–43
air quality, 11, 142–43, 186
air resistance, 170
air traffic control, 138–39, 140–41
air turbulence, 144

TABLET FLIGHT

Flight logs and flight bags, containing information about the plane and its route, have long accompanied pilots on their journeys. But they are increasingly becoming an aviation relic.

Numerous airlines have embraced the iPad as a replacement for the heavy logs and maps that pilots traditionally carry. Under American Airlines' "electronic flight bag" program, 8,000 iPads were distributed to pilots—replacing an estimated 24 million pages of printed material.

American is not alone. JetBlue recently provided custom-equipped iPads to its pilots, and both the U.S. Marine Corps and U.S. Air Force have begun to replace military maps with iPads for their own purposes.

AIS (automated identification system), 49
Alcántra Bridge, Spain, 112
Aldrin, Edwin "Buzz," 166
Aloha Airlines, 146
alphabet flags, 51
alternator, 79
American Bureau of Shipping, 72
amphibious transport docks, 25
anchorages, 67
anchors, 66
angle of attack, 128, 133
Apollo space program, 170
APT (advanced passenger train), 107
aqueducts, 113
arch bridges, 112, 113, 114
Archimedes, 21
Arlanda Airport, Sweden, 140
Armstrong, Neil, 166
artificial reefs, 111
asbestos, 73
asphalt, 86, 87, 153
astrolabes, 38, 39
astronauts:
  life in space, 178–79
  space suits, 177, 181
  transporting, 192
AT/B (articulated tug barge), 61
automation, 186, 188–89
automobiles, 78–81, 99
  anatomy of, 78–79
  brakes, 79, 94
  cooling systems, 78
  crash testing, 97
  crumple zones, 96
  driverless, 188–89
  electric, 94, 95, 186, 191
  electrical systems, 78
  emissions, 79
  engine, 78
  exhaust system, 79
  flex cars, 190
  hybrid, 17, 94, 186
  milestones, 14–15, 16, 17
  noiseless, 94
  safety features, 96–97
  steering, 78
  transmissions, 79
autopilot, 135, 188
azimuth thrusters, 29
azipods, 29

B

backscatter imaging, 162, 163
ballast, 21, 100, 101
ballast tanks, 27
ballast water, 34
balloons, hot air, 12, 126
Barbary pirates, 70
barges, 13, 21, 61
bascule bridges, 115
batteries, 26, 78, 79, 94, 95
beacons, 37, 42, 44–45, 46, 53, 136

Beckton Desalination Plant, London, 33
Bell, Alexander Graham, 23
Bellamy, "Black Sam," 70
Benz, Karl, 15, 76
Benz motorwagen, 14, 76
Bessemer steel process, 98
bicycle tires, 77
bilge, rounded, 22
biofuels, 190
bitts, 67
Blackbeard (pirate), 70
black box, 147
blimps, 131
Boeing aircraft:
  314 flying boat, 15
  601 geosynchronous satellites, 175
  707 jet, 142
  747 jet, 16, 127, 131
  787 Dreamliner, 127
  B-2 stealth bomber, 127
  B-29 Superfortress, 16, 127
  B-52 Stratofortress, 127
bogies, 100, 108
bollards, 67
borescopes, 158
Borman, Frank, 178
bow thrusters, 29
boxcars, 105
brake systems:
  airplanes, 155, 165
  automobiles, 79, 94
  railroads, 104, 106
  trucks, 83
bridges, 112–19
  arch, 112, 113, 114
  cantilever, 114, 115
  collapses, 116–17, 118
  compression and tension in, 114
  construction, 116–17
  curling, 115
  decks, 117, 118
  designs, 114–15
  fixed, 114
  inspections, 118–19
  iron, 112, 113
  maintenance, 119
  movable, 115
  painting, 119
  safety, 118–19
  suspension, 114–15, 116–17
  tower construction, 116
Brooklyn Bridge, 112, 113
BTM (barrier transfer machine), 91
bubbler system, 62
Buick coupe, 16
bulbous bows, 21, 22
bulk carriers, 24
buoyancy, 21, 27
buoys, 46–47
buoy tenders, 47
buses, 85

C

cable cars, 108, 109
cables, 114, 117
cable-stayed bridges, 114
Cabo Branco Lighthouse, Brazil, 45
Cadillac Eldorado, 16
caissons, 116
cameras, traffic control, 89
Canada, and space, 176
Canadian Ice Service (CIS), 57
canal locks, 65
canals, 13, 64–65, 76, 186
canoes, dugout, 20, 21
cantilever bridges, 114, 115
Cape Hatteras Light, U.S., 44
caravels (carracks), 20
carbon monoxide, 79
car carriers, 84
car ferries, 60
cargo, movement of:
  by air, 151, 159
  by canal, 76
  containerization, 16, 24, 31, 65, 68–69, 70–71, 73, 104, 105
  costs of, 11
  by rail, 68, 69, 104, 105
  by ship, 16, 59, 65, 68–69
  by truck, 15, 16, 68, 69, 84
car pooling, 93
cars, *see* automobiles; railroads
catalytic converters, 79
catamarans, 22
catapults, 131
cathedral hulls, 22
cat's eyes, 96, 97
catwalks, 117
Cayley, George, 126
Cayley glider, 13
celestial navigation, 38, 39
Celtic chariots, 77
cement mixers, 84
center beam bulkhead, 105
Central Intelligence Agency, 72
centrifugal force, 107
Cerro Tololo Inter-American Observatory, Chile, 183
Chapultepec aqueduct, Mexico, 113
chariot wheels, 77
Chesapeake Bay Bridge-Tunnel, 113
Chicago, Bloomingdale Trail, 111
China:
  Grand Canal, 13
  high-speed rail, 106
  junks, 20, 21
  missile test in, 182
  space program of, 166
chined (deep-V) hulls, 22
chip logs, 38
chronometers, 39
cities:
  rail systems, 98, 108–9
  traffic control, 93
Clipper ships, *Flying Cloud*, 13
coal-mine refuse, 87
Coast Guard, U.S., 22–23, 67
Coast Guard cutters, 25, 60
COFC (container on flat car), 105
cold cats, 131
Columbia space shuttle, 167, 171
Columbus, Christopher, 20, 38, 40
commuter rail, 109
compactors, 100
compass:
  gyroscopic, 39, 135
  magnetic, 38, 39, 135
compressed natural gas, 186, 191
Concorde, 127, 186
Conlin, Jonathan, 11
containerization, 16, 24, 31, 65, 68–69, 70–71, 73, 104, 105
continental shelf, 48
continuous flow intersections, 90
continuous span bridges, 114
controllable pitch propellers, 29
Cook, Captain James, 39
corvettes (ships), 25
Cospas/Sarsat, 53
Costa Concordia, running aground of, 37
covered hoppers, 105
cranes, 61, 69, 72, 101, 110, 117, 176, 189
crankshaft, 78
crossover tracks, 102
cross-staffs, 39
crosswalks, 88
cruise ships, 24, 31, 33
curling bridge, London, 115
Curtis-Schmidt telescope, 183
cylinders, 78, 79, 83

SEGUE TO SEGWAY

Segway Personal Transporters (PTs), while hardly omnipresent, have found a home in the world of transportation. An estimated 1,500 police and security departments now use them as a tool for community and facility policing.

The Segway is driven by shifting weight forward or backward. Gyroscopic and fluid-based leveling sensors detect the change in the center of its mass and then rebalance the load while adjusting the vehicle's speed accordingly.

The Segway can go 24 miles (38 km) on a single battery charge and reach a top speed of 12 mph (20 km/h). At a cost of roughly one cent per mile, it's among the cheapest forms of powered movement on the market.

COCKPIT CHANGES

Why is the cabin of an airplane called a cockpit? Curiously enough, the word's origins were on land and sea—not in the air.

A "cockpit" is a pit in the ground used to hold fighting cocks—a sport that dates back thousands of years and remains popular today in certain parts of the world. Suggesting a crowded and poorly lit space, the word was used in the eighteenth century to refer to certain offensive lower-deck spaces on sailing ships.

By the nineteenth century, the word was used generally for the steering area, or well, of a ship. A century later, the word took flight and today is used to refer to the steering area of an aircraft.

D

davit-launched lifeboat, 36
dead reckoning, 38–39, 40
dead water, 55
deep-V (chined) hulls, 22
De Havilland Comet, 16, 127, 142
Denver International Airport, 149, 152, 161
derailment, 110
Derwent steam engine, 13
desalination, 32–33
Detroit, People Mover, 108
dhows, 20
diesel engines, 28, 105
diesel fuel, 79, 80, 81, 83, 99
diligence (stagecoach), 11
diners, 111
Diolkos wagonway, 98
DIRECTV satellites, 175
dirigibles, 13
Disney Company, 108
displaced left-turn intersections, 90
dive boats, 60
dogs, detector, 71, 162, 163
double-stack trains, 104, 105
drawbridges, 115
dredges, 63
drones, 17, 189
dry dock, 72–73
DSC (digital selective calling), 53
DSN (deep space network), 173
Dubai International Airport, 149
dump scows, 63
Dunlop, John, 77

E

Earhart, Amelia, 126, 127
earthquakes, and bridges, 118
eddy current brakes, 106
Edinburgh Airport, Scotland, 140
Edmund Fitzgerald, sinking of, 55
Egyptians:
  chariot wheels, 77
  roads of, 86–87
  ship construction, 20, 21
Eisenhower, Dwight D., 92
electrolysis, 178
embankments, 86
Emma Maersk (containership), 31
energy security, 11
English Channel, 11, 23
Enoshima Lighthouse, Japan, 45
Enterprise, 30
environmental impact, 11, 77, 79, 186, 190–91
EPIRB (emergency position indicating radio beacon), 37, 53
Erie Canal, 13, 64, 65
Estonia, sinking of, 37
European Union, and space, 166, 167, 175, 176
evaporator, 32
exclusive economic zone, 48
expansion joints, 117

F

FAA (Federal Aviation Administration), 138, 162
FedEx, 159
ferries, 60
fifth wheel, 82
fireboats, 60
fishing vessels, 25
fixed-block system, 103
flash-butt welding, 100
flat-bottomed boats, 22
flight, 126–47
  directional, 133
  holding patterns, 137, 141
  by instruments, 134, 136, 137
  milestones, 127
  Next Gen, 139
  rules, 136, 137
  see also airplanes
floating cranes, 61
fly ash, 87
foam, high-expansion, 165
Ford, Henry, 15, 76
Forth Bridge, Scotland, 112, 119
Fort Worth Alliance Airport, Texas, 141
fossil fuels, 11, 80, 186, 187, 190
free-fall lifeboat, 36
Fresnel lens, 45, 154
fuel cell technology, 186, 191
fuel storage tanks, 157
fuel systems, 79, 186
Fulton, Robert, 12
future, 186–93

G

Gagarin, Yuri, 166
gantry cranes, 72, 101
garbage trucks, 84
gasoline, 80–81
gas stations, 80, 81
gas turbines, 20, 28
George Washington Bridge, New York, 45
Germany, rail signals in, 102
Giffard dirigible, 13
"glasphalt," 87
Glenn, John, 166
gliders, 13, 126
Glomar Explorer, 72
GMDSS (global maritime distress and safety system), 52
gnomic projection, 43
Goddard, Robert H., 16, 166
Golden Gate Bridge, 113, 117
Google car, 17, 188–89
GPS (global positioning system), 39, 40, 41, 136, 138, 174, 175
graving dock, 72
great circle routes, 43
gridlock, 90
ground-penetrating radar, 123
ground tackle, 66
gyroscopic compass, 39, 135

H

halon systems, 146, 164
harbors, 60–61
Harland & Wolff, Belfast, 72
Haystack radar, 183
Heathrow Airport, London, 141, 149
heavy-lift ships, 25
heavy metals, 73
helicopters, 71, 132–33
Henry VII, king of England, 72
highway systems, 76
  interchanges, 90–91
  interstate, 92
  numbering, 92
  speed limits, 97
Hindenburg, 126
hinge pipe beams, 118
Holland Tunnel, New York, 113
Homer, *The Odyssey*, 55
Hong Kong, port of, 59
hopper barges, 61
hopper cars, 104–5
hopper dredges, 63
horse-drawn carriages, 85
horse-drawn railways, 98
horsepower, 79
hospital ships, 25
hovercraft, 20, 23
HOVs (high-occupancy vehicles), 93
Hubble telescope, 173
Hughes Company, 72
Humber Bridge, England, 118
hump yards, 104
hurricanes, 57, 144
hydraulic dredges, 63
hydrocarbons, 79, 80
hydrofoils, 20, 23
hydrographic survey ships, 43
hydrostatic technology, 63
hypoxia, 181

I

ice, mapping, 57
icebergs, 56–57
icebreakers, 20, 28, 57, 62
ice sheets and shelves, 56
ice tongues, 56
incineration, of waste, 34
inclinometer, 134
India, space program of, 166
Industrial Revolution, 12, 14, 26, 76, 112, 120
infrared thermography, 118, 119
infrastructure, 13
Inmarsat, 52
internal combustion engines, 14–15, 20, 78–79, 94, 186, 190
International Maritime Organization, 50, 190
International Space Station, 167, 172, 173, 174, 176, 178, 192
interstate highway system, 92
Iron Bridge, England, 113
iron bridges, 112, 113

J

Japan:
  monorail system, 109
  and space, 166, 176
Jeffrey's Hook Light, U.S., 45
Jersey barriers, 96, 97
jet fuel, 157
JFK Airport, New York, 149, 159
Johnson Space Center, Texas, 172
Jolly Roger flags, 70
jughandle turns, 90

K

kerosene, 80, 157
keystones, 112
kingpin, 82
Knarraros Lighthouse, Iceland, 44
knots, 67
Kõpu lighthouse, Estonia, 44

L

Laerdal Tunnel, Norway, 112
LaGuardia Airport, New York, 148
land:
  behind the wheel, 76–97
  bridges and tunnels, 112–23
  rails, 98–111

JETPACKS DOWN UNDER

Personal human flight has been a dream since Leonardo da Vinci's day. But the idea of a jetpack, worn to launch humans into the air, may finally be coming of age.

A New Zealand company recently received a permit from the country's civil aviation agency to test a 132-pound jetpack in two rural areas. Though the manned tests are limited to 20 and 25 feet above ground and water, respectively, a test with a dummy pilot reached 5,000 feet before deploying its return parachute.

The jetpack is designed primarily for recreational use. Nevertheless, the project is of interest to the military and might have applications for civilian search-and-rescue operations as well.

Lanterna of Genoa, Italy, 44
laser dozers, 100
LCAC (landing craft air-cushioned), 61
LCS (littoral combat ship), 61
lead and heavy metals, 73
Lenin (icebreaker), 62
Leonardo da Vinci, 126, 132, 187
letters of marque, 70
Liberty ships, 16
lifeboats, 22–23, 36–37, 60
life preservers, 36–37
light curtains, 93
lighthouses, 44–45
lightning, 145
light-rail systems, 108
Lindbergh, Charles, 126
Little Red Lighthouse, New York, 45
Lloyd's Register, 72
LNG (liquefied natural gas) carriers, 24
locomotives, 12, 13, 98, 105
London, central zone, 93
London Bridge, 113
London Underground, 14, 108
Long Beach, port of, 58
longshoremen, 59
LORAN (long-range navigation system), 40
Los Angeles, Port of, 58
Los Angeles International Airport, 141
low boys, 84

M

Macquarie Lighthouse, Australia, 44
Madrid-Barajas Airport, Spain, 141
maelstroms, 55
Maersk Alabama, 70
magnetic compass, 38, 39, 135
magnetic levitation (maglev), 99, 106, 107, 187
magnetic north pole, 39
magnetometers, 118, 162
mail trains, 99
Malacca, Straits of, 70
maps, 43
Marconi, Guglielmo, 52
Marine Broadcasting Offenses Act (1967), 52
maritime containers, 16
maritime travel, 50
Mark V boats, 60
Mars, missions to, 166, 193
masts, 20, 21
McLean, Malcom, 16, 68
Memphis, Tennessee, hub, 159
mercator projection, 43
Mersey Railway Tunnel, 113
Michigan lefts, 90
microbursts, 144
Milan, congestion pricing in, 93
Mir space station, 167, 172, 176, 181
missiles, 170
Model T Fords, 15, 76
Monitor, 14
monorails, 108, 109, 187
Mont Blanc, France, 122
Montgolfier brothers, 126
moon, race to, 186, 187
motorist clubs, 77
motor vehicle, first patent for, 76
moving-block system, 103

N

naphtha, 80
NASA, 167, 173, 183, 192
National Hurricane Center, 144
National/Naval Ice Center, U.S., 57
NATO, 48
navigation, 38–57
  airplane, 135, 136–37
  astrolabes, 38, 39
  avoiding the elements, 56–57
  buoys, 46–47
  celestial, 38, 39
  dead reckoning, 38–39, 40
  and depth, 43
  electronic, 40–41
  GPS, 39, 40, 41, 136, 138, 174, 175
  international standards, 148
  lighthouses, 44–45
  magnetic compass, 38, 39, 135
  nautical charts, 42, 43
  origins of, 38
  traffic at sea, 49–51
  underwater, 40, 48
NAVSAT, 39, 40
Newton, Isaac, 169
New York:
  emergency Hudson River landing, 146
  harbor, 60
  High Line, 111
  subways, 14, 108, 110
New York Central, 110
New York Times, 166
Nixon, Richard M., 162
Normandie, 35
norovirus, 33
Norwegian Cruise Lines, *Crown Princess*, 33
nuclear fission, 28

O

Oakland International Airport, 140
octane ratings, 81
oil, refining, 80–81
Oresund Bridge, 113
oxygen masks, 143
ozone layer, 190

P

Panama Canal, 17, 64–65
Panama Railroad, 99
Panamax vessels, 17, 65
pantographs, 108
Paris, Promenade Plantée, 111
Paris Metro, 14, 108
passports, 162
PCBs, 73
peace symbol, 51
pedicabs, 85
people movers, 85, 108–9
personal electronic devices, 147
petroleum, 80–81
Peugeot car, 14
Phoenician warships, 21
pigtail, truck, 83
pilings, 118
pilot boats, 60
pirate radio, 52
pirates, 70
pistons, 78, 79
planetary gear set, 79
pneumatic drills, 120
police boats, 60
pollution, 11
Polynesians, ship builders, 20
Ponte Vecchio, Florence, 113
pontoons, 115
Pont Saint-Bénezet/Pont d'Avignon, France, 113
ports, 58–59
  containerships in, 68–69
  dredging, 63
  security in, 70–71
  traffic in, 49
Port Said Light, Egypt, 44
Postal Service, U.S., 15, 99, 126, 159
pozzolana, 112
Prius (Toyota), 17, 94
privateers, 70
Project Jennifer, 72
propellers:
  airplane, 130, 131
  ships, 20, 21, 28, 29
Puente del Alamillo, Spain, 113

R

rack-and-pinion steering, 78
Rackham, "Calico Jack," 70
racons (radar transponder beacons), 46
radar, 39, 123, 138–39, 140, 189
radiation portal monitors, 71
radio waves, 39
rail cars, 99, 105
  recycling, 111
railroads, 76, 98–111, 186
  accidents, 110
  boxcars, 105
  dead man's switch, 110
  double-stack trains, 104, 105
  driverless, 99, 189
  electrified, 98
  elevated, 14
  freight, 104–5
  high-speed, 11, 15, 16, 99, 106–7
  hump yards, 104
  light-rail, 108
  marshaling or classification yards, 104
  milestones, 12–16
  safety, 110
  signals, 89, 99, 102, 103, 110
  tilting trains, 107
  transcontinental, 14, 98, 99
  tunnels, 113
  underground, 14
railroad tracks:
  crossings, 110
  gauges, 100
  inspections of, 100
  laying, 100–101
  stabilizers, 101
  steel, 98, 99
  switches, 102, 104
  ties ("sleepers"), 100
  velvet/welded, 100, 106
recirculating-ball steering, 78
recycling, 87, 111
reefer (refrigerated truck/boxcar), 68, 84, 105
reefs, artificial, 111
reverse osmosis, 33
RFID (radio frequency identification), 162
rhumb line, 43
Richmond International Airport, 141
rickshaws (*tuk-tuks*), 85
road railers, 105
roads:
  anatomy of, 86
  and bridges, 117
  driving right/left, 77
  highway systems, 76, 90–91
  intersections, 90–91
  lane shifters, 91
  paving, 86–87
  Roman, 13, 76, 86
  safety features, 96–97
  stop signs, 90
  striping, 87
  toll roads, 76, 93, 189
rockets, 16–17, 166–67, 168–69, 174, 187
rogue waves, 55
rollers, 76
Roman roads, 13, 76, 86
  and bridges, 112, 114
Roosevelt Island, New York, 109
ro-ro (roll-on/roll-off) vessels, 24
rotaries, 90, 91
Rotterdam, port of, 59
roundabouts, 90
ROVs (remotely operated vehicles), 71
Royal Caribbean, *Allure of the Seas*, 31
rudders, 20, 21, 29
rumble strips, 96, 97
Russia:
  icebreakers, 28, 62
  and space, 16, 166–67, 170, 172, 176, 192–93

S

Saarinen, Eero, 149
sails, 20, 21
St. Anthony Falls Bridge, 118
St. Elmo's Fire, 145
San Francisco, cable cars, 109
San Francisco International Airport, 140
San Francisco-Oakland Bay Bridge, 118
San Jose International Airport, CA, 140
satellites, 174–75
  cleaning space debris, 183
  collision of, 182
  communication, 52
  navigation, 39, 40, 41, 53, 174, 175
  orbits of, 175
school bus, 85
scurvy, 33
sea:
  closer to shore, 58–73
  communication at sea, 51–53
  life at sea, 20–37
  navigation, 38–57
  rules of the road, 50
  safety at sea, 70–71

RAILS AFLOAT

When does a railroad car move by water? When it's on a car float.

Car floats are barges with rail tracks that move freight trains from one side of a body of water to the other. Railcars move from traditional tracks over a float bridge onto the car float, which is powered by tugs to move to its destination.

Now a rare sight, car floats once littered New York Harbor: at their height, 323 of them crossed the Hudson between New Jersey and New York City. While only one service remains today, vestiges of the past can be seen in float-bridge remains celebrated in waterfront plazas and parks.

sea *(cont.)*
territorial waters, 48
traffic, 49–51
waste disposal, 34
waves, 54–55
see also ships
SEAL delivery vehicles, 60
seat belts, 96
seawater, undrinkable, 32
Seikan Tunnel, Japan, 112
semaphore, 51, 89, 102, 103
sensors, 88
Severn Tunnel, 113
sewage treatment, 34
sextants, 39
Shanghai, port of, 58
shear length beams, 118
Shepard, Alan, 166
shingles, in paving, 87
Shinkansen (bullet train), Japan, 16, 106
ship-breaking industry, 73
ships:
abandon ship, 36–37
anchoring, 66, 67
bridge-to-bridge radio, 52
bulbous bow, 21, 22
bulkheads, 21, 35
bunkering, 30
buoyancy, 21, 27
crews, 31
docking plans, 73
fire on, 35
flooding on, 35
fueling, 30
harbor vessels, 60–61
hull shapes, 20, 21, 22–23
hydrographic survey ships, 43
inventions, 21
iron, 20, 21
lightships, 44
lights of, 50
maintenance of, 72–73
moorings, 66–67
mooring systems, 67
nuclear-powered, 20, 28, 62
port security vessels, 71
resistance, 23
sinking, 37
SOS distress call, 53
speed of, 186
stability of, 21, 23
steamships, 12, 14, 20, 21, 28
stopping, 66–67
types of, 24–25
under-way replenishment, 30
watertight doors, 35
wooden, 20
shipwrecks, 166
shock absorbers, 78
Singapore:
congestion pricing in, 93
port of, 59, 60
Sirius radio networks, 175
skewed propellers, 29
Skylab, 166, 167, 170, 176, 179
slip switches, 102
slurry, 63
Smithsonian Institution, *A Method of Reaching Extreme Altitudes*, 166
smoke detectors, 164
smuggling, 70
SOLAS (Safety of Life at Sea) convention (1914), 37
Somali pirates, 70
sonar, 26, 43
Sopwith Camel, 127
Sopwith Schneider flying boat, 148
SOS distress call, 53
Soyuz spacecraft, 166–67, 170–71, 172, 193
space, 166–83
animals in, 167
commercial flights to, 192
communication in, 173, 174–75
crew seats, 171
dangers in, 180–81
debris in, 182–83
decompression in, 181
exploration, 17, 166–67
and fire, 181
food, 178, 179
fungi in, 181
launch, 168–69
life in, 178–79
milestones, 166–67
mission control, 172–73
oxygen/air, 178–79
radiation in, 180
return from, 170–71
satellites, 174–75
solid waste in, 178
transporter vehicles, 168
walking in, 166, 177
water recovery system, 178–79
weightlessness in, 180
Space Exploration Technologies (SpaceX), 192–93
space race, 192–93
SpaceShipOne, 167
space shuttle, 17, 171, 179
space suits, 177, 181
spandrel-braced bridges, 114
sparkplugs, 79
Sperry, Lawrence, 135
Spirit of St. Louis, 127
sprinkler systems, 35
Sputnik, 16
stagecoaches, 11, 12
steam engines, 12, 13, 14, 98
steamships, 12, 14, 20, 21, 28
steering systems, auto, 78
Steinke hoods, 37
Stephenson, George, 98
Stevens, John, 13
stiletto (warship), 61
Stockholm, congestion pricing in, 93
stone arches, 112, 113, 114
stop signs, 90
streetcars, 85, 98
submarines, 20, 21, 25, 26–27
basic design, 26
blowing the tanks, 27
coordinating, 48
steering, 26
submerging, 27
submersibles:
dry docks, 73
ROVs, 71
subways, 14, 98, 100, 108–9, 110
Suez Canal, 64, 65
Sullenberger, Chesley "Sully," 146
supercavitating propellers, 29
supertankers, 73, 80
surveillance/research ships, 25
survival immersion suits, 37
suspension bridges, 114–15, 116–17
suspension systems, auto, 78
sustainability, 190–91
Suvarnabhumi Airport, Thailand, 140
SWATHs, 22
Swift, Hildegarde, *The Little Red Lighthouse and the Great Gray Bridge*, 45
swing bridges, 115
Sydney Airport, Australia, 140

T

Tacoma-Narrows Bridge "Galloping Gertie," 116–17
tamper/tamping machine, 101
tank cars (railroad), 105
tanker ships, 16, 24, 31, 73, 80
tanker trucks, 84
tanktainers, 68
taxis, 85
Taylor, David, 21

SCHOOL BUS YELLOW

Every day, an estimated 480,000 buses transport over 25 million children to school in the United States. Nearly all of those buses—and many in Canada as well—are painted one color: school bus yellow.

Such uniformity can be traced to one man, a professor at Teachers College, Columbia, named Frank Cyr, and to a conference he held to establish school bus standards in 1939. In addition to strict construction and safety standards, conference attendees agreed upon the use of black lettering on what has come to be known as school bus yellow—as it was thought to be easiest to see in the early morning.

TBM (tunnel boring machine), 120–21
TGV, France, 106
thermals, 144
third rail, 108, 109
Thomas Point Lighthouse, U.S., 44
Three Gorges Dam, China, 65
Thresher (submarine), 27
thruster propulsion system, 47
tires, pneumatic, 14, 77
Titanic, 15, 20, 31, 37, 52
torque, 79, 132
Torre de Hercules, Spain, 44
track stabilizer, 101
traffic circles, 90, 91
traffic lights, 88–89
traffic separation schemes, 49
trains, *see* railroads
trams/trolleys, 85, 98
Trans-Australian Railway, 99
Trans-Canada railway, 99
Transcontinental Railroad, 14, 98, 99
transmission, 79
transponders, 93
transportation:
  automation in, 186, 188–89
  cost of, 186
  impact of, 11
  milestones, 12–17
  sustainability in, 190–91
Trans-Siberian Railway, 99
trash disposal unit (TDU), 34
trash skimmers, 60
treadles, 93
trimarans, 22, 61
trucks, 15, 69, 82–85
  aerodynamics, 82
  and containerization, 68, 69
  diesel fuel for, 79, 83
  fifth wheel, 82
  pigtail, 83
  steering, 78
  tractor trailers, 82–83, 104
tsunamis, 54
tugboats, 61
tuk-tuks (rickshaws), 85
tunnels, 112–13, 120–23
  cut-and-cover, 120, 121
  maintenance, 123
  safety, 122
  ventilation, 121, 122, 123
Turtle (submarine), 21

HYBRIDS AT SEA

Hybrid vehicles are no longer confined to the roads. In a novel historical twist, ships with "SkySails" are now being piloted on the world's waterways.

The SkySail consists of a large kite and a series of controls that allow it to billow out high above a ship (330–980 ft; 100–300 m). The wind power captured by the sail complements the work of traditional fuel-based engines, leading to an estimated 10 to 15 percent fuel savings.

To date, the technology has been piloted on containerships and fishing vessels. Plans are afoot for it to be employed on large yachts as well.

U

undersea waste disposal, 34
underwater navigation, 40, 48
UPS, 159
U.S. Airways, 146

V

V-22 Osprey, 133
Vancouver SkyTrain, 108
Vanderbilt, Cornelius, 110
variable pitch propellers, 29
variometer, 135
Verne, Jules, *Twenty Thousand Leagues Under the Sea*, 26
Verrazano-Narrows Bridge, 117
vertical lift bridges, 115
VHF radio, 136
Vikings, 20
vitamin C, 33
Voith Schneider propellers, 29
VOR stations, 136, 137
Vostock rocket, 16
VTS (vessel traffic system), 49

W

warships, 21, 25, 61
waste disposal:
  airplane toilets, 143, 156
  bin liners, 68
  incineration, 34
  sewage treatment, 34
  solid, in space, 178
  trash skimmers, 60
  undersea, 34
water:
  potable, 32–33
  saline concentrations in, 32
water depths, 43
waterfront, 58, 59
water spouts, 55
Watt, James, 79
wavelength, 54
wave-making resistance, 22, 23
waves, 54–55
  internal, 55
  rogue, 55
  tsunamis, 54
Weather Reconnaissance Squadron, U.S. Air Force Reserve, 144
wheel, history of, 76–77
Whittle, Sir Frank, 157
Wilkinson, John, 21
wind shear, 144
wind turbines, 191
World's Fair (1939), 188
World War II, aviation in, 126, 157
Wright brothers, 15, 126, 127, 148
Wright Flyer, 15

Y

Yokohama Marine Tower, Japan, 45

Z

zebra crossings, 88
Zeppelin, Count Ferdinand von, 126
Zeppelins, 15, 126
zipper machines, 91